Low-temperature Technologies

Edited by Tatiana Morosuk and Muhammad Sultan

Published in London, United Kingdom

IntechOpen

Supporting open minds since 2005

Low-temperature Technologies
http://dx.doi.org/10.5772/intechopen.78524
Edited by Tatiana Morosuk and Muhammad Sultan

Contributors
Abraham Engeda, Ahmed Hegazy, Lazaro Canizalez Davalos, Oleg Pakhomov, Alexander Baranov, Alexander Fedorov, Andrey Zaitsev, Vladimir Ivanov, Ruslan Polyakov, Nencho Stanev Deliiski, Natalia Yordanova Tumbarkova, Tatiana Morosuk, Stefanie Tesch, George Tsatsaronis, Amira A-Abdalall, Anthony Adeyanju, Shaimaa Seyam, Muhammad Sultan, Hassan Niaz, Takahiko Miyazaki, Eraldo Santos, Carlisson Azevedo, Caio Macedo, Lana Baia, Yves Alexandrino, Carla Azevedo, Raíssa Araújo, Sílvia Gomes, Esam Elsarrag, Phillip Davies, Opus Igobo, Mohamed Elnaggar, Mohammed Alnahhal, Ravinder Kumar

First published in London, United Kingdom, 2020 by IntechOpen
IntechOpen is the global imprint of INTECHOPEN LIMITED, registered in England and Wales, registration number: 11086078, 7th floor, 10 Lower Thames Street, London, EC3R 6AF, United Kingdom
Printed in Croatia

British Library Cataloguing-in-Publication Data
A catalogue record for this book is available from the British Library

Additional hard and PDF copies can be obtained from orders@intechopen.com

Low-temperature Technologies
Edited by Tatiana Morosuk and Muhammad Sultan
p. cm.
Print ISBN 978-1-83880-666-8
Online ISBN 978-1-83880-667-5
eBook (PDF) ISBN 978-1-83880-668-2

For EU product safety concerns:
IN TECH d.o.o., Prolaz Marije Krucifikse Kozulić 3, 51000 Rijeka, Croatia, info@intechopen.com or visit our website at intechopen.com.

Meet the editors

Professor Tatiana Morosuk (Tetyana Morozyuk) is the Head of the Department "Energy-Based Methods for Refrigeration Systems" at the Technische Universität Berlin, Germany. She studied refrigeration engineering in the Odessa State Academy of Refrigeration, Ukraine, and received her diploma in 1990. She received her Ph.D. in 1994 and Professorship in 2001, all in the Ukraine. Professor Morosuk has over twenty years teaching experience in the fields of refrigeration, energy engineering, and applied thermodynamics. She is associated with several scientific organizations as well as many international energy-related conferences and recognized international journals. She serves as an associate editor for the following international journals: "International Journal of Energy and Environmental Engineering" (Springer), "International Journal of Natural Gas Science and Engineering" (Elsevier), "International Journal of Energy Research" (Wiley), "Energies" (MDPI), "Entropy" (MDPI), and "Journal of Energy Resources Technology" (ASME). Professor Morosuk's areas of scientific activities include the application of energy-based methods to the improvement of the thermodynamic, economic, environmental performance of different power generation systems, refrigeration/cryogenic systems and chemical plants. Particular attention is given to hydrogen economy, systems associated with the liquefaction of natural gas and the regasification of LNG, alternative refrigeration processes for sustainable industrial and commercial applications, and smart energy supply and use in industrial parks, including innovative concepts of liquid air energy storage. She is the author or co-author of 7 books and 12 book chapters as well as more than 300 publications and 10 patents. Fourteen Ph.D. theses and more than 100 master theses have been successfully completed under her supervision/co-supervision. Prof. Morosuk has vast administrative experience being the study dean of four international master's programs in Germany, and the Head of the Energy Engineering Department at TU Berlin Campus El Gouna.

Dr Muhammad Sultan is a Assistant Professor at Bahauddin Zakariya University, Pakistan. He received his Bachelor degree in agricultural engineering (2008) with distinctions and his Masters degree in environmental engineering (2010) from the University of Agriculture, Faisalabad, Pakistan. Muhammad Sultan was awarded prestigious Monbukagakusho and JASSO fellowships from the Japanese government for his Ph.D. (2015) and postdoctoral studies (2017), respectively, in Energy & Environmental Engineering from Kyushu University, Japan. He also completed postdoctoral studies (2019) as a Canadian Queen Elizabeth II Advance Scholar in mechatronic systems engineering at Simon Fraser University, Canada). He has authored/co-authored more than 100 journal and conference articles. He has published 2 book chapters and edited/co-edited 2 books. He is serving as Regional Editor for the Evergreen Journal of Kyushu University. His research interests include HVAC, desiccant air-conditioning, evaporative cooling, adsorption cooling, energy recovery ventilator, adsorption heat pump, Maisotsenko cycle (M-cycle), wastewater, and agriculture and livestock applications.

Contents

Preface XI

Chapter 1 1
Central Air Conditioning: Systems and Applications
by Mohamed Elnaggar and Mohammed Alnahhal

Chapter 2 21
Investigation of Desiccant and Evaporative Cooling Systems for Animal Air-Conditioning
by Muhammad Sultan, Hassan Niaz and Takahiko Miyazaki

Chapter 3 39
Solar Pond Driven Air Conditioning Using Seawater Bitterns and $MgCl_2$ as the Desiccant Source
by Esam Elsarrag, Opubo N. Igobo and Philip A. Davies

Chapter 4 59
Theoretical and Experimental Analysis of a Thermoelectric Air-conditioning System
by Adeyanju Anthony Ademola

Chapter 5 77
An Experimental Study of Synthetics and Natural Refrigerants Gases
by Eraldo Cruz dos Santos, Carlisson Azevedo, Caio Macêdo, Carla Azevedo, Raíssa Araújo, Sílvia Soares Gomes, Lana Baia and Yves Alexandrino

Chapter 6 101
Effect of Nanoparticles on Performance Characteristics of Refrigeration Cycle
by Ravinder Kumar

Chapter 7 113
Modeling of the Energy for Bound Water Freezing in Logs Subjected to Refrigeration
by Nencho Deliiski and Natalia Tumbarkova

Chapter 8 133
Technique and Technology of Whole-Body Cryotherapy (WBC)
by Alexander Baranov, Oleg Pakhomov, Alexander Fedorov, Vladimir Ivanov, Andrew Zaitsev and Ruslan Polyakov

Chapter 9 159
Comparative Evaluation of Cryogenic Air Separation Units from the Exergetic and Economic Points of View
by Stefanie Tesch, Tatiana Morosuk and George Tsatsaronis

Chapter 10 179
Impact of Air-Conditioning Filters on Microbial Growth and Indoor Air Pollution
by Amira Hassan Al-abdalall, Sarah Abdullah Al-dakheel and Hmidah Abdulhadi Al-Abkari

Chapter 11 207
Designing Spiral Plate Heat Exchangers to Extend Its Service and Enhance the Thermal and Hydraulic Performance
by Lázaro Canizalez Dávalos, Edilberto Murrieta Luna, Mario Alberto Rodríguez Ángeles and Víctor J. Cruz Delgado

Chapter 12 223
Air Cooling in Steam Plant Condenser Using Refrigeration System for Improving Vacuum Pump Performance
by Ahmed Hegazy and Abraham Engeda

Chapter 13 245
Energy and Exergy Analysis of Refrigeration Systems
by Shaimaa Seyam

Preface

The low-temperature market is growing and expanding. It plays an essential and growing role in the global economy. A wide range of low temperatures are required almost in all branches of everyday life and industries, including the food and beverage industries, air conditioning (comfort, technological, technical, IT equipment), water and "dry" ice, winter sports, chemical industries, freeze-drying (for food, pharmaceutical, biological applications, hair care industries, etc.); metallurgy, mechanical industries, machinery, civil engineering, liquefaction of natural gas, etc. The total number of low-temperature systems recently in operation over all the world is roughly 3 billion. Such a development must be sustainable, with limited impact on the environment and reasonable energy consumption (https://iifiir.org/en).

The scientific publications related to refrigeration cover the directions of new and/or advanced technologies for application, thermodynamic cycles, working fluids (one-component and mixtures), heat and mass transfer processes, materials etc. The methods for the analysis and optimization are adjusted for in the refrigeration machines because of the processes below/crossing the temperature of the environment. The initial idea of this book was to provide the reader with a comprehensive overview of the latest developments and perspectives of research in the very wide spectrum of refrigeration: feasibility of new technologies and improvements in existing technologies; energetic, economic, and environmental effectiveness of machines and processes; different optimization methods; and many other aspects related to low temperatures.

This book, "Low-temperature Technologies", consists of 13 chapters. These chapters cover a wide range of low-temperature applications from the temperature slightly below the environment (air conditioning) down to cryogenics.

The energy consumption for industrial, commercial, and domestic heat, ventilation, and air conditioning has become a very important issue in the energy balance of many countries. Therefore, continuous updating to the latest developments helps the researchers and practical engineers in their work as well as the students for the education process. Chapter 1 discusses the central air condition systems and their applications: all-air, all-water, and air-water systems. The topic of a special air conditioning system and its new application is addressed in Chapter 2. The desiccant-based systems have been evaluated: standalone and assisted by the M-cycle. The recommendations for the optimal applications have been formulated. Desiccant-based air conditioning systems are also the focus of Chapter 3. A solar-based technology, i.e. solar pond driven air conditioning systems, have been investigated theoretically and experimentally. Chapter 4 reports the evaluation of the micro thermoelectric air conditioning system. The aim of this work is to validate the theoretical model using the experimental data. The results obtained from experimental study have been reported in Chapter 5 as well. The research interest of the authors is synthetics and natural refrigerants gases usually used for air conditioning systems in Brazil.

The research in the field of nanorefrigerants is one of new developments for improving the performance of refrigeration systems. Chapter 6 summarizes the information about behavior of different nanoparticles in vapor-compression refrigeration machines.

Within Chapter 7, the authors suggest an approach for modeling the total thermal energy needed for freezing the bound water in logs subjected to refrigeration. The approach maximally considers the physics of the freezing process of the bound water in wood.

Chapter 8 addresses the application of low-temperature technologies for medical application. Heat transfer aspects have been discussed for the technologies applied to whole-body cryotherapy.

"Cryogenics" is the key word for Chapter 9. The authors reported the evaluation of air separation units. An exergetic analysis has been applied in order to identify the thermodynamic inefficiencies and the processes that cause them. The economic characteristics of these systems have been also reported.

The remaining chapters discuss the performance of the equipment. Chapter 10 is dedicated to the evaluation of impacts of air-conditioning filters on microbial growth and indoor air pollution. The design of the spiral plate heat exchangers is the topic of Chapter 11. The authors focus on the thermal and hydraulic performance of such a heat exchanger. Computational fluid dynamics is performed to validate the thermal and hydraulic method. Particular attention is given to the maintenance of the spiral plate heat exchangers. Within Chapter 12, the influence of low temperatures on the performance of the vacuum pumps for steam power plants using air cooling is discussed.

Chapter 13 is dedicated to the overview of the energy and exergy analyses applied to the different types of refrigeration systems.

We hope that the readers will find this book to be interesting and useful for their professional activities.

Tatiana Morosuk (Editor)
Professor,
Technische Universität Berlin,
Germany

Muhammad Sultan (Co-editor)
Assistant Professor,
Bahauddin Zakariya University,
Pakistan

Chapter 1

Central Air Conditioning: Systems and Applications

Mohamed Elnaggar and Mohammed Alnahhal

Abstract

It became evident nowadays that modernization influences domestic and commercial HVAC industry, and thus high technological and energy-efficient central air conditioning systems are demanded. Therefore, the selection of proper type of central air conditioning system is a crucial target in the construction industry as improper selection can maximize initial and/or running costs of the system and decreases the human comfort and indoor air quality levels. In fact, a pre-assessment of the construction type and budget available is required for selecting the proper type of central air conditioning system. Therefore, there is a continuous need for an updated material in the literature that reviews the central air conditioning systems and applications, which is the motivation of the present chapter. The present chapter reviews the central air conditioning systems and applications. Specifically, all-air systems, all-water systems, and air-water systems are discussed. In addition, all provided systems are further explored through several developed schematic diagrams enabling the identification of their various components and the understanding of their working principles. It is may be of interest to note that this chapter is suitable for undergraduate level students in the fields of HVAC and R, mechanical, and construction engineering.

Keywords: central air conditioning, all-air systems, all-water systems, air-water systems, air handling unit

1. Introduction

In central air conditioning, air, water, or both are used as working fluids to produce the required heating and/or cooling, and therefore based on working fluids, central air conditioning systems can be classified into three groups [1–5], namely:

1. All-air systems: in these systems air is used as working fluid for heating and/or cooling purposes.

2. All-water (hydronic) systems: in these systems water is used as working fluid for heating and/or cooling purposes.

3. Air-water systems: in these systems both air and water are used as working fluids for producing heating and cooling purposes.

It is may be of interest to note that each type of the central air conditioning systems has several systems of sometimes different configurations, and the use of any system depends on its advantages and disadvantages [1–5]. The following sections will provide descriptions of all types of central air conditioning systems.

2. Classifications of central air conditioning systems

2.1 All-air systems

In these systems air alone is used as working fluid to produce cooling or heating in air-conditioned zones; besides that air is responsible for controlling the zones' humidity level and provide the required ventilations to air-conditioned zones. In addition, in all-air systems, air is used for aromatizing purposes. Therefore, only air as working fluid is responsible for providing comfort, i.e., cooling, heating, controlling of humidity and ventilation odor, and thus these systems are called all-air systems [1–5].

2.1.1 Air handling unit (AHU)

Air handling unit can be considered as the heart of all-air systems since cooling and heating take place in the air handling unit. It also mixes the outside air after being purified with the return air, then the necessary psychometric processes are carried out. Air conditioner is then expelled or withdrawn to the place to be air-conditioned. These units are used for capacities exceeding 100,000 CFM (50 m^3/s) air. The main components of the air handling unit are shown in **Figure 1** [1–6].

The main components of the air handling unit shown in **Figure 1** are described below:

- **Supply fan**. Centrifugal fan type is used to provide the conditioned air to various zones.

- **Fan motor**. Electric motor is used to provide the rotating motion to the supply fan.

- **Cooling coil**. Coil placed in AHU where cold water from chiller is circulating in medium- and large-capacity AHU or expanded refrigerant in small-capacity AHU.

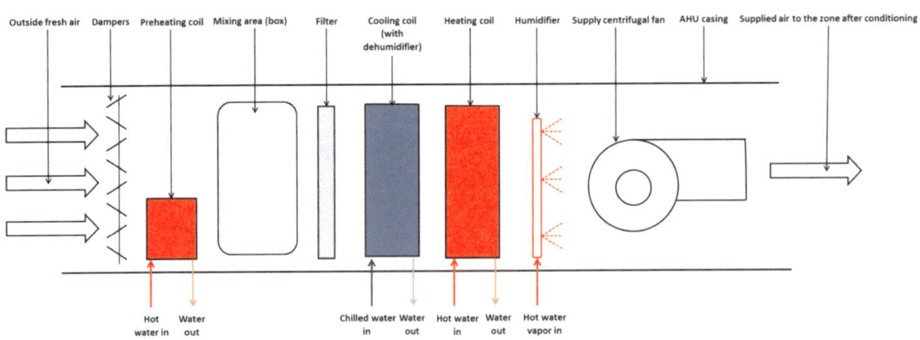

Figure 1.
A schematic diagram of an air handling unit with its main components.

- **Filters**. Filters or strainers are placed at the early air path in AHU. Filter type used may depend on application type.

- **Mixing box**. It is place where fresh air is mixed with zone return air or with fresh conditioned air. Mixing processes are performed to obtain the desired air temperature and humidity or to maintain energy-efficient performance.

- **Dampers**. Dampers are used to control the amount and direction of air before or after conditioning is performed.

- **Heating coil**. Coil placed in AHU where hot liquid or vapor water from boilers is circulating.

- **Preheating coil**. Preheating coil is placed at AHU entrance before cooling and heating coils. The task of preheating coil in hot days is to reduce the entering fresh air's relative humidity, thus preventing possibly the condensation of water vapor on cooling coil, hence preventing frost formation on the cold coil. In addition, preheater coil will prevent the freezing of water inside the coils in cold days.

- **Humidifier**. It is a system which is responsible for increasing the humidity in the conditioned zone. Humidifiers are usually used in cold days where maintaining hot climate is desired; however this will accompany low-humidity levels, and thus the use of humidifier becomes essential to maintain the comfort edge. Humidifier can provide humidity either as hot vapor or water spray. The first one is more preferable particularly in healthcare applications as hot vapor will prevent the growth of biological organisms such as bacteria or algae besides hot vapor compared to the water spray. Besides that hot vapor is more preferable as provided humidity will have higher temperature and thus will not reduce the conditioned hot air provided to the zones.

- **Centrifugal pumps**. Centrifugal pumps are used for cooling and heating air processes. They are used to maintain pumping cycle of hot water from boilers to the heating or preheating coils and back to boiler and/or pumping cycle of cold water from the chiller or cooling tower to cooling coil in AHU and back to the chiller or cooling tower.

- **Control systems**. Control systems can vary from simple control system to advanced control system that use the latest technologies such programmable logic controllers. Usually controllers are used to control the temperatures and humidity of the supply air to the zones. It also controls the damper systems in the AHU. Moreover, advanced control systems can even control the fan rotation and thus the rate of air supplied to the zone based on the required temperature in the zone and zone exit damper. In such systems, pressure sensor connected to the control system will be placed in the air duct, and as attaining the required temperature in the zone, dampers will get closer increasing the duct-sensed pressure through which the fan connected to the control system will reduce its speed and thus maintain energy-efficient performance [7, 8].

- **Casing**. Casing is a kind of AHU cover that includes all the above AHU components.

It is may be of interest to note that the configuration of air handling unit can differ slightly in design and components according mainly to the type of application and AHU capacity (e.g., healthcare buildings or other), but also initial and running costs can affect the selection of various AHU components. In addition, air handinbased on various classificationsg units can be classified based on the structure and based on the location where it is placed [2]. The following sections will demonstrate all types of air handling units based on various classifications.

2.1.1.1 Classification of air handling unit based on structure

A. Horizontal air handling unit (horizontal AHU)

The horizontal air handling units place the supply air fan, cooling coil, heating coil, and humidifier in the casing horizontally (see **Figure 2**). This design requires a large floor area as AHU components are large in size. Horizontal AHU type are usually of high capacity and are usually placed in plant room, whereas small horizontal AHU type can be placed carefully on the roof [2].

B. Vertical air handling unit (vertical AHU)

In vertical air handling units, the centrifugal fan is placed in a position above the cooling coil, heating coil, and humidifier as shown in **Figure 3**. Vertical air handling units are small in size and are low-capacity units, and thus they are installed in a small floor area of plant room [2].

2.1.1.2 Classification of air handling unit based on location of installation

A. Internal air handling unit (internal AHU)

These air handling units are normally installed indoor in a plant room. This type of air handling unit is called simply air handling unit [1].

B. Fresh air handling unit (fresh AHU)

In this type, air handling units are installed in the outdoor environment such as on the building roof. This type of air handling unit is called fresh air handling unit. Fresh air handling units are designed to withstand climatic variations. In addition,

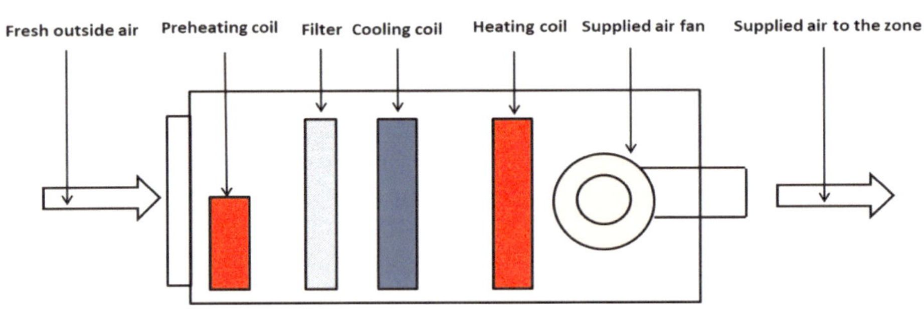

Figure 2.
A schematic diagram of horizontal air handling unit. Redrawn with modification from [2].

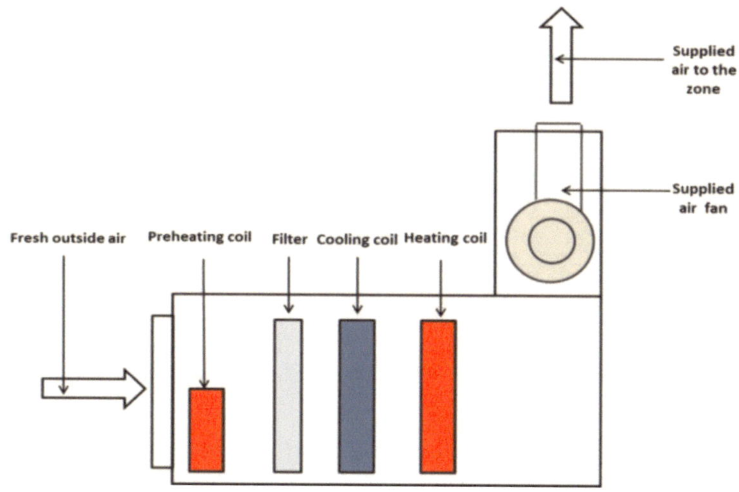

Figure 3.
A schematic diagram of vertical air handling unit. Redrawn with modification from [2].

safety measures must be taken before installing fresh air handling units on roofs due to their large weights [2].

2.1.1.3 Classification of air handling unit based on supply air fan placement

A. Blow-through units

In the blow-through units, supply air fan forces or pushes the air through the cooling coil (see **Figure 4**) to reduce the increase of supplied air temperature due to friction, and thus air can be cooled before being supplied to the various zones [2].

B. Draw-through units

In this type, the supply fan is placed after cooling coil (see **Figure 5**), heating coil, filter, and humidifier, and thus air is pulled by the supply fan. This system is commonly used as filters and coils which require small air speed and larger ducts

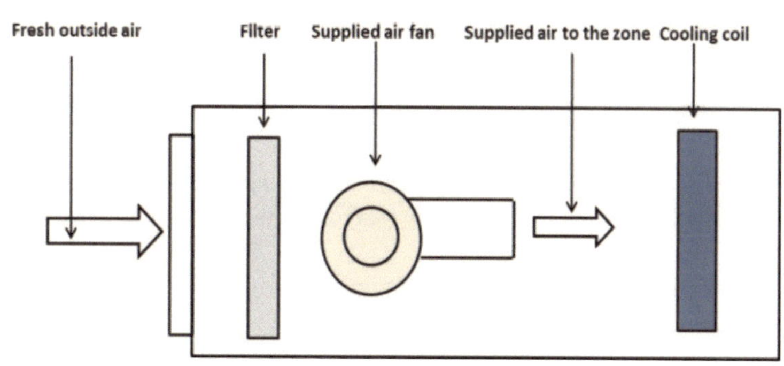

Figure 4.
A schematic diagram of blow-through type of air handling unit. Redrawn with modification from [2].

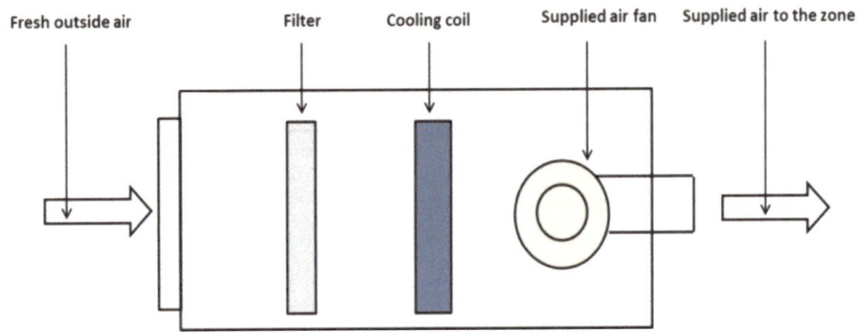

Figure 5.
A schematic diagram of draw-through type of air handling unit. Redrawn with modification from [2].

than large speeds and small ducts through the fan. The only disadvantage of this system is that the fan sound can travel with supply air to the conditioned zones [2].

2.1.2 Types of all-air systems

2.1.2.1 Conventional systems

A. Classification of conventional systems

These systems can be classified into two groups [1–5], namely:

i. Fixed supplied air volume flow rate with variable supplied air temperature

In these systems the rate of air flow remains constant, and the zone-required temperature is obtained by varying the supplied air temperature.

i. Variable supplied air volume flow rate with fixed supplied air temperature

These systems maintain fixed supplied air temperature; however the zone-required temperature is maintained by varying the volume flow rate of supplied air.

B. Advantages of conventional systems

1. Simplicity as the components are of simple configuration and can be found separated from each other

2. Low initial cost since the system has simple configuration

3. Low running cost as ventilation, return air, and side passage flows can be used, which can maintain energy-efficient performance

4. Quite operation as air handing unit including the fan is placed in plant room away from zone

5. Centralized maintenance as air handling unit is placed in the plant room

2.1.2.1.1 Single-duct conventional systems

Single-duct systems are generally used for conditioning stores, offices, and industries. A schematic diagram of a single-duct system is shown in **Figure 6**, which is fixed supplied air volume flow rate and variable supplied air temperature type, and thus the required temperature in the zone is attained by the adjustment of heating and cooling coil flow rates and thus their temperature. On the contrary, the required temperature in conditioned zone can also be attained by varying supplied volume flow rate and holding fixed supplied air temperature (**Figure 7**) which is accomplished by simply placing dampers in the single duct; thus supplied air volume rate can be varied, e.g., in summer days, higher supplied air volume rates are provided to the zone to reach required temperature.

It is worth mentioning that single-duct system is used in both single and multi-zones. Return air can be mixed with the fresh air in the mixing box in appropriate ratio. This will provide energy-efficient running of the system; however care must be taken on the quality of return air as air may have higher percentages of humidity levels [1–5, 8]. This configuration can be seen in **Figure 6** and **Figure 7**; however, in **Figure 7** variable supplied air volume flow rate and fixed supplied air temperature type has new arrangement called side passage flow (green lines) where the air once cooled is directed immediately to the mixing box to accelerate the cooling process.

2.1.2.1.2 Multi-duct conventional systems

Characteristics the characteristics of multi-duct systems are similar to single-duct systems except that two or more ducts can be used for providing conditioned air to two or more zones. The usage of several ducts in these systems provides flexibilities as long as the required condition can be varied according to the requirement in each zone. These systems are used where control of temperature and humidity in a building zone is required [1–5]. **Figure 8** shows two-duct system of variable supplied volume flow rates with fixed supplied air temperature.

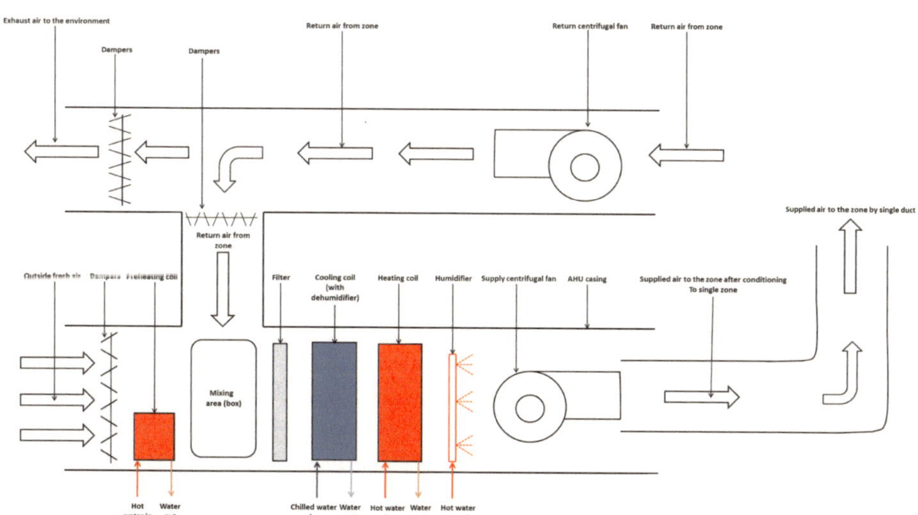

Figure 6.
A schematic diagram of single-duct system (fixed supplied air volume flow rate and variable supplied air temperature).

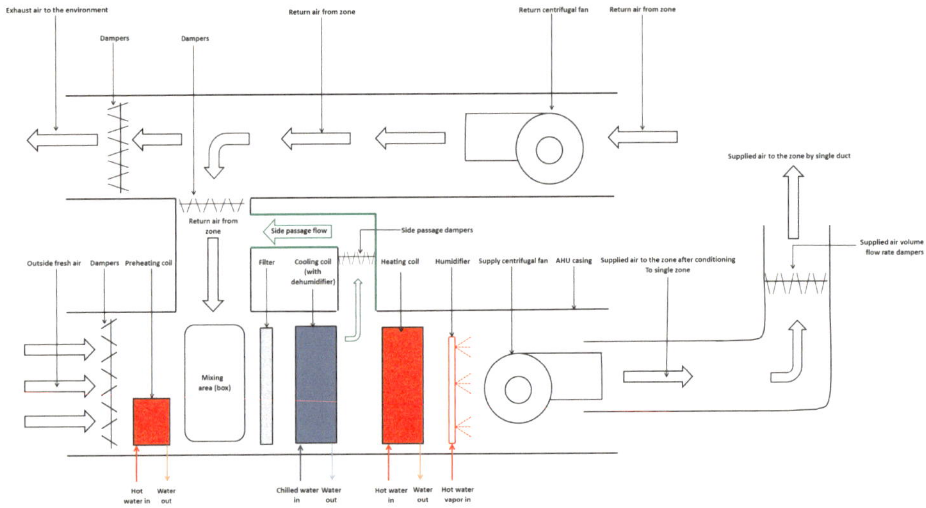

Figure 7.
A schematic diagram of single-duct system (variable supplied air volume flow rate and fixed supplied air temperature).

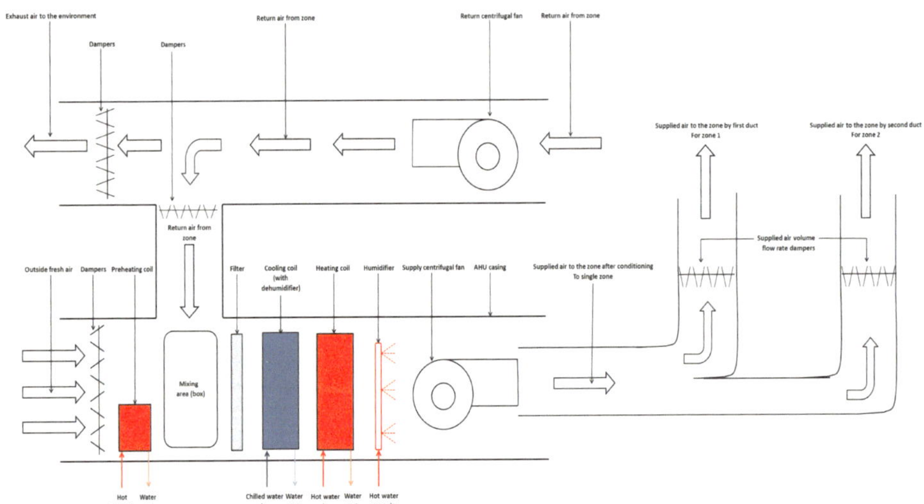

Figure 8.
A schematic diagram of multi-duct system (variable supplied air volume flow rate and fixed supplied air temperature).

2.1.2.2 Reheat systems

Reheat systems are used in applications of variable loads. In these systems air will be cooled to the zone's lowest required temperature, and then air will be circulated to all zones, and the required temperatures of various zones can be obtained by reheating the supplied air. Electric heater or hot water can be used as reheaters which are located in the terminal units (see **Figure 9**) of conditioned zones [1–5].

In addition, reheaters can be used for reducing the levels of humidity in cold temperature zones by increasing supplied air temperatures. Usually zone thermostat will be used to control the reheater according to required conditions.

DOI: http://dx.doi.org/10.5772/intechopen.89455

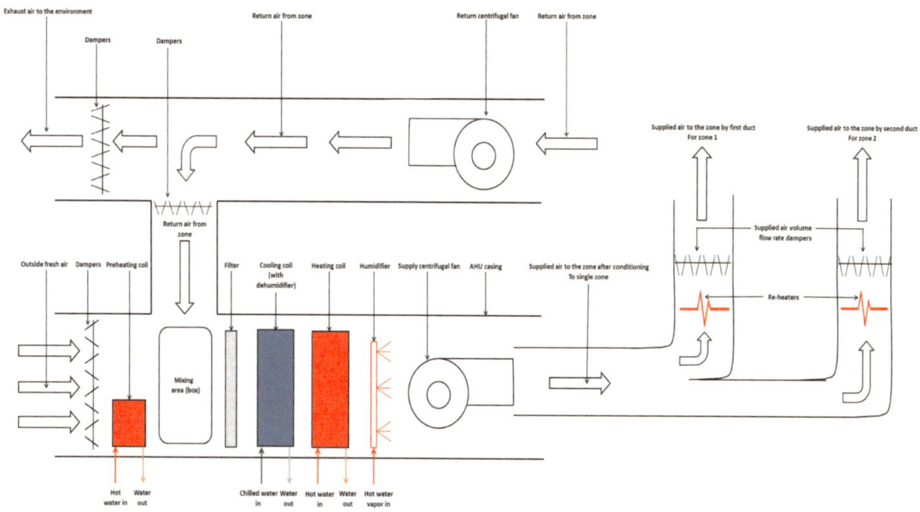

Figure 9.
A schematic diagram of a reheat system.

2.1.2.3 Constant volume induction system

In these systems, an induction unit is used either for ventilation or heating purposes. The induction unit usually contains hot water coil, and it induces the zone's air to provide further heating or air ventilation (**Figure 10**). Single or several induction units can be used based on the number of conditioned zones, and hot water is provided to them from boilers through a water circuit which return the water again for reheating. Primary air is usually provided from the air handling unit which is usually responsible for the latent and sensible thermal loads and provides the desired humidity level. In addition, primary air will be provided to induction unit. Induction units are used to meet the increase heating loads and thus can provide comfort quicker with lower running cost [1–5].

Advantages of induction units

1. Possibility of controlling room temperature as every room with induction unit can be considered as zone

2. Simple air duct design

3. Centralized supplied primary air

4. Control system simplicity

5. Economic running

6. Possibility of controlling the air ventilation and odor

7. Quite operation as fans are away from the room

2.1.2.4 Multi-zone unit systems

In multi-zone unit systems, the cooling and heating coils are placed parallel to each other where an amount of air supplied with a constant temperature is

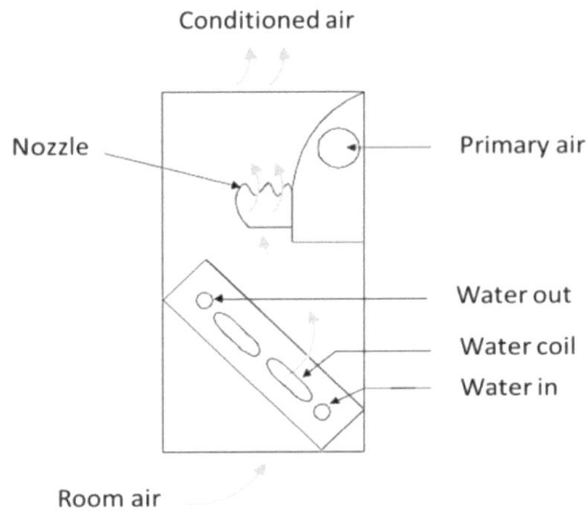

Figure 10.
A schematic diagram of an induction unit. Readapted from [5].

maintained. In these systems, hot air and cold air are mixed in required proportions, and thus the supplied air of fixed temperature and fixed volume will be provided to conditioned zone [1–5]. **Figure 11** shows a multi-zone unit system that supplies air separately to three different zones as dampers can be used.

This system is used and suitable for the following applications, namely:

i. Buildings that contain a number of small and large zones in which separate temperature control is desired as in schools and offices.

ii. Areas in building of different directions and different internal loads such as buildings, e.g., banks

iii. Buildings that have interior zones of varying sizes as in radio or television studios

Advantages of multi-zone unit systems

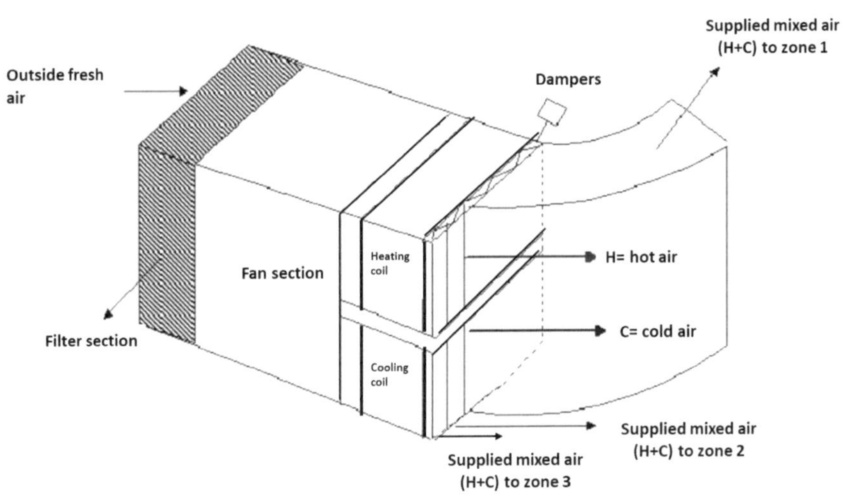

Figure 11.
A schematic diagram of multi-zone unit system. Readapted from [2].

1. These systems allow separate control of temperature in zones or places that are considered as separate area as supplied air is provided at the desire temperature.

2. It is simple to have the smallest unit size. These systems can either be assembled in plant location or in the factory and fit all requirements.

3. They are easy to switch operation from cold to hot in summer and winter seasons and vice versa as this can be simply done manually from air handling unit in plant room.

4. These systems allow easy air distribution and balance. Using only one air duct with various exits and outlets makes the balance process easier.

5. Centralized refrigeration equipment as air handling unit is used.

6. Centralized maintenance as all air handling unit is placed in the plant room.

7. Low-cost operation.

8. Quiet operation.

2.1.2.5 Dual-duct systems

Dual-duct systems allow separate control of temperatures in conditioned places and zones. Temperature control is achieved by supplying the mixing box with air from hot air duct and cold air duct; that is hot air and cold air are mixed in mixing box in proper required proportion based on the zone thermostat, and then air can be supplied to the zone to maintain required zone temperatures. These systems are commonly used in multiroom buildings such as offices, hotels, apartments, hospitals, schools, and large laboratories. **Figure 12** shows a dual-duct system that supplies different zones [1–5].

Advantages of dual-duct systems

1. Provide separate control in the temperatures in each zone as cold and hot air presence at the same time allows rapid change in temperatures.

2. Dual-duct systems can be found in the smaller size as the number of served zone by the central system is reduced, whereas the supply air is maintained through the mixing box which contains cold and hot air just at the each zone.

3. Easy switching from hot to cold modes and vice versa. This is accomplished by the zone or place thermostat which is adjusted once a year.

4. The refrigeration equipment and boilers are placed in one place, and thus electricity, water, and sewage services are located only in plant room but not in the building parts.

5. Centralized maintenance and service are accomplished.

6. Centralized outdoor air inlets. This will ensure no winds or rain are likely to enter from outdoor environment.

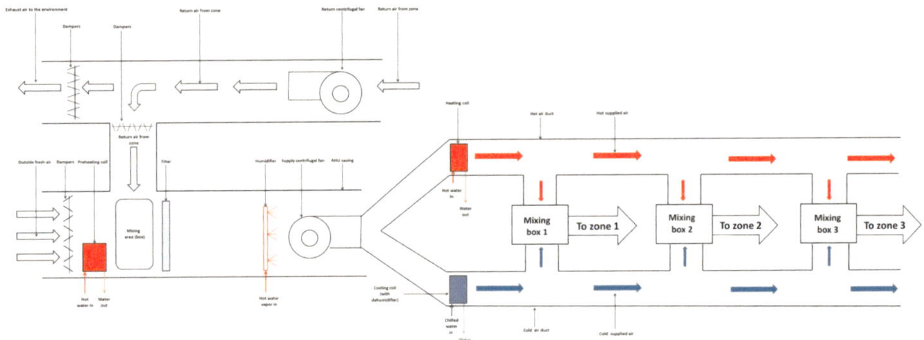

Figure 12.
A schematic diagram of dual-duct system. Redrawn with modification from [1].

7. Higher filter efficiencies can be attained.

8. Low-cost operation can be obtained with these systems.

9. These systems are operating quietly as the refrigeration machines and fan are placed away from zones.

10. These systems have flexible air duct system designs. The choice of medium and high air speeds is possible on an economic basis and according to the requirements of the building.

Disadvantages of the dual-duct systems

1. The use of separate ducts increases the initial cost compared to other systems.

2. Precise control needs a large handling unit, and as a result the total cost of the system will be enlarged.

3. The dual-duct systems are considered inefficient energy systems, and currently these systems are not recommended.

2.1.2.6 Variable air volume (VAV) systems

Variable air volume systems can vary the thermal loads according to zone thermal load variations. One advantage of these systems is that both the initial cost and the operating cost are low because the air volume requires simple control within 20% of the air outlets. These systems are used with fixed thermal loads throughout the year. Applications of these systems can be found in commercial stores, office buildings, hotels, hospitals, housing, and schools. **Figure 13** shows various common terminal units of variable air volume [1–5].

2.2 All-water systems

In all-water systems, water is used as a working fluid for providing heating and cooling. Water is pumped to the fan coil unit located in the zone which is to be conditioned. And then the fan coil unit will use the zone air or sometimes outdoor air for cooling or heating purposes. If outdoor air is used with the fan coil units, then separate air duct system will be introduced to the structure of the building [1–5].

Figure 13.
Pictures of some common variable air volume terminal units. Readapted from [2].

2.2.1 Fan coil unit

Fan coil units are used in all-water central air conditioning. Cold and/or hot water from central chillers and boilers flows through the unit coils. The air temperature is controlled by controlling the flow rate of water through the coil via control valves, e.g., solenoid valves. Fan coil units are cheap in price and widely used in hotels, office buildings, and medical centers. **Figure 14** shows a schematic diagram of a fan coil unit [1–5].

Advantages of fan coil units

1. Low cost

2. Do not need ducts

3. Does not occupy much space

4. Easy to install

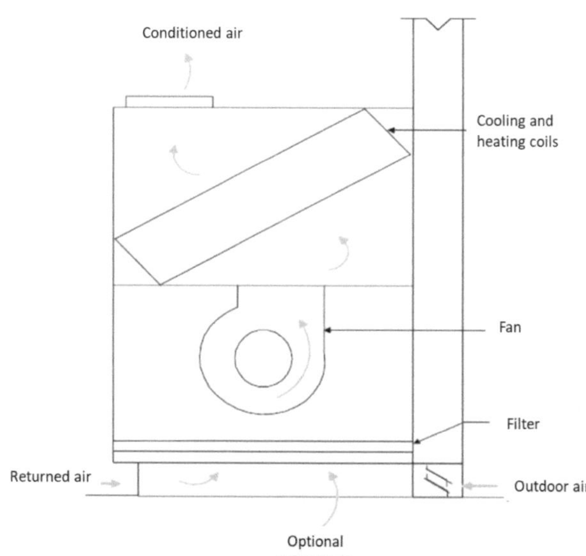

Figure 14.
A schematic diagram of a fan coil unit. Readapted from [5].

Disadvantages of fan coil units

1. Do not provide good control of room air humidity.

2. Require maintenance within the air-conditioned places.

3. Provide suitable medium for bacteria grow in water pipes.

4. The ventilation of the rooms is affected by the speed of wind and rain and insect leakage through the wall openings and cracks which are associated with optional air ventilation duct systems.

2.2.2 Types of all water systems

The all water systems are classified based on water pipe connections to the fan coil units, into two types [1–5], namely:

2.2.2.1 Single piping systems

A. Single piping system (reversed return)

In this system, there are two pipes, one pipe to feed either the cold or hot water to the fan coil unit and another pipe for returning the water back to the chiller or boiler (see **Figure 15**). In these systems the supplied water is in counterflow with the return water.

B. Single piping system (direct return)

In this system, there are two pipes, one pipe to feed either the cold or hot water to the fan coil unit and another pipe for returning the water back to the chiller or boiler (see **Figure 16**). In these systems the supplied water is flowing in a pipe which is connected directly to the chiller or boiler, where the same pipes receive the return water from each fan coil unit.

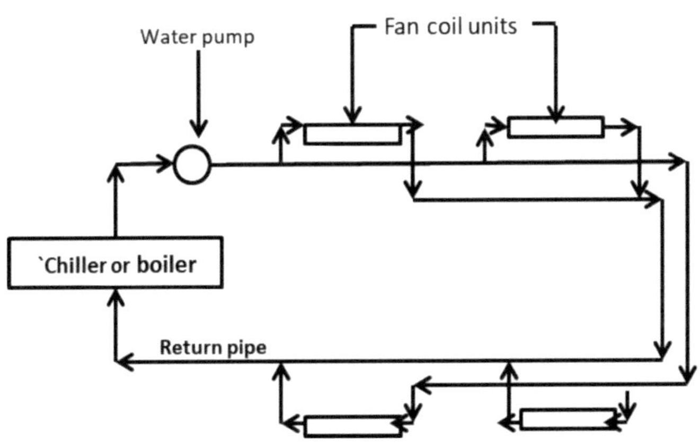

Figure 15.
A schematic diagram of a fan coil unit, two-pipe system of reversed return. Redrawn from [2].

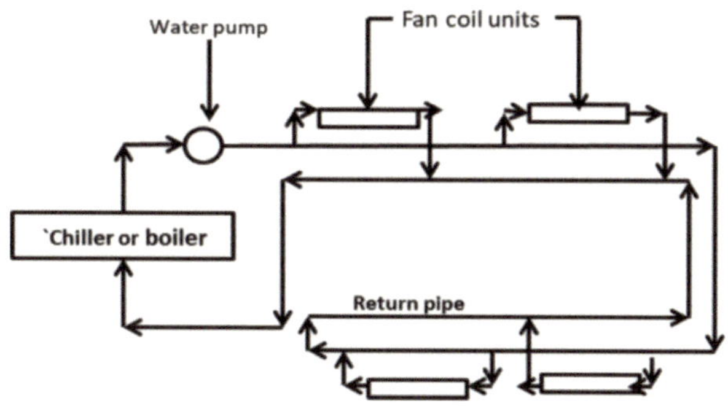

Figure 16.
A schematic diagram of a fan coil unit, two-pipe system of direct return. Redrawn with modification from [2].

2.2.2.2 Multi-piping systems

A. Three-pipe systems

In these systems, there are two pipes for providing hot and cold water to the fan coil unit and one pipe for return water (three-pipe system). **Figure 17** shows a three-pipe all-water-type system.

B. Four-pipe systems

In these systems, there are two pipes for providing hot and cold water to the fan coil unit and two pipes for return, one for cold water return and one for hot water return (four-pipe system). These systems are considered as the best system as the use of two return pipes can maintain efficient energy performance. **Figure 18** shows a four-pipe all-water-type system.

2.3 Air-water systems

In these systems both air and water produce heating and cooling effects. Usually induction or fan coil units can be used in air-water systems. The following section will describe the induction unit systems and fan coil system [1–5].

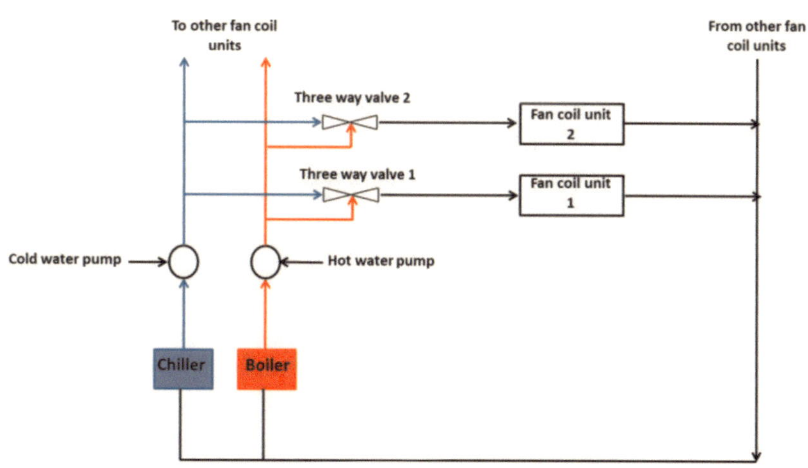

Figure 17.
A schematic diagram of a fan coil unit of three-pipe system. Redrawn with modification from [2].

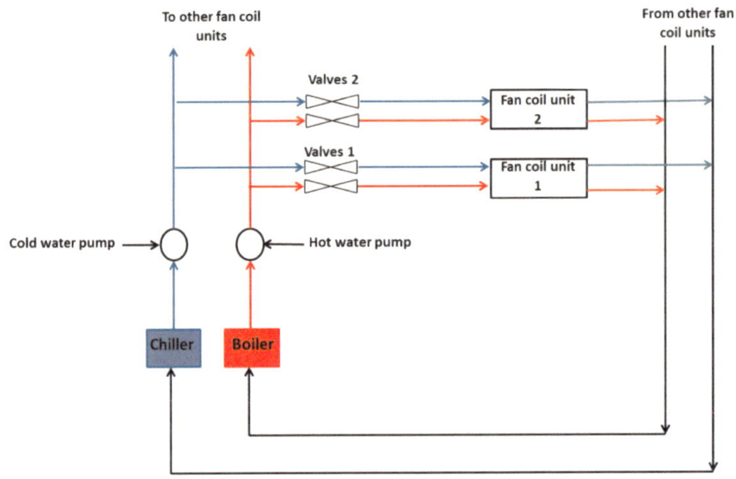

Figure 18.
A schematic diagram of a fan coil unit of four-pipe system. Redrawn with modification from [2].

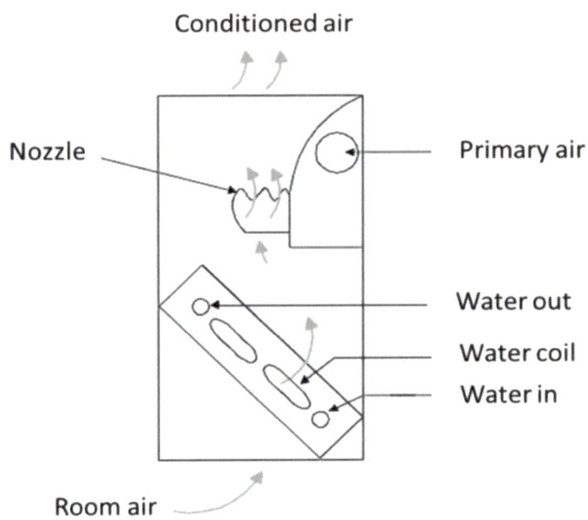

Figure 19.
A schematic diagram of an induction unit system. Readapted from [5].

2.3.1 Induction unit system

Induction unit systems in air-water central air condition are used in building of multi-surrounding rooms such as offices, hotels, hospital patient rooms, as well as apartments. These systems are used where higher thermal loads are present. In addition, these systems are suitable where some rooms require cooling, while the next other rooms require heating. Besides that, these systems are suitable for buildings such as skyscrapers where the spaces are limited. **Figure 19** shows a schematic view of induction unit system which uses primary air and external air. Return air can be used if all primary air is greater than the minimum ventilation requirements [1–5].

2.3.2 Primary air fan coil systems

The working principle of primary air fan coil systems is very similar to the induction unit system. The main difference is the use of fan. These systems are

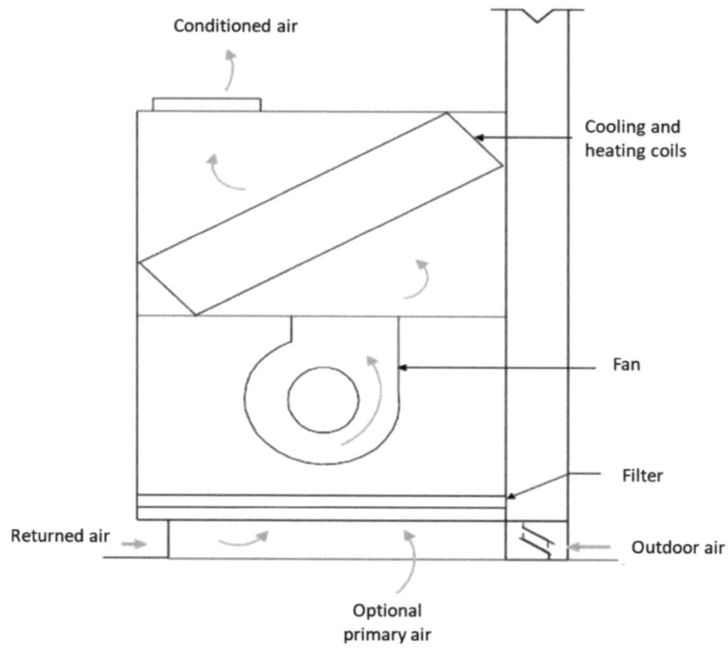

Figure 20.
A schematic diagram of a fan coil unit system. Readapted with modification from [5].

generally used in multiroom buildings such as hotels, hospitals, and apartments where the operating mode can be switched to work in cold season.

Advantages of these systems over induction units are the quite operation and the means to control required condition as fan speed can be usually controlled. But the initial cost of fan coil unit which is higher than induction units makes the induction units more preferred. **Figure 20** shows a schematic diagram of a fan coil unit with its components [1–5].

3. Conclusion

This chapter reviews the types of central air conditioning systems. Specifically, the types and applications of all-air, all-water, and air-water systems are provided. In addition, this chapter gives further insight to all systems through several developed schematic diagrams as various components and working principles of these diagrams and systems can be identified and figured out, respectively. Moreover, the differences among the given systems can be easily distinguished through this chapter.

Author details

Mohamed Elnaggar[1*] and Mohammed Alnahhal[2]

1 Laboratory of Refrigeration and Air conditioning, Department of Engineering Professions, Palestine Technical College, Deir El-Balah, Gaza Strip, Palestine

2 Department of Refrigeration, Air conditioning and Heating Engineering, Israa University, Gaza, Gaza Strip, Palestine

*Address all correspondence to: mohdhn@yahoo.com

References

[1] Pita EG. Air Conditioning Principles and Systems: An Energy Approach. 4th ed. England: Pearson; 2002

[2] Air conditioning systems and equipment (in Arabic), General Organization for Vocational and Technical Training, General Department of Curriculum Design and Development, Kingdom of Saudi Arabia; 2008

[3] Owen M. Heating, Ventilating, and Air-Conditioning Systems and Equipment. USA: ASHRAE; 2012

[4] Seyam S. Types of HVAC systems, In: HVAC System. England: IntechOpen; 2018

[5] McDowall R. Fundamentals of HVAC Systems. England: Elsevier; 2006

[6] Carrier A. Air Handling Unit Manual 39 HQ. Turkey: Trane; 2015

[7] Levenhagen J. HVAC Control System Design Diagrams. USA: McGraw Hill Companies Inc.; 1999

[8] Chua KJ, Chou SK, Yang WM, Yan J. Achieving better energy efficient air conditioning - a review of technologies and strategies. Applied Energy. 2013; **104**:87-104

Chapter 2

Investigation of Desiccant and Evaporative Cooling Systems for Animal Air-Conditioning

Muhammad Sultan, Hassan Niaz and Takahiko Miyazaki

Abstract

Productivity of livestock animals particularly sheep, goats, dairy, and beef cattle are usually affected due to high thermal/heat (sensible and latent) stresses, particularly in the developing countries. Different types of heating, ventilation, and air-conditioning (HVAC) systems are used worldwide depending upon the ambient air conditions to achieve the animals' thermal comfort. In this chapter, few low-cost options for the air-conditioning system and for farm building designs are discussed. Desiccant-based two air-conditioning systems are considered i.e., standalone desiccant air-conditioning (D-AC) and M-cycle assisted D-AC (M-DAC) system. The feasibility of both systems is thermodynamically checked for climatic conditions of Multan, Pakistan. Daily- basis data of ambient and processed air from both systems are analyzed for the thermal comfort of Holstein Friesian cows. Temperature humidity index (THI) is calculated to investigate the thermal heat stress conditions. Results showed that the D-AC system can be used efficiently in the humid climatic conditions with relatively moderate-to-low temperatures. On the other hand, the M-DAC system can be used in humid climatic conditions with relatively high-temperature conditions. It is important to mention that the typical direct evaporative cooling systems can be obviously low-cost options in case of dry climatic conditions.

Keywords: desiccant, evaporative cooling, air-conditioning, animal, M-cycle, thermal comfort

1. Introduction

Air-conditioning (AC) is a basic need for thermal comfort of humans as well as for animals. Living space per capita in case of human is decreasing due to population increase; therefore, it is difficult to provide huge space to animals with high natural ventilation. Thus, compact size farms with good thermal comfort conditions are needed. For this purpose, many air-conditioning systems are being used globally. The conventional AC systems use almost half of the total energy which is a huge amount of primary energy. So, energy efficient AC systems are principally required for animals particularly for developing countries. In case of Pakistan, animals contribute about 70–75% of the total agricultural GDP share [1] which is huge number. Livestock is sometime neglected area of research in developing countries e.g., Pakistan, Bangladesh, India, African countries etc. Therefore, the benefit from the livestock sector of such countries can be significantly increased by providing low-cost sustainable farm technologies [2]. Animals' thermal comfort includes the mechanisms of metabolism rate, skin heat transfer, rate of respiration, genetic factor and nature of feed, etc., [3–5]. That's why, it is important to air condition the space for animals to enhance the output products e.g., milk, fertility, meat and other related products etc., [2, 4, 6].

Figure 1 shows the illustration of the different heat transfer phenomena between environment and animals and the psychrometric thermal comfort zones for different animal species [7]. Most commonly heat transfer phenomena are considered which govern the fundamental equations for load calculations. Similarly, comfort zones elaborate the limit of temperature and relative humidity required for the ideal growth of different species [5, 8].

This chapter focuses on ideal heat transfer and comfort zones for the different species of animals. A set of equations is used for the measurement of the heat load

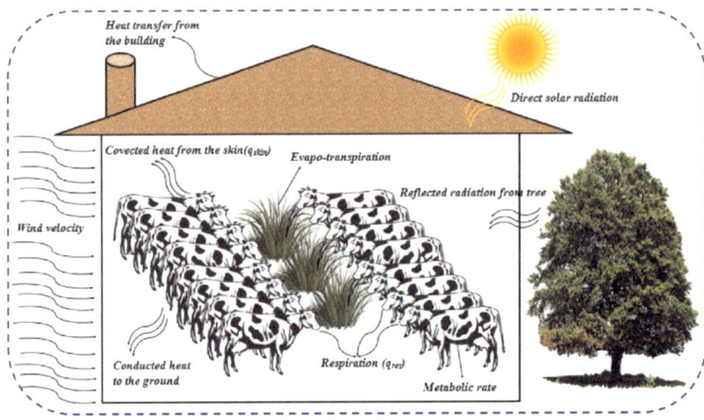

A

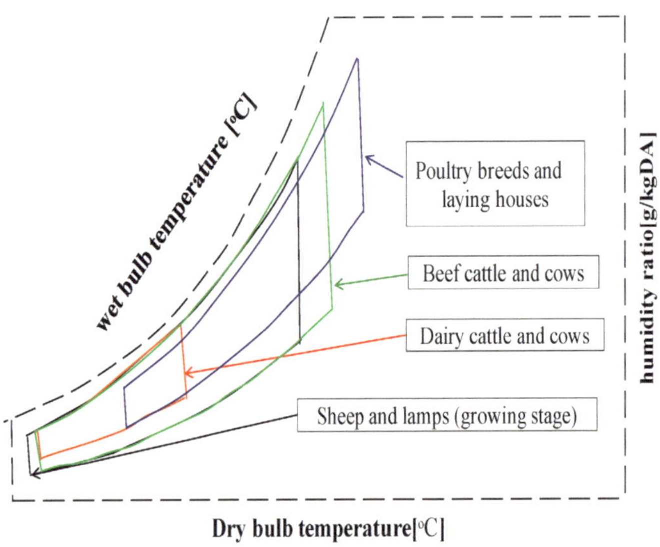

B

Figure 1.
Simplified illustration of: (A) animals heat transfer phenomena, and (B) psychrometric thermal comfort zones for different animals.

calculation for farm animals (Holstein Friesian Cow and Poultry application). Evaporative cooling and desiccant based air-conditioning systems are discussed according to the thermal comfort requirements for the animals. Jurinak model [9–13] is used for the evaluation of desiccant block and simplified correlations are used for performance evaluation of heat exchanger [14] and M-cycle cooling system [5]. Building designs and associated key factors affecting are also discussed for the subjected application. Moreover, the feasibility of systems is checked for the climatic conditions of Multan, Pakistan. Some other authors [15–18] also evaluate for animal applications the different AC systems (desiccant based and evaporative cooling-based systems) for different ventilation arrangements.

The novelty of this book chapter is to introduce the methods to calculate the heat load calculations for Holstein Friesian cows and poultry applications for the climatic conditions of Multan, Pakistan. The AC systems (standalone and combined) proposed for animal air-conditioning are not used for discussed applications before for the climatic conditions of Multan, Pakistan.

2. Ventilation rate and building designs for animal housings

2.1 Significance of ventilation rate

Ventilation is one of the important techniques used to control the thermal stresses in case of animals. Lots of studies have showed the significance of ventilation rate for achieving the ideal thermal comfort, and thereby, ventilation rate is considered an essential parameter for animal air-conditioning [5, 15, 16–19]. For example, **Figure 2** shows the factors affecting the selection of ventilation rate [20]. The suffocation or level of O_2/CO_2 is usually controlled by the air flow rate which must be designed precisely depending upon the nature of application. For livestock applications, ventilation is highly required to avoid the heat stress condition especially dairy cattle [17–20]. Dairy products, meat production, and other livestock applications need enough oxygen for longer storage period especially at commercial level. The ventilation rate depends upon the nature of the application and the climatic condition of the area for which the AC is needed [5, 21, 22]. It is considered the key parameter to measure and to create the thermal comfort.

2.2 Climate control and farm building designs

Advanced animals' AC options are limited in developing countries due to high initial and maintenance cost. Small farmers/stake holders cannot afford much cost. In addition, building designs are not very efficient in terms of providing thermal comfort to the farm animals [5, 7, 23]. In many developing countries including Pakistan, open side walls covered with sieve like material (natural ventilation) are used for cross flow of air with a flat roof [24]. Natural ventilation is good in maintaining the temperature, humidity and suffocation in cattle barns especially [23]. Instead, mechanical ventilation (active or passive) avoids the excess amount of carbon dioxide and ammonia in the air which can cause diseases and growth rate declination. Furthermore, the building designs can be further modified with respect to ventilation requirements and thermal comfort requirements [5, 7, 8, 23]. In this regard, researchers use different types of ventilation orientations [3, 5, 23, 25].

First type of ventilation is base ventilation as shown in **Figure 3a**. Negative pressure is created by lateral fans inside the house ejecting the exhaust air outside (brown arrows). Supply clean air (green arrows) is entered through the windows on the opposite wall that creates ventilation flow in the cross direction that guarantees

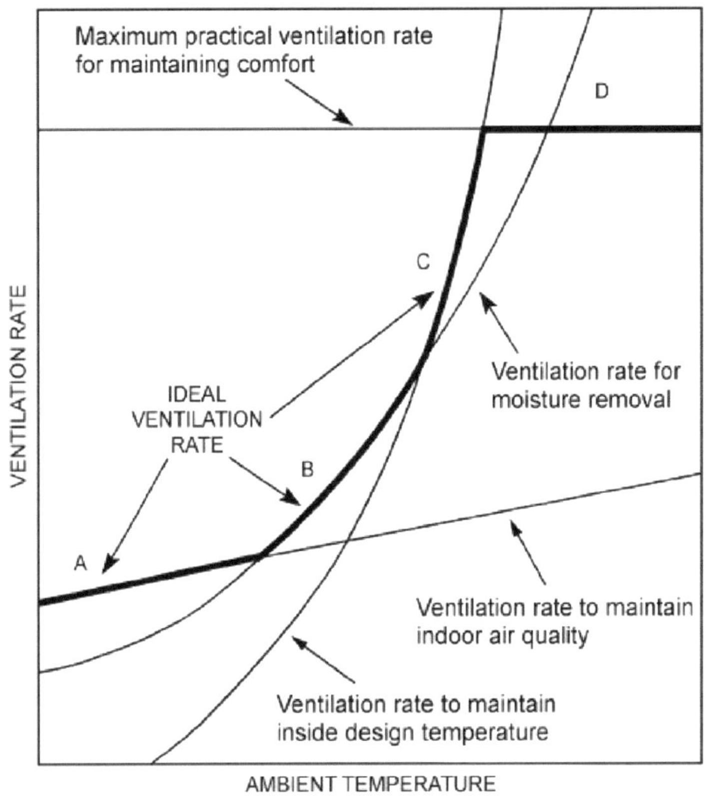

Figure 2.
Selection of ventilation rate for livestock application [20].

indoor air quality control. Base ventilation in rectangular shaped farm sheds which are used with fans at one end and the holes at roof and opposite side of the building which allows the cross flow of the air. Base flow is normally used to control indoor air quality for the circulation of the air which reduces the contaminants of harmful gases in the air [3, 25]. The second type of the ventilation is tunnel ventilation as shown in **Figure 3b**. This is due to the horizontal movement of air in the farm shed. There are greater number of fans used than that of base ventilation. It can be seen that the hot air is ejected through the fans placed at the end of the house outside (red arrows), that creates a negative pressure inside the building. Due to the pressure difference, fresh outdoor air is entered (blue arrows) through the inlets decreasing the indoor air temperature. In this type of ventilation, air temperature is decreased due to the removal of thermal emission of the animals and wind-chill effect is produced [3, 7, 14, 26]. Air speed does not exceed 3 ms^{-1} in this type of ventilation, otherwise thermal discomfort occurs. When tunnel ventilation is not enough to provide thermal comfort in the farm sheds, evaporative cooling systems are additionally used. The evaporative cooling (EC) shown in the **Figure 3c** converts sensible heat into latent which is carried out by evaporative pads. It can be seen that the negative pressure is created inside the house by the fans which lets the hot external air pass through the wet evaporative pads that decreases the air temperature (red/blue arrows). The indoor air temperature is decreased and is expelled by the tunnel ventilation fans (red arrows). During the warmest periods, ambient air passes through the evaporative pads where air is cooled due to evaporation of the water [3, 7, 8, 25]. This decreases the outlet air and consequently, indoor air temperature also decreases. Cooling pads are configured on the larger side of the

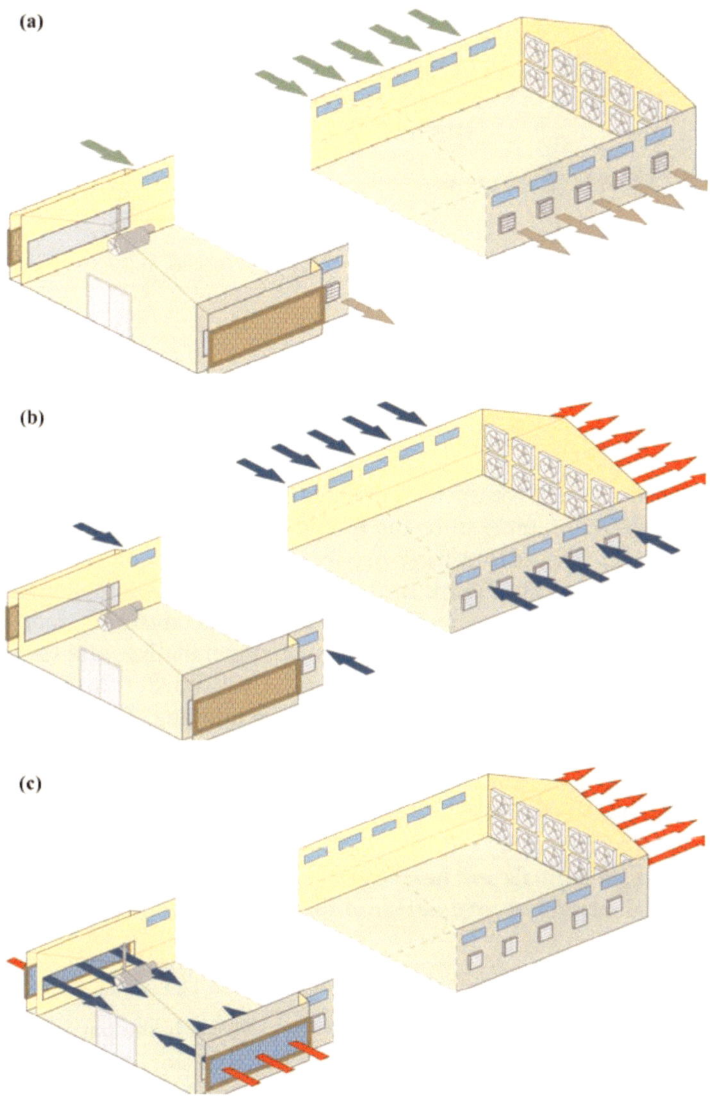

Figure 3.
Illustration of operating principle for: (a) base ventilation, (b) tunnel ventilation, and (c) evaporative cooling [3].

building wall opposite to the tunnel ventilation fans which allows the passage of cooled air due to negative pressure [3].

The side view of the evaporative pads used in the farm sheds is shown in **Figure 4a** and the schematic diagram of the evaporative cooling system as shown in the **Figure 4b**. Evaporative cooling (EC) is a system of water vapor evaporating into air that cools the air by water evaporation. In a study [7, 16], authors used honeycomb like structure for the evaporation supportive channel. In that channel, water is supplied from upside down with the help of a sprinkler. Air is passed through that honeycomb structure from one side to another side. The fan is used for the controlled flow of air in the system. There is a cylinder of water that collects water whereas a water pump is used to flow the water from the tank to the sprinkler as shown in **Figure 4b**. This can be a low-cost method to attain thermal comfort conditions in controlled sheds in hot climatic conditions except in rainy season (monsoon season).

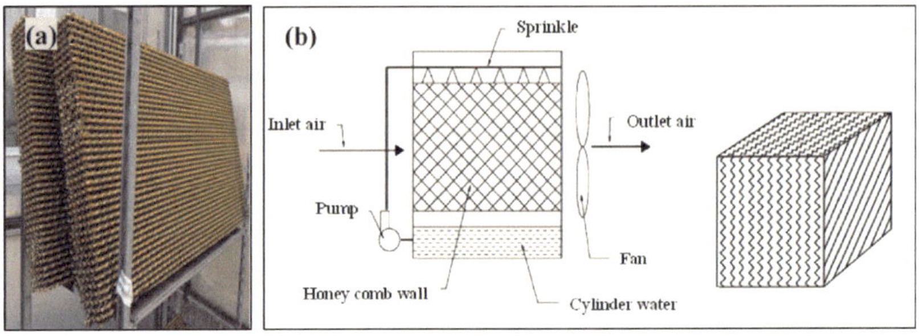

Figure 4.
A view of: (a) cellulose evaporative pads [3]; and (b) schematic diagram of evaporative cooling system [7].

3. Methodology: animals' thermal comfort

Farm animals' thermal comfort is estimated for the viewpoint of optimum productivity. Different temperature and humidity values are required for different types of animals [7]. The heat transfer between farm animals and surrounding environment is expressed in **Figure 1a**. The heat transfer phenomenon between animal and environment is associated with conduction, convection, radiation, evaporation, evapotranspiration, wind velocity and metabolism rate (sensible and latent energy/heat transfer through animal skin). Heat transfer through building is also considerable while designing the animals' farm building. Temperature humidity index (THI) is one of the key parameters to measure the environmental condition for the farm animals. The thermal threshold values for Holstein Friesian cows are 72 for thermal neutral region and above than 72 for heat stress region [6]. The total heat load is calculated by following Eqs. (1)–(9). Total amount of heat is the sum of heat load for animals and heat load for building as shown in Eq. (8). Eq. (1) is used for the calculation of total heat load for animals [14] as follows:

$$Q_a = q_{skin} + q_{res} + S \tag{1}$$

where, Q_a is total amount of heat for animals (kW), q is the partial heat load for the different sections (kW), and S is the amount of heat stored in the body (for ideal case: S = 0). The subscripts "a" and "res" denote animals and respiration, respectively. Heat load for respiration is calculated by the Eqs. (2) and (3) [14, 19] as given by:

$$q_{skin} = S_A M_R \tag{2}$$

$$S_A = 0.147\ W^{0.57} \tag{3}$$

where, W is the weight of animal (kg), S_A is the surface area of skin (m^2), and M_R denotes the metabolism rate (met). The heat of respiration which is the sum of evaporation and convection heat loss is calculated by the Eqs. (4)–(6) [14].

$$q_{res} = C_{res} + E_{res} \tag{4}$$

$$C_{res} = [0.0014\ M_R(34 - T_a)] \tag{5}$$

$$E_{res} = [0.0173\ M_R(5.87 - P_a)] \tag{6}$$

where, C is the convection heat loss from respiration [$W/(h.m^2)$], E is the evaporation heat loss from respiration [$W/(h.m^2)$], and M_R is metabolic rate (met). The terms T_a and P_a represents ambient air temperature (°C) and vapor pressure (kPa), respectively. Total amount of heat transfer through buildings is calculated by Eq. (7) as given by:

$$Q_b = U\,A\,\Delta T \tag{7}$$

where, Q_b is the total heat load for buildings (kW), U is overall heat transfer coefficient for building [W/(m^2.K)], A is the area of building (m^2), and ΔT is the temperature difference (°C or K). Total amount of heat required for the thermal comfort of farm animals is calculated by Eq. (8) as follows:

$$Q = Q_a + Q_b \tag{8}$$

where, Q is the total amount of heat load (kW). Temperature humidity index (THI) is important parameter which is used to measure the environment condition i.e., whether the region is in heat stress condition or in thermal neutral condition. THI is calculated by the Eq. (9) [4, 6] as given below:

$$\mathrm{THI} = (1.8\,\mathrm{T} + 32) - [(0.55 - 0.0055\,\mathrm{RH})(1.8\,\mathrm{T} - 26)] \tag{9}$$

where, THI is the temperature humidity index [−], T is the temperature (°C), and RH is the relative humidity (%). As the THI is dependent on temperature and humidity, therefore, typical values of THI can also be found from the literature as given by reference [6]. A region is considered thermal neutral region when THI < 68, and the region is thermally stable region when THI = 68–72. The region is moderate heat stress region when THI = 72–80 and the region is severe heat stress region when THI > 80.

Similarly, to cows and cattle, THI is also calculated for poultry birds [23, 27]. THI equation for poultry applications is given by Eq. (10) as reported in Ref. [27]:

$$\mathrm{THI} = 0.6\mathrm{T}_{db} + 0.4\mathrm{T}_{wb} \tag{10}$$

where, T is the temperature (°C). The subscripts "db" and "wb" denote dry-bulb and wet-bulb, respectively. The interior dry-bulb and wet-bulb temperatures can be calculated according to the Eqs. (11) and (12). For the air-conditioned space by evaporative cooler, interior dry-bulb temperature and humidity ratio is calculated from sensible and latent heat equations for the building as follows:

$$t_{i,\,db} = t_{o,\,db} + \frac{Q_s(\mathrm{age}, t_{o,\,db})\,n_{birds} m_{birds} - \beta\,\dot{m}_w\,h_{fg}}{\dot{m}_a c_p} \tag{11}$$

$$W_i = W_o + \frac{Q_L(\mathrm{age}, t_{i,\,db})\,m_{birds} - \beta\,\dot{m}_w\,h_{fg}}{\dot{m}_a h_{fg}} \tag{12}$$

where, t is the temperature (°C), Q is the heat production (W), n_{birds} is the number of birds, and m_{birds} is the mass per bird (kg). The term $\beta\,\dot{m}_w$ represents the mass flow rate of the moisture in the air and h_{fg} denotes latent heat of vaporization i.e., nearly 2.43 MJ kg^{-1} at 30°C. The term C_p shows specific heat of air i.e., 1006 J kg^{-1} K^{-1}. The subscripts "i," "o," "db," "wb," "S," "L," "a," and "w" denote inside, outside, dry-bulb, wet-bulb, sensible, latent, air, and water, respectively.

4. Proposed air-conditioning systems

There are many innovative modern AC technologies that are used globally in addition to vapor compressor-based systems. These systems mainly use the conception of evaporative cooling adsorption cooling and desiccant AC. These systems are not explored extensively in developing countries. Direct evaporative cooling system (swamp cooler) is most common system used worldwide wherever dry climatic conditions exists. Thermal comfort and/or temperature/humidity control are becoming more popular and demanding day by day in agriculture sector particularly for product storage, post-harvest processing, farm animals' buildings, as well as transportation of dairy, meat and food products [1, 15, 28, 29]. Thermal comfort for agricultural products and livestock requires cooling (temperature control) as well as humidity control and ventilation. These requirements change with the change in application and climatic conditions from one to another. Therefore, above mentioned thermally driven AC systems can be used for this purpose [5, 7, 8, 15].

4.1 Standalone desiccant air-conditioning (D-AC) system

Standalone desiccant air-conditioning (D-AC) system consists of a desiccant unit (wheel or block) mostly with the addition of heat-exchanger (HX). There is no cooling device principally used in the D-AC system. The D-AC system is used for the humidity control. It is relatively more efficient in the humid regions where humidity control is primarily required. It may also feasible for the applications in which humidity control is mainly concerned (i.e., storage of onion and leafy vegetables) [8, 11, 15, 26].

The schematic diagram of the D-AC system is shown in **Figure 5a**. The standalone D-AC system consists of two desiccant blocks (DB-1 and DB-2) used for the dehumidification of air. The dehumidified (relatively warmer) air is passed through the HX where the temperature of the air becomes equal to the ambient air (ideally). After that, the cooled air may be used for the desired application. The total process is known as the air dehumidification cycle. However, in regeneration cycle, the desiccant unit is regenerated by passing the hot air. Therefore, a heating unit is used D-AC system to heat up the air. Then the heated air is passed through the desiccant unit to remove its moisture in order to be used for cyclic process. The heating unit usually uses thermal energy sources i.e., waste heat, biogas, biomass, direct thermal energy (from solar or likewise), geothermal energy, etc. The regeneration depends upon the material type, ambient air conditions and other factors related to the material properties. Thus, regenerations temperature may be changed with the change in these factors.

A set of equations given by Jurinak model [9–13, 30] is used for performance evaluation of the desiccant unit.

$$F_{1,\,ip} = \frac{\varphi_1}{(T_{ip} + 273.15)^{1.49}} + \tau_1 \left(\frac{w_{ip}}{1000}\right)^{\gamma_1} \tag{13}$$

$$F_{2,\,ip} = \frac{(T_{ip} + 273.15)^{1.49}}{\varphi_2} - \tau_2 \left(\frac{w_{ip}}{1000}\right)^{\gamma_2} \tag{14}$$

$$\eta_{F1} = \frac{F_{1,\,2} - F_{1,\,1}}{F_{1,\,8} - F_{1,\,1}} \tag{15}$$

$$\eta_{F2} = \frac{F_{2,\,2} - F_{2,\,1}}{F_{2,\,8} - F_{2,\,1}} \tag{16}$$

where, F_1 and F_2 are combined potentials [−], and η_{F1}, η_{F2} are the efficiencies of the combined potentials. The term "*ip*" indicates the state of air in the system (1, 2, and 8). Eqs. (13)–(16) are used for the performance evaluation of the

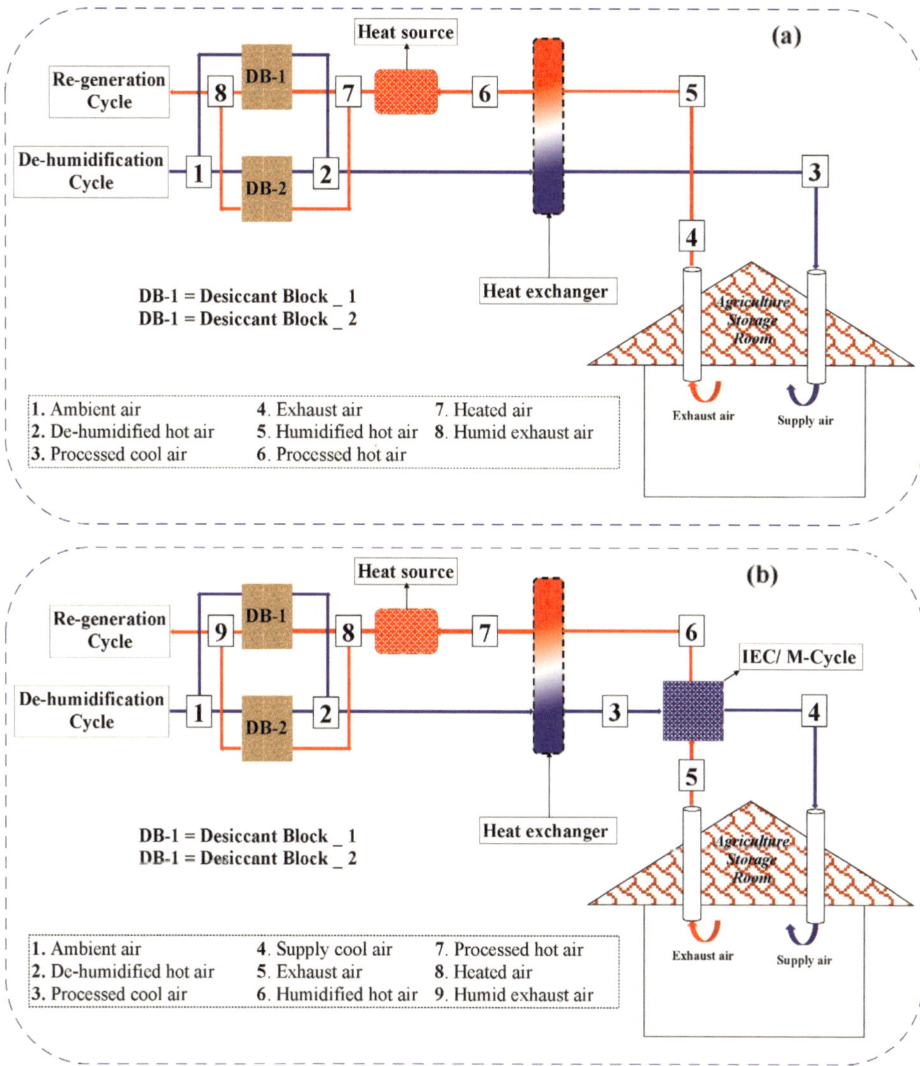

Figure 5.
Schematic diagram for: (a) standalone desiccant AC (D-AC) system and (b) M-cycle assisted desiccant AC (M-DAC) system.

desiccant unit with the help of MS excel solver. The typical values for the efficiencies are taken as: $(\eta_{F1}, \eta_{F2}) = (0.05, 0.95)$ for the high performance of desiccant unit as reported by [8–13]. The values for the constants are given in **Table 1**.

The performance of the desiccant unit was evaluated by given equations and also can be validated for the different climatic conditions (especially for developing countries). The processed air through the desiccant unit has higher temperature and low relative humidity as comparison with the ambient air conditions. The hot air passes through the heat exchanger where the temperature of the air decreases, but the humidity ratio remains constant. The temperature of the air is calculated by Eq. (17) [14].

$$T_3 = T_{2,db} - \varepsilon_{HX}\left(T_{2,db} - T_{1,db}\right) \tag{17}$$

where, "ε" is the efficiency of the HX whose value is taken as 0.90. The T_3 is the air temperature (°C) which has to be calculated. The T_2 is the temperature (°C) of

Parameter	Value	Parameter	Value
φ_1 [−]	−2865	φ_2 [−]	6360
τ_1 [−]	4.344	τ_2 [−]	1.127
γ_1 [−]	0.8624	γ_2 [−]	0.07969

Table 1.
Constants of Jurinak model for the evaluation of desiccant unit.

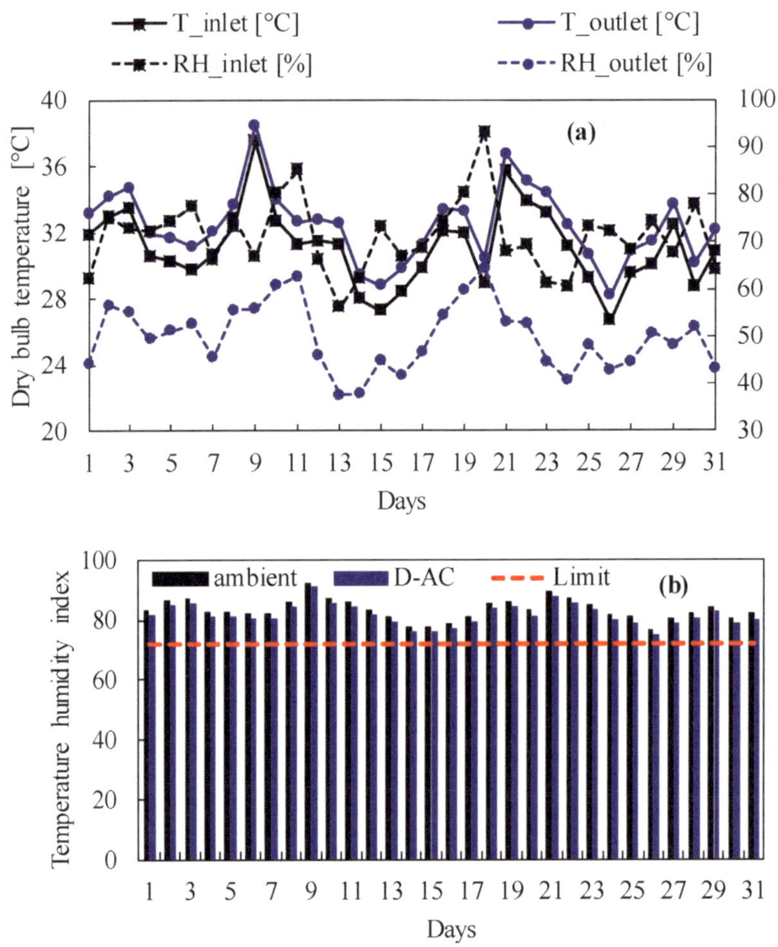

Figure 6.
Results of D-AC system for July (daily-basis) for climatic conditions of Multan, Pakistan: (a) temperature-humidity profile and (b) THI profile.

the dehumidified air after passing through the desiccant and the T_1 is the ambient air temperature (°C).

Figure 6 shows the results of the standalone D-AC system for July (daily based) according to climatic conditions of Multan, Pakistan. The data of temperature and humidity is the average data of 20 years obtained from the METRONOME software. The D-AC is evaluated in terms of system analysis and also THI analysis which elucidate the D-AC system's applicability for animal air-conditioning. **Figure 6a** shows the difference between ambient air temperature and relative humidity with the product air temperature and relative humidity. It is clear that the product air temperature is higher than ambient air temperature due to latent heat of adsorption [15, 29, 30]. However, relative humidity of product air from the D-AC system is lower than ambient air relative humidity due to adsorption of water by desiccant material. Feasibility of the standalone D-AC system for animal air-conditioning is checked by the THI analysis which clarifies the system's feasibility for desired application. **Figure 6b** shows the THI values of ambient and product air, as well as the permissible limit of heat stress. The permissible limit of heat stress for cows is 72. The THI of the ambient air as well as product air is higher than the permissible limit for the whole month. So, it may not be suitable for the ambient conditions of Multan for the month of July.

4.2 M-cycle assisted desiccant air-conditioning (M-DAC) system

The schematic diagram of Maisotsenko cycle (M-cycle) assisted desiccant air conditioning (M-DAC) system is shown in **Figure 5b**. It consists of desiccant unit with the addition of HX and an exclusive M-cycle unit. The M-cycle cooling device lowers down the temperature of the processed air from the HX. Therefore, this system can be used for the humidity and temperature control. This can be efficient in hot and humid climatic regions where temperature and humidity control are essential. It may also be efficiently feasible for the applications in which humidity and temperature control is concerned (i.e., agricultural storage and livestock applications) even in hot and humid climatic conditions [5, 11, 16, 26]. The system consists of two desiccant blocks (DB-1 and DB-2) used for the air dehumidification and regeneration purposes. The dehumidified air is passed through the heat-exchanger where the temperature of the air becomes nearly equal to the ambient air. The air is further passed through the M-cycle unit where air is further cooled up to the desired temperature and humidity conditions to be used for desired application. The desiccant unit is regenerated by hot air and therefore a heating unit is principally required in this system which is supposed to be operated on low-cost thermal energy options. The heated air is passed through the desiccant unit to remove moisture from the desiccant material for cyclic usage.

It is important to mention that the M-cycle is an advanced indirect evaporative cooling conception that cools down the temperature of the working air up to the dew point by capturing energy from the air step by step as the humidity of the system remains constant [5, 16, 31, 32]. M-cycle is well-known in the air-conditioning field due to its working range for the dew point evaporative cooling. The details can be found from authors' previous work as reported in [16]. Many researchers used different structures (channels) and materials for the manufacturing of the M-cycle channels as well as flow arrangements [8, 16, 31, 32]. In one-way configuration of M-cycle unit, the ambient air is passed through the dry channels (in cross flow direction to the wet channel) and then part of this air is mixed into the wet channel. The cyclic process brings the product air temperature to the dew point (theoretically) of the ambient air temperature. In another way configuration of M-cycle unit, the wet channel is sandwiched between two dry channels. Process air is passed firstly through one of the dry channels followed by the wet channel. In the cyclic process, this will lower the temperature of the product air passing through the other dry channel up to the ambient air wet bulb (finally approaches to the dew point) temperature. In authors' previous work [5], a simplified correlation was developed for performance evaluation of the M-cycle unit as given by following Eqs. (18) and (19):

$$T_o = A_1 + B_1(T_i) + C_1(H_{spc}) \tag{18}$$

$$Q = A_2 + B_2(T_i) + C_2(H_{spc}) \tag{19}$$

where T_i and T_o represent the air temperature (°C) at inlet and outlet of M-cycle channels, respectively. The term H_{spc} represents the specific humidity (g/kg-DA) and Q represents the specific cooling capacity. The terms A, B, and C are the constants for simplified correlations, and optimized values for these constants are given in **Table 2**.

Figure 7 shows the daily based analysis of M-cycle based D-AC system for the month of July for Multan, Pakistan. The data of temperature and humidity is the average data of 20 years obtained from the METRONOME software. The M-DAC is

Constant	Value	Constant	Value	Constant	Value
A_1 [−]	6.70	B_1 [−]	0.2630	C_1 [−]	0.5298
A_2 [−]	−5.48	B_2 [−]	0.7317	C_2 [−]	−0.5946

Table 2.
Numerical values of the constants of the simplified M-cycle correlations [5].

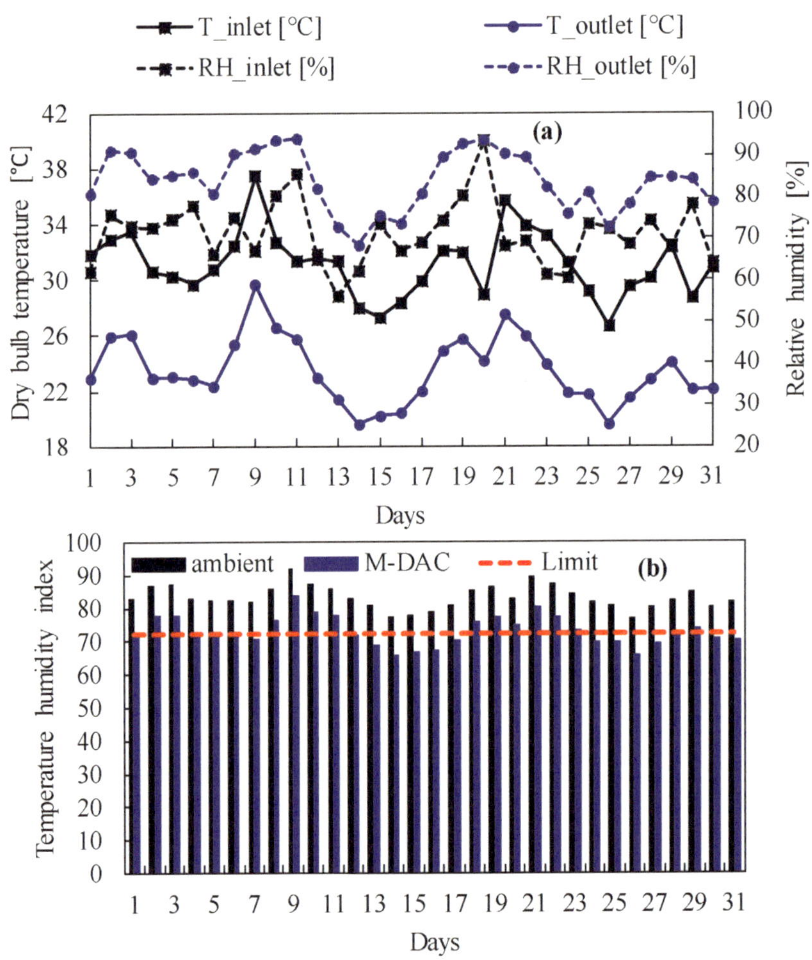

Figure 7.
Results of M-DAC system for July (daily-basis) for climatic conditions of Multan, Pakistan: (a) temperature-humidity profile and (b) THI profile.

evaluated in terms of system analysis and THI analysis which elucidate the M-DAC system's applicability for animal air-conditioning. **Figure 7a** shows the difference between ambient air temperature and relative humidity with the product air temperature and relative humidity. It is clear from the figure that the product air temperature is lower than ambient air temperature due to water evaporation [30, 33–35]. However, relative humidity of product air is higher than ambient air relative humidity which is required for animal air-conditioning.

Thereby, feasibility of the M-cycle based D-AC system for animal air-conditioning is checked by THI calculated from Eq. 9. **Figure 7b** shows the THI values of ambient air and product air, and permissible limit of heat stress is also shown in the figure. The permissible limit of heat stress for cows is 72. The figure clearly indicates that the THI of the ambient air is higher than the permissible limit for the whole month, while the THI of the product air is lower than permissible limit in most of the days. So, it may be feasible for the ambient conditions of Multan for the month of July (greater relative humidity and temperature).

The psychrometric representation of the product air of proposed AC systems according to the temperature and relative humidity values is shown in the **Figure 8** with the help of different markers. The values are obtained for the ambient conditions of Multan, Pakistan for the month of July. The black color line box shows the thermal comfort zone for animals. The circular marker shows the ambient air conditions, triangular marker shows the D-AC system's air conditions and diamond shaped marker shows M-DAC system's air conditions. The ambient conditions and D-AC output conditions are not able to provide the thermal comfort for animals. However, M-DAC output conditions are relatively suitable; therefore, this system can be applicable for animal air-conditioning subjected to the conditions.

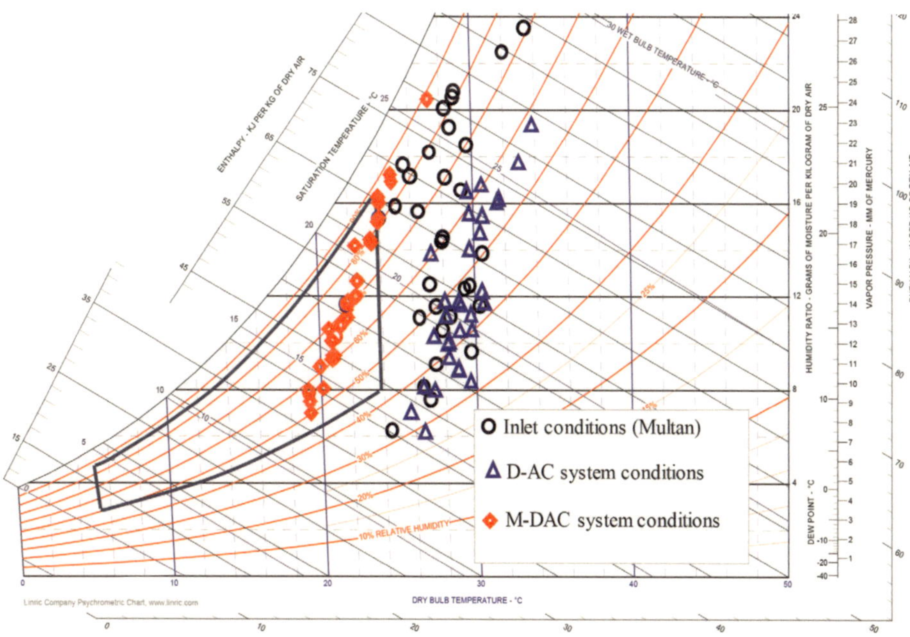

Figure 8.
Psychrometric plot of results identifying the feasibility of D-AC and M-DAC systems. Black line marked closed region shows animal air conditioning zone.

5. Conclusions

This chapter reviews the fundamental requirements for animals' air conditioning (AC) particularly for developing countries. The significance of ventilation rate and the associated animal housing designs are discussed from the view-point of energy-efficient climatic control systems. The study explores the literature related to temperature humidity index for Holstein Friesian cows. The importance of evaporative cooling systems is obvious in developing countries; however, this chapter highlights its usage/integration with the thermally-driven desiccant based open cycle systems. In this regard, the study focuses the climatic conditions of Multan, Pakistan. The performance of two kinds of evaporative and desiccant based AC systems is checked for the subjected application. Results indicate that the performance of the systems vary with the climatic conditions and systems' operation. The standalone desiccant AC (D-AC) system nicely dehumidifies the air which increases the temperature as well due to adsorption heat. However, the dry warm air is unable to provide the optimum thermal comfort conditions for the animals for most of the climatic conditions. However, M-cycle assisted desiccant AC (M-DAC) system dehumidifies the air on one side whereas it also control the sensible AC load on the other side. Thus, it makes it feasible to provide thermal comfort conditions for animals for most of the climatic conditions. Thus, it is energy efficient to use the evaporative cooling options in humid climates if integrated with desiccant units. The D-AC system should be used only for humid climatic conditions where latent load control (dehumidification) is mainly required. The D-AC system is not found effective for the summer conditions of Multan, Pakistan. However, it may be used for drying purposes as per authors' previous work. The M-DAC system is good to use in all kind of humid climatic conditions including the hotter regions. It has been found feasible for thermal comfort of cows for summer conditions including rainy season for Multan, Pakistan.

Conflict of interest

The authors declare no conflict of interest.

Author details

Muhammad Sultan[1*], Hassan Niaz[1] and Takahiko Miyazaki[2]

1 Department of Agricultural Engineering, Bahauddin Zakariya University, Multan, Pakistan

2 Faculty of Engineering Sciences, Kyushu University, Fukuoka, Japan

*Address all correspondence to: muhammadsultan@bzu.edu.pk; sultan@kyudai.jp

References

[1] Ministry of Finance, Government of Pakistan. [Online]. Available from: http://www.finance.gov.pk/ [Accessed: 11 February 2019]

[2] De Rensis F, Lopez-Gatius F, García-Ispierto I, Morini G, Scaramuzzi R. Causes of declining fertility in dairy cows during the warm season. Theriogenology. 2017;**91**: 145-153

[3] Costantino A, Fabrizio E, Ghiggini A, Bariani M. Climate control in broiler houses: A thermal model for the calculation of the energy use and indoor environmental conditions. Energy and Buildings. 2018;**169**:110-126

[4] Karimi M, Ghorbani G, Kargar S, Drackley J. Late-gestation heat stress abatement on performance and behavior of Holstein dairy cows. Journal of Dairy Science. 2015;**98**(10):6865-6875

[5] Sultan M, Miyazaki T, Mahmood MH, Khan ZM. Solar assisted evaporative cooling based passive air-conditioning system for agricultural and livestock applications. Journal of Engineering Science and Technology. 2018;**13**(3):693-703

[6] Palacio S, Bergeron R, Lachance S, Vasseur E. The effects of providing portable shade at pasture on dairy cow behavior and physiology. Journal of Dairy Science. 2015;**98**(9):6085-6093

[7] Sultan M, Miyazaki T. Energy-efficient air-conditioning systems for nonhuman applications. Refrigeration. Janeza Trdine: InTech Open; 2017:97-117. DOI: 10.5772/intechopen.68865

[8] Niaz H, Sultan M, Miyazaki T, Khan ZM. Study on desiccant air-conditioning system for livestock application in Pakistan. In: 2nd International Conference on Energy Conservation and Efficiency (ICECE, 2018); 2018. pp. 19-22

[9] Panaras G, Mathioulakis E, Belessiotis V. Proposal of a control strategy for desiccant air-conditioning systems. Energy. 2011;**36**(9):5666-5676

[10] Panaras G, Mathioulakis E, Belessiotis V. Solid desiccant air-conditioning systems—Design parameters. Energy. 2011;**36**(5): 2399-2406

[11] Panaras G, Mathioulakis E, Belessiotis V. Achievable working range for solid all-desiccant air-conditioning systems under specific space comfort requirements. Energy and Buildings. 2007;**39**(9):1055-1060

[12] Panaras G, Mathioulakis E, Belessiotis V, Kyriakis N. Theoretical and experimental investigation of the performance of a desiccant air-conditioning system. Renewable Energy. 2010;**35**(7):1368-1375

[13] Panaras G, Mathioulakis E, Belessiotis V, Kyriakis N. Experimental validation of a simplified approach for a desiccant wheel model. Energy and Buildings. 2010;**42**(10): 1719-1725

[14] ASHRAE. Handbook—Fundamentals. Atlanta, GA, USA: American Society of Heating, Refrigerating and Air-conditioning Engineers. Inc.; 2017

[15] Sultan M, El-Sharkawy II, Miyazaki T, Saha BB, Koyama S. An overview of solid desiccant dehumidification and air conditioning systems. Renewable and Sustainable Energy Reviews. 2015;**46**:16-29

[16] Mahmood MH, Sultan M, Miyazaki T, Koyama S, Maisotsenko VS.

Overview of the Maisotsenko cycle—A way towards dew point evaporative cooling. Renewable and Sustainable Energy Reviews. 2016;**66**:537-555

[17] Corrales-hernández A et al. Physiological responses in horses, donkeys and mules sold at livestock markets. International Journal of Veterinary Science and Medicine. 2018; **6**:97-102

[18] Southall RG. An assessment of the potential of supply-side ventilation demand control to regulate natural ventilation flow patterns and reduce domestic space heating consumption. Energy and Buildings. 2018;**168**:201-214

[19] Wang X, Zhang G, Choi C. Effect of airflow speed and direction on convective heat transfer of standing and reclining cows. Biosystems Engineering. 2018;**167**:87-98

[20] ASHRAE. Environmental control for animals and plants. Handbook—Fundamentals. Atlanta, GA, USA: American Society of Heating, Refrigerating and Air-Conditioning Engineers. Inc.; 2005

[21] Deng Y, Feng Z, Fang J, Cao SJ. Impact of ventilation rates on indoor thermal comfort and energy efficiency of ground-source heat pump system. Sustainable Cities and Society. 2018;**37**: 154-163

[22] Mukhtar A, Ng KC, Yusoff MZ. Design optimization for ventilation shafts of naturally-ventilated underground shelters for improvement of ventilation rate and thermal comfort. Renewable Energy. 2018;**115**:183-198

[23] Abid M, Wajid HA, Khan ND, Akhtar S, Zahid MA, Usman M. Optimization of ventilation system for existing environmentally controlled poultry sheds in Pakistan. World Applied Sciences Journal. 2013;**24**(9): 1221-1233

[24] Rong L, Bjerg B, Batzanas T, Zhang G. Mechanisms of natural ventilation in livestock buildings: Perspectives on past achievements and future challenges. Biosystems Engineering. 2016;**151**:200-217

[25] Baniyounes AM, Ghadi YY, Rasul MG, Khan MMK. An overview of solar assisted air conditioning in Queensland's subtropical regions, Australia. Renewable and Sustainable Energy Reviews. 2013;**26**:781-804

[26] Pandelidis D, Anisimov S, Worek WM, Drąg P. Analysis of different applications of Maisotsenko cycle heat exchanger in the desiccant air conditioning systems. Energy and Buildings. 2017;**140**:154-170

[27] Gates RS, Zhang H, Colliver DG, Overhults DG. Regional variation in temperature humidity index for poultry housing. Transactions of ASAE. 1995; **38**(1):197-205

[28] Qi R, Tian C, Shao S. Experimental investigation on possibility of electro-osmotic regeneration for solid desiccant. Applied Energy. 2010;**87**(7): 2266-2272

[29] Sultan M, Miyazaki T, Koyama S. Optimization of adsorption isotherm types for desiccant air-conditioning applications. Renewable Energy. 2018; **121**:441-450

[30] Sultan M, Miyazaki T, Koyama S, Khan ZM. Performance evaluation of hydrophilic organic polymer sorbents for desiccant air-conditioning applications. Adsorption Science and Technology. 2018;**36**(1–2):311-326

[31] Pandelidis D, Anisimov S, Worek W, Drąg P. Numerical analysis of a desiccant system with cross-flow Maisotsenko cycle heat and mass exchanger. Energy and Buildings. 2016; **123**:136-150

[32] Pandelidis D, Anisimov S, Worek W, Drąg P. Comparison of desiccant air conditioning systems with different indirect evaporative air coolers. Energy Conversion and Management. 2016;**117**:375-392

[33] Sultan M, Miyazaki T, Saha BB, Koyama S. Steady-state investigation of water vapor adsorption for thermally driven adsorption based greenhouse air-conditioning system. Renewable Energy. 2016;**86**:785-795

[34] Sultan M, El-Sharkawy II, Miyazaki T, Saha BB, Koyama S, Maruyama T, et al. Water vapor sorption kinetics of polymer based sorbents: Theory and experiments. Applied Thermal Engineering. 2016; **106**:192-202

[35] Sultan M, Miyazaki T, Saha BB, Koyama S, Kil HS, Nakabayashi K, et al. Adsorption of difluoromethane (HFC-32) onto phenol resin based adsorbent: Theory and experiments. International Journal of Heat and Mass Transfer. 2018;**127**:348-356

Chapter 3

Solar Pond Driven Air Conditioning Using Seawater Bitterns and $MgCl_2$ as the Desiccant Source

Esam Elsarrag, Opubo N. Igobo and Philip A. Davies

Abstract

Solar energy is used for a wide range of applications such as electricity production, desalination, cooling, heating, etc. Solar-based technologies are widely spread and increasingly studied in the industry. This theoretical and experimental study focuses on solar ponds as a desiccant and low-grade energy source. A thermal model has been developed for a salinity gradient solar pond (SGSP) with a non-convective zone split into 10 sub-zones. A solar pond was constructed and used as a case study for the validation of the predictive model capabilities. The dimensional characteristics of the pond, as well as the solar radiation intensity and ambient temperature data obtained from the meteorological data, were used to produce the solar pond's zone thermal behaviour data. With regards to the thermal behaviour measurements obtained from the solar pond, the predicted data were found to be higher. There is a significant difference between the real-world and meteorological data obtained, the difference between the predicted and real-world pond temperature data was also attributed to the fact that the actual absorbed solar radiation was reduced due to wall shading effect, turbidity and insufficient duration of operation of the pond. In the following year, the stored heat from the previous summer would be expected to improve thermal storage values obtained partially.

Keywords: solar energy, solar pond, desiccant cooling, carbon emissions, humid climates

1. Introduction

The continuous increase in energy demand raises the need for alternative energy sources. Energy consumption from air conditioning will continue to increase which will raise the need for innovative solutions in many industrial sectors. One of the significant sources alternative widely and increasingly studied in the industry is solar-based technologies. Solar energy is used for a wide range of applications such as electricity production, desalination, cooling, heating, etc. Solar ponds are relatively simple and yet effective thermal storage [1].

A solar pond consists of a body of salty water which collects solar energy and converts it for thermal storage. A solar pond can be either convective or non-convective. The principle of convective solar ponds largely depends on the water's surface being covered by an insulating material to store the system's collected heat; the most commonly used solar pond of this type is the shallow solar pond. Non-convective solar ponds, however, operate by limiting the process of natural convection by the use of a water collector or storage medium.

In this study, a salinity gradient solar pond (SGSP) is designed and constructed. The SGSP is a large, low-cost solar-thermal energy collection and storage system which consists of a large body of saltwater (with salinity gradient) such that solar energy incident on the pond is partially transmitted to the bottom of the pond where a portion (20–30%) of it is absorbed.

A typical SGSP consists of three regions: the upper-convective zone (UCZ), the non-convective zone (UCZ) and the lower-convective zone (LCZ). The UCZ is the topmost layer of the solar pond and is a relatively thin layer (usually 0.1–0.5 m) which contains almost no salinity (about 0–5% concentration). The NCZ is the middle region (of about 0.7–1.5 m thickness) and has an increasing concentration (salinity gradient) relative to the UCZ, and it also acts as insulation on the LCZ, because convection motion in the NCZ is ideally suppressed if the concentration gradient is sufficiently large. The LCZ is the layer in which the salt concentration is the greatest (about 26%), and there is no concentration gradient in it, as depicted in **Figure 1** [3].

Resultantly, large amounts of heat can potentially be stored in these systems [4]. The utilisation of solar ponds as energy carriers was first conceived by Tabor [5, 6] when observing the natural phenomenon in a Hungarian lake at the turn of the twentieth century. The studies proposed the possibility of simulating the natural phenomenon experimentally for energy production. The authors laid the all-important foundations of current solar pond research conducted to date. However, the work of Tabor [6] was not developed as far as its potential would allow. Nevertheless, in recent decades, with the current global search for alternative energy, as aforementioned, solar ponds have gained a substantial increase in interest in academia and industry [1, 3, 7].

A wide variety of models have been developed to investigate the energy properties and potential capabilities of solar ponds. Rabl and Nielsen [4] first examined the possibility of utilising the thermal energy from solar ponds for space heating, by deriving a set of formulae specific to a certain type of salt gradient pond. The pond,

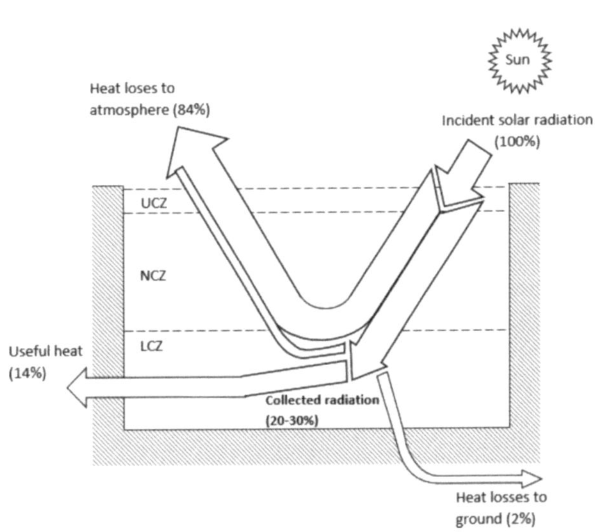

Figure 1.
SGSP solar radiation distribution [2].

in essence, consisted of two homogeneous layers: an upper convecting layer (a lower salt content) and a lower non-convecting layer which acts as the thermal storage part of the pond.

The authors considered the varying temperature ranges that a standard-sized dwelling would encounter seasonally in varying locations and climates. They subsequently reported that the use of a solar pond would undoubtedly provide adequate heating at prices, competitive with those of conventional heating solutions, for the various climates and geographic locations they considered; including the Arctic Circle. Due to its simplified nature, the model developed by Rabl and Nielsen [4] revived solar pond research and had subsequently been used to study the prospects of solar ponds in many regions and climates [8].

Kooi [9] developed an analytical model to study operating characteristics, such as its temperature distributions and energy fluxes, on the assumption that the non-convective zone temperature was equal to the ambient wet-bulb temperature. The model would allow the analysis of the performance of three-layer solar pond systems to be conducted in a similar calculation model to the Hottel-Whillier-Bliss equations [10] used for steady-state flat plate solar energy collectors.

With validation, established models such as that of Rabl and Nielsen [4] deemed the steady-state salt gradient of a solar pond system to be similar to that of a flat plate collector. Kooi [11] also modified the formulae developed by Rabl and Nielsen [4] to account for reflected radiation. He assumed that, since most solar ponds would operate near the solubility limit, this would naturally increase the reflectivity of the solar pond floor. Kooi concluded that the increase of reflectivity, in fact, reduced the efficiency of the steady-state system and that avoidance of supersaturation would be the key for an economical solar pond system.

Other authors also studied the influence of a solar pond's physical properties on its thermal storage efficiency. Wang and Seyed-Yagoobi [12] developed the equations reported by Kooi [11] to investigate the influence of the water's clarity and salt concentrations on the penetration of solar radiation underwater. The authors used turbidity as a parameter for the solar pond's water clarity. The authors reported that solar radiation did not affect the penetration of solar radiation underwater. However, the clarity of the water was found to be imperative, as the turbidity was observed to affect solar radiation penetration with increasing depths.

Karakilcik et al. [13] assessed the performance effect of the presence of shade on each of the solar pond zones. The authors reported a major influence on the solar pond's efficiency caused by the solar pond's shading effect. Another potential performance effect was proposed by Jaefarzadeh and Akbarzadeh [14]. The authors suggested that wind-induced mixing could affect the salinity gradient required for an effectively operating solar pond. The use of floating rings on the surface of the pond would help to mitigate such effects and thus improve performance year-round.

With a great amount of progress focussed on the analysis of the efficiency, performance and adverse effect mitigation, the salt gradient solar pond has been shown by many accounts to be a very promising technology for energy storage that can be adapted to many climates and geographical locations. Elsarrag et al. [15] reviewed the possibility of supplying the necessary energy required for the regeneration stage of a liquid desiccant cooling system using a salt gradient solar pond. The authors considered different solar pond system configurations, designs and solute materials which would be suitable for implementations in a potential solar-powered desiccant cooling system. Sayer et al. [16] researched the feasibility and performance gel pond and compared with the salinity gradient solar pond for low temperature applications. Amro and Yusuf [17] conducted a theoretical study of using solar ponds for seawater desalination in Qatar.

This study aims to investigate theoretically and validate a salt gradient solar pond experimentally as a desiccant and energy source in a hot-humid climate. The model is comprised of energy balances of the pond (including each salt gradient, pond wall, and surface area), saltwater thermo-physical properties and soil temperature.

2. Modelling of a salinity gradient solar pond

The behaviour of a solar pond, like any other solar-thermal collector, is majorly influenced by its geographical location. The pond of interest for the title-study has been constructed and has been analysed for more than 10 months in a hot-humid climate. Initially, the weather data for the location, as obtained from the NASA Atmospheric Science Data Centre at 25.2867°N and 51.5333°E (the geographical location of the salinity gradient solar pond of interest) is given in **Table 1**.

The temperature profile needs to be determined to characterise the thermal behaviour of the pond. The temperature varies with depth (and time). The temperature profile of the solar pond can be obtained from an energy balance of the solar pond. The general energy balance equation is in the form:

$$\begin{pmatrix} \textit{rate ofchange} \\ \textit{ofenergy content} \\ \textit{of elemental layer} \end{pmatrix} = \begin{pmatrix} \textit{rate ofheat} \\ \textit{flow} \\ \textit{into the element} \\ \textit{layer} \end{pmatrix} + \begin{pmatrix} \textit{rate of heat} \\ \textit{generation in} \\ \textit{the elemental} \\ \textit{layer} \end{pmatrix} - \begin{pmatrix} \textit{rate of heat} \\ \textit{flow} \\ \textit{out from the} \\ \textit{elemental layer} \end{pmatrix}$$

Month	Insolation (kWh/m^2/day)	$T_{amb, av}$ (°C)	V (m/s)	RH (%)
Jan	3.42	19.5	4.11	52.5
Feb	4.25	20.1	4.71	51.8
Mar	4.88	22.5	4.44	51.4
Apr	5.84	26.7	4.12	47.3
May	6.92	31.4	4.52	42.2
Jun	7.4	33.7	4.75	41.7
Jul	7.01	35.2	4.35	42
Aug	6.57	35.4	4.28	43.5
Sep	5.84	33.4	3.83	44.8
Oct	4.84	30	3.5	47.8
Nov	3.78	25.9	3.52	50.2
Dec	3.2	21.9	4.05	52.9

Table 1.
Metrological data for Doha.

Assuming an initial pre-stable ideal state, convection is ideally suppressed in the pond, thus heat flow is primarily by conduction. Thus, the energy balance can be expressed in terms of the one-dimensional heat conduction equation in differential form as:

$$\rho C_p \frac{\partial T}{\partial t} = \frac{\partial}{\partial z}\left(k\frac{\partial T}{\partial z}\right) + \dot{g}(z,t) - L(z,t) \quad (1)$$

where the thermo-physical properties (density, thermal conductivity and specific heat capacity) of the saltwater vary with temperature and concentration.

For example, for NaCl pond, the following correlations are widely employed [18]:

$$k = 0.5553 - 0.0000813S + 0.0008(T - 20) \quad (2)$$

$$\rho = 998 + 0.65C - 0.4(T - 20) \quad (3)$$

$$Cp = 4180 + 4.396C + 0.0048S^2 \quad (4)$$

3. Energy analysis of the solar pond

The temperature profile of the solar pond can be obtained from an energy balance of the different zones of the solar pond. With the assumptions that [2]:

- The temperature variation in the horizontal direction is assumed negligible. Thus the temperature and concentration distribution can be considered one-dimensional.
- The three zones of the pond (UCZ, NCZ and LCZ) are considered distinct enough to have a clear fixed boundary.
- The bottom surface of the pond is assumed appropriately blackened; as such the radiation reaching the LCZ is completely absorbed by the saltwater and the pond's bottom.
- Due to the presence of convection in the UCZ and the LCZ, the temperature and concentration in these zones are considered uniformly constant; such that they can be treated as single cells with a thickness zu and zl, respectively.
- The temperature varies with depth in the NCZ, and as such, in applying the energy balance, this part of the pond can be divided into several imaginary layers, i of thickness Δz each.
- The pond is considered very large. Thus, the side effects such as convection current at the wall can be ignored [19].

3.1 Upper-convective zone (UCZ)

Due to convection in the UCZ, it can be treated as having a uniform temperature. The heat balance equation for the UCZ can be given as:

$$Q_{UCZ} = Q_{NU} + Q_{solar} - Q_U \quad (5)$$

$$Q_U = Q_{Uc} + Q_{Ur} + Q_{Ue} + Q_{Us} \tag{6}$$

The solar radiation intensity, I, at a given layer (depth) in the pond can be obtained as a fraction of the radiation that penetrates the pond's surface.

The solar radiation in the pond decays exponentially with depth

$$I_z = \tau I_o \tag{7}$$

The fraction (τ) varies with the depth (z), and can be expressed as:

$$\tau = 0.36 - 0.08\ln(z) \tag{8}$$

While, the solar radiation that penetrates the pond's surface can be expressed in terms of the incident radiation on the pond's surface; taking into consideration that not all the incident rays penetrate (refracted) at the surface, as some are reflected back.

$$I_o = \beta I \tag{9}$$

$$\beta = 1 - \frac{1}{2}\left[\frac{\sin^2(\theta_i - \theta_r)}{\sin^2(\theta_i + \theta_r)} + \frac{\tan^2(\theta_i - \theta_r)}{\tan^2(\theta_i + \theta_r)}\right] \tag{10}$$

are related to the refractive index, n (=1.33 for water) as:

$$\sin\theta_i = n sin\theta_r \tag{11}$$

The angle of incidence can be obtained from:

$$\cos\theta_i = cos\delta cos\theta cos\omega + sin\delta sin\theta \tag{12}$$

$$\delta = 23.45\sin\left(\frac{360(284+N)}{365.25}\right) = 23.45\sin\left(\frac{360(N-80)}{370}\right) \tag{13}$$

$$\omega = \frac{2\pi(h-12)}{24}\ \text{(in rad)} \tag{14}$$

$$\omega = \frac{360(h-12)}{24}\ \text{(in degree)} \tag{15}$$

The convective heat loss rate from the pond's surface, $\dot{Q}_{Uc}$ is:

$$\dot{Q}_{Uc} = Ah_c(T_u - T_a) \tag{16}$$

where the convective heat transfer coefficient can be obtained as:

$$h_c = 5.7 + 3.8V \tag{17}$$

The heat loss rate due to radiation $\dot{Q}_{Ur}$ is

$$\dot{Q}_{Ur} = \varepsilon\sigma A\left(T_u{}^4 - T_{sky}{}^4\right) \tag{18}$$

The sky temperature T_{sky} can be determined as:

$$T_{sky} = T_a - \left(0.55 + 0.061\sqrt{P_w}\right)^{0.25} \tag{19}$$

The evaporative heat loss rate $\dot{Q}_{Ue}$ is obtained as:

$$\dot{Q}_{Ue} = \frac{\lambda h_c (P_u - P_w) A}{1.6 C_{p_a} P_{atm}} \tag{20}$$

$$P_u = \exp\left(18.403 - \frac{3885}{T_u + 230}\right) \tag{21}$$

$$P_w = R_H \exp\left(18.403 - \frac{3885}{T_a + 230}\right) \tag{22}$$

$$C_{p_a} = 1.005 + 1.82Rs \tag{23}$$

The heat loss through the sidewall $\dot{Q}_{Us}$ can be obtained as:

$$\dot{Q}_{Us} = C_w A_{wU} (T_u - T_{GU}) \tag{24}$$

$$C_w = \frac{1}{R_w} = \frac{k_p k_c}{S_p k_c + S_c k_p} \tag{25}$$

The solar energy absorbed by the zone can be obtained as the difference between the radiation entering the zone and the radiation leaving the zone.

With the foregoing, the energy balance equation for the UCZ can be written in differential form as:

$$\rho C_p \frac{\partial T_U}{\partial t} = \frac{\partial}{\partial z}\left(k \frac{\partial T}{\partial z}\right) + \frac{\partial I_{(z,t)}}{\partial z} - \dot{q}_u \tag{26}$$

$$z_1 A \rho C_p \frac{\partial T_U}{\partial l} = kA \left.\frac{\partial T}{\partial z}\right|_{z=z_1} + \beta A_{(z_1)} I\left(\tau_{(0,t)} - \tau_{(z_1,t)}\right) - \dot{Q}_u \tag{27}$$

Thus, the heat balance in non-differential form can be written as:

$$z_1 A \rho_u C_{p_u} \frac{T_{U(t+1)} - T_{U(t)}}{\Delta t} = \frac{k_{\frac{1}{2}} A_{\frac{1}{2}} \left(T_{(1,t)} - T_{U(t)}\right)}{\frac{\Delta z_1}{2}} + \beta A_{e(u)} I\left(\tau_{(0,t)} - \tau_{(z_1,t)}\right) - \dot{Q}_{u(t)} \tag{28}$$

Thus, the UCZ layer temperature can be obtained as:

$$T_{U(t+1)} = T_{U(t)} + \frac{\Delta t}{z_u A_u \rho_u C_{p_u}} \left\{ \frac{k_{\frac{1}{2}} A_{\frac{1}{2}} \left(T_{(1,t)} - T_{U(t)}\right)}{\frac{\Delta z_1}{2}} + \beta A_{e(u)} I\left(\tau_{(0,t)} - \tau_{(z_1,t)}\right) - \dot{Q}_{u(t)} \right\} \tag{29}$$

3.2 Non-convective zone (NCZ)

The NCZ is assumed to be divided into several imaginary layers i, with the first and last layers having a boundary with the UCZ and LCZ denoted as 1 and f, respectively; thus, the temperature in the first layer can be denoted as T_1 and in the last layer as T_f.

Here, the heat balance for the NCZ can be given as:

$$Q_{(i,t+1)} = Q_{(i+1,t)} + Q_{(solar)} - Q_{(i-1,t)} - Q_{s(i,t)} \tag{30}$$

Thus, the energy balance can be written in differential form as:

$$A\rho_i C_{p_i}\frac{\partial T}{\partial t} = \frac{\partial}{\partial z}\left(kA\frac{\partial T}{\partial z}\right)_{z_{i+1}} + \beta I\frac{\partial\left(A_{e(z)}\tau_{(z)}\right)}{\partial z} - \frac{\partial}{\partial z}\left(kA\frac{\partial T}{\partial z}\right)_{z_{i-1}} - C_w\frac{\partial A_{w(z)}}{\partial z}\left(T_{(i,t)} - T_G\right) \tag{31}$$

Or,

$$\Delta z_i A_i \rho_i C_{p_i}\frac{\partial T}{\partial t} = kA\left.\frac{\partial T}{\partial z}\right|_{z=z_{i+1}} + \beta A_{e(i)} I\left(\tau_{\left(i-\frac{1}{2},t\right)} - \tau_{\left(i+\frac{1}{2},t\right)}\right) - kA\left.\frac{\partial T}{\partial z}\right|_{z=z_{i-1}} - C_w A_{w(i)}\left(T_{(i,t)} - T_G\right) \tag{32}$$

And in non-differential form:

$$\Delta z_i A_i \rho_i C_{p_i}\frac{T_{(i,t+1)} - T_{(i,t)}}{\Delta t} = \frac{k_{\left(i+\frac{1}{2}\right)}A_{\left(i+\frac{1}{2}\right)}\left(T_{(i+1,t)} - T_{(i,t)}\right)}{\Delta z_i} + \beta A_{e(i)} I\left(\tau_{\left(i-\frac{1}{2},t\right)} - \tau_{\left(i+\frac{1}{2},t\right)}\right) - \frac{k_{\left(i-\frac{1}{2}\right)}A_{\left(i-\frac{1}{2}\right)}\left(T_{(i,t)} - T_{(i-1,t)}\right)}{\Delta z_i} - C_w A_{w(i)}\left(T_{(i,t)} - T_G\right) \tag{33}$$

Hence, the temperature of a layer of the NCZ can be expressed as:

$$T_{(i,t+1)} = T_{(i,t)} + \frac{\Delta t}{\Delta z_i A_i \rho_i C_{p_i}}\left\{\frac{k_{\left(i+\frac{1}{2}\right)}A_{\left(i+\frac{1}{2}\right)}\left(T_{(i+1,t)} - T_{(i,t)}\right)}{\Delta z} + \beta A_{e(i)} I\left(\tau_{\left(i-\frac{1}{2},t\right)} - \tau_{\left(i+\frac{1}{2},t\right)}\right) - \frac{k_{\left(i-\frac{1}{2}\right)}A_{\left(i-\frac{1}{2}\right)}\left(T_{(i,t)} - T_{(i-1,t)}\right)}{\Delta z} - C_w A_{w(i)}\left(T_{(i,t)} - T_{G(i)}\right)\right\} \tag{34}$$

3.3 Lower-convective zone (LCZ)

Using the same procedure that outlined previously, the heat balance equation for the LCZ can be expressed as:

$$Q_L = Q_{(solar)} - Q_{L\to N} - Q_{s(i,t)} - Q_G - Q_{ext} \tag{35}$$

$$z_l A_l \rho_l C_{p_l}\frac{\partial T_l}{\partial t} = \beta A_{e(z_3)} I\left(\tau_{(z_3,t)}\right) - kA\left.\frac{\partial T}{\partial z}\right|_{z=z_2} - C_w A_{w(z_l)}\left(T_{(l,t)} - T_G\right) - C_{gw}A_{z_3}\left(T_{(l,t)} - T_G\right) - Q_{ext} \tag{36}$$

$$z_l A_l \rho_l C_{p_l}\frac{T_{(l,t+1)} - T_{(l,t)}}{\Delta t} = \beta A_{e(z_3)} I\left(\tau_{(z_3,t)}\right) - \frac{k_l A_{z_2}\left(T_{(l,t)} - T_{(f,t)}\right)}{\frac{\Delta z}{2}} - C_w A_{w(z_l)}\left(T_{(l,t)} - T_G\right) - C_{gw}A_{z_3}\left(T_{(l,t)} - T_G\right) - Q_{ext} \tag{37}$$

Thus, the temperature of the LCZ can be given as:

$$T_{(l,t+1)} = T_{(l,t)} + \frac{\Delta t}{z_l A_l \rho_l C_{p_l}}\left\{\beta A_{e(z_3)} I\left(\tau_{(z_3,t)}\right) - \frac{k_l A_{z_2}\left(T_{(l,t)} - T_{(f,t)}\right)}{\frac{\Delta z}{2}} - C_w A_{w(z_l)}\left(T_{(l,t)} - T_G\right) - C_{wg}A_{z_3}\left(T_{(l,t)} - T_G\right) - Q_{ext}\right\} \tag{38}$$

3.4 Dimensions of the solar pond

The dimensional characteristics of each zone of the solar pond would need to be initially ascertained to obtain the temperature predictions of each of the pond's zones. The geometry of the pond could be characterised as shown in **Figure 2**. Here, the relevant cross-sectional surface areas (required for heat transfer) are deduce in relation to the dimensional references of the zones.

3.4.1 *Surface areas of the LCZ*

The cross-sectional area of the LCZ, A_l may be calculated at mid-plane of the zone (as shown in **Figure 2**).

The average area is:

$$A_l = x_l \cdot w_l \tag{39}$$

where the average length of the pond at the level (measured at mid-level):

$$x_l = X - 2\left(\frac{z_2 + \frac{z_l}{2}}{tan\vartheta}\right) \tag{40}$$

Similarly, the average width of the pond can be given as:

$$w_l = W - 2\left(\frac{z_2 + \frac{z_l}{2}}{tan\varphi}\right) \tag{41}$$

Thus, the area is:

$$A_l = \left(X - 2\left(\frac{z_2 + \frac{z}{2}}{tan\vartheta}\right)\right) \cdot \left(W - 2\left(\frac{z_2 + \frac{z_l}{2}}{tan\varphi}\right)\right) \tag{42}$$

Similarly,

$$A_{z_2} = \left(X - 2\left(\frac{z_2}{tan\vartheta}\right)\right) \cdot \left(W - 2\left(\frac{z_2}{tan\varphi}\right)\right) \tag{43}$$

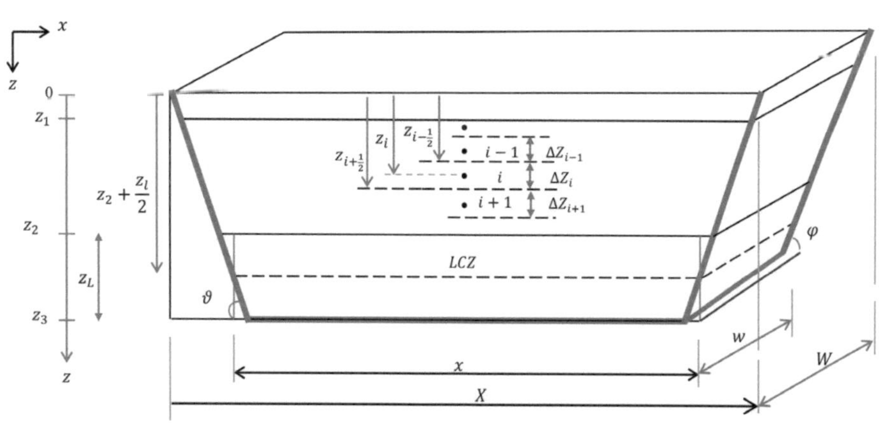

Figure 2
The dimensional characteristics of the solar pond.

Side wall area, $A_{w(z_l)}$,

$$A_{w(z_l)} = 2A_{w(z_l),W} + 2A_{w(z_l),X} \tag{44}$$

The area of one side (along the width) is:

$$A_{w(z_l),W} = \frac{1}{2}\left[\left(W - 2\left(\frac{z_2}{tan\varphi}\right)\right) + \left(W - 2\left(\frac{z_3}{tan\varphi}\right)\right)\right]\frac{z_l}{sin\vartheta} \tag{45}$$

Hence,

$$A_{w(z_l),W} = \frac{z_l}{sin\vartheta}\left[W - \frac{z_2}{tan\varphi} - \frac{z_3}{tan\varphi}\right] \tag{46}$$

Similarly, the area of an adjacent side along the length is:

$$A_{w(z_l),X} = \frac{z_l}{sin\varphi}\left[X - \frac{z_2}{tan\vartheta} - \frac{z_3}{tan\vartheta}\right] \tag{47}$$

Thus, the total area of the four side walls of the LCZ is:

$$A_{w(z_l)} = 2\frac{z_l}{sin\vartheta}\left[W - \frac{z_2}{tan\varphi} - \frac{z_3}{tan\varphi}\right] + 2\frac{z_l}{sin\varphi}\left[X - \frac{z_2}{tan\vartheta} - \frac{z_3}{tan\vartheta}\right] \tag{48}$$

3.4.2 Surface areas of the UCZ

Following the same procedure, the area for the UCZ can be deduced to be:

$$A_u = \left(X - 2\left(\frac{0 + \frac{z_u}{2}}{tan\vartheta}\right)\right) \cdot \left(W - 2\left(\frac{0 + \frac{z_u}{2}}{tan\varphi}\right)\right) \tag{49}$$

$$A_u = \left(X - \frac{z_u}{tan\vartheta}\right) \cdot \left(W - \frac{z_u}{tan\varphi}\right)$$

Similarly, the total area of the four side walls for the UCZ can be obtained as:

$$A_{w(z_u)} = 2\frac{z_u}{sin\vartheta}\left[W - \frac{0}{tan\varphi} - \frac{z_1}{tan\varphi}\right] + 2\frac{z_u}{sin\varphi}\left[X - \frac{0}{tan\vartheta} - \frac{z_1}{tan\vartheta}\right] \tag{50}$$

$$A_{w(z_u)} = 2\frac{z_u}{sin\vartheta}\left[W - \frac{z_1}{tan\varphi}\right] + 2\frac{z_u}{sin\varphi}\left[X - \frac{z_1}{tan\vartheta}\right] \tag{51}$$

3.4.3 Surface areas of the NCZ

$$A_i = \left(X - 2\frac{z_i}{tan\vartheta}\right) \cdot \left(W - 2\frac{z_i}{tan\varphi}\right) \tag{52}$$

$$A_{i+\frac{1}{2}} = \left(X - 2\frac{z_{i+\frac{1}{2}}}{tan\vartheta}\right) \cdot \left(W - 2\frac{z_{i+\frac{1}{2}}}{tan\varphi}\right) \tag{53}$$

$$A_{i-\frac{1}{2}} = \left(X - 2\frac{z_{i-\frac{1}{2}}}{tan\vartheta}\right) \cdot \left(W - 2\frac{z_{i-\frac{1}{2}}}{tan\varphi}\right) \tag{54}$$

The total surface area of the four side walls of an elemental layer in the NCZ is:

$$A_{w(\Delta z_i)} = 2\frac{\Delta z_i}{sin\vartheta}\left[W - \frac{z_{i-\frac{1}{2}}}{tan\varphi} - \frac{z_{i+\frac{1}{2}}}{tan\varphi}\right] + 2\frac{\Delta z_i}{sin\varphi}\left[X - \frac{z_{i-\frac{1}{2}}}{tan\vartheta} - \frac{z_{i+\frac{1}{2}}}{tan\vartheta}\right] \quad (55)$$

where

$$z_i = z_1 + (i-1)\Delta z + \frac{\Delta z}{2} \quad (56)$$

$$z_{i-\frac{1}{2}} = z_1 + (i-1)\Delta z \quad (57)$$

$$z_{i+\frac{1}{2}} = z_1 + i\Delta z \quad (58)$$

3.5. Soil temperature

Although the earth surface temperature at the location can easily be obtained alongside the metrological data; the temperature of the soil varies with depth, as such requires to be calculated separately as a function of depth and time.

The annual variation of the average soil temperature at different depths can be obtained as [20]:

$$T_g(z,t) = \overline{T}_g + A_0 e^{\frac{-z}{d}} \sin\left[\omega(t - t_0) - \frac{z}{d}\right] \quad (59)$$

$$A_o = \left(T_{g(0)\,\max} - T_{g(0)\,\min}\right)/2 \quad (60)$$

$$d = \sqrt{2D_h/\omega} \quad (61)$$

$$\omega = \frac{2\pi}{365} \quad (62)$$

4. Calculation procedure

The solar radiation intensity and ambient temperature data were obtained from the meteorological data for Qatar. The calculation was initialised (at time t = 0) by setting the temperature of the various layers of the pond to be equal to the ambient temperature; while setting the initial salinities of the UCZ and LCZ as equal to 2 and 26%, respectively (i.e. assuming that the pond was initially stabilised artificially).

In the sequence of calculation, the parameters—heat transfer coefficients and the properties of the liquid (in the different layers)—are first determined by the initial (ambient) temperature. Then, the obtained liquid properties are employed together with the solar radiation to determine the temperatures of the different layers of the pond at time interval Δt. The temperature of any layer at a time interval is determined with the liquid properties previously obtained with the preceding temperature of that layer. Using the same procedure, the temperatures of the layers for any selected time interval or time of the day can be calculated. In the simulation, the thickness of UCZ, NCZ and LCZ were taken to be 0.2, 1.3 and 0.5 m, respectively.

5. Modelling results

Results were computed by performing energy balances throughout each layer and sub-layer of the pond to achieve accurate temperatures. As the middle NCZ layer has the greatest depth and in turn has the most changes in salinity and density, it had to be divided into more layers. Each layer, as they have different densities,

Month	T_{atm}	T_{UCZ}	T_{LCZ}
January	19.50	19.50	42.36
February	20.10	20.10	51.69
March	22.50	22.50	58.07
April	26.70	26.70	65.63
May	31.40	31.40	74.51
June	33.70	33.70	79.48
July	35.20	35.20	78.65
August	35.40	35.40	76.23
September	33.40	33.40	71.31
October	30.00	30.00	63.98
November	25.90	25.90	55.58
December	21.90	21.90	49.51

Table 2.
Calculated temperatures of the top (UCZ) and bottom (LCZ) layers.

Month	T_{i1}	T_{i2}	T_{i3}	T_{i4}	T_{i5}
January	21.13	23.74	25.96	27.95	29.82
February	22.31	25.95	29.12	32.00	34.73
March	24.98	29.09	32.68	35.94	39.03
April	29.43	33.91	37.78	41.30	44.64
May	34.44	39.38	43.64	47.50	51.17
June	36.93	42.19	46.74	50.87	54.79
July	38.26	43.26	47.59	51.54	55.30
August	38.26	42.96	47.04	50.76	54.29
September	36.05	40.42	44.24	47.73	51.04
October	32.35	36.29	39.75	42.93	45.95
November	27.93	31.38	34.44	37.27	39.96
December	23.78	26.99	29.86	32.51	35.03

Table 3.
Calculated temperatures of the upper half of the middle (NCZ) layer of the salinity gradient solar pond, divided into 10 sub-layers.

loses heat to the layers immediately below and above dissimilarly. A solar pond operating in the geographical area of interest is predicted to reach a maximum temperature of approximately 79.5°C in the summer months with lows of 40°C in the winter. **Tables 2–4** show the diverse temperatures expected with the months of the year based on the energy balance models discussed in the previous section.

6. Empirical validation

Following the predictions obtained from the calculation model, empirical validation would need to be conducted on the solar pond constructed. Thus, the model calculations reported in this study were simultaneously compared and validated against the temperature gradient data gathered from a salinity gradient solar pond constructed as aforementioned. The solar pond of interest was constructed with the dimensions shown in **Figure 3**.

Month	T_{i6}	T_{i7}	T_{i8}	T_{i9}	T_{i10}
January	31.68	33.61	35.71	38.08	40.80
February	37.42	40.17	43.08	46.24	49.74
March	42.08	45.20	48.48	52.02	55.92
April	47.95	51.35	54.96	58.89	63.24
May	54.82	58.59	62.62	67.02	71.86
June	58.69	62.71	66.99	71.63	76.72
July	59.02	62.84	66.90	71.28	76.06
August	57.79	61.39	65.19	69.30	73.79
September	54.32	57.66	61.18	64.95	69.07
October	48.92	51.93	55.07	58.40	62.02
November	42.60	45.24	47.96	50.83	53.92
December	37.49	39.96	42.48	45.13	47.98

Table 4.
Calculated temperatures of the lower half of the middle (NCZ) layer of the salinity gradient solar pond, divided into 10 sub-layers.

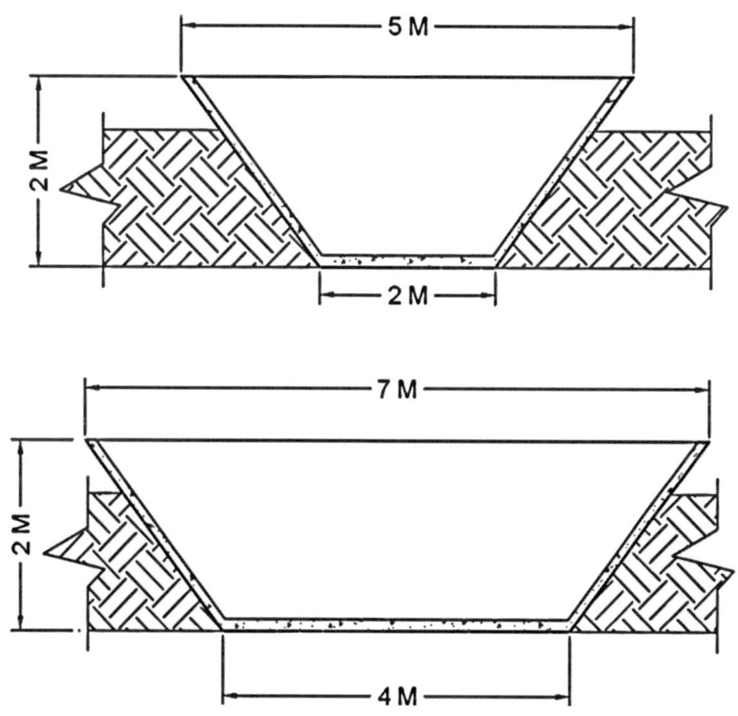

Figure 3
The dimension of the side view (top) and front view (bottom) of the salinity gradient solar pond used for empirical validation of the developed calculation model. The dimensions detailed were used to produce the thermal predictions for the salinity gradient zones of the pond.

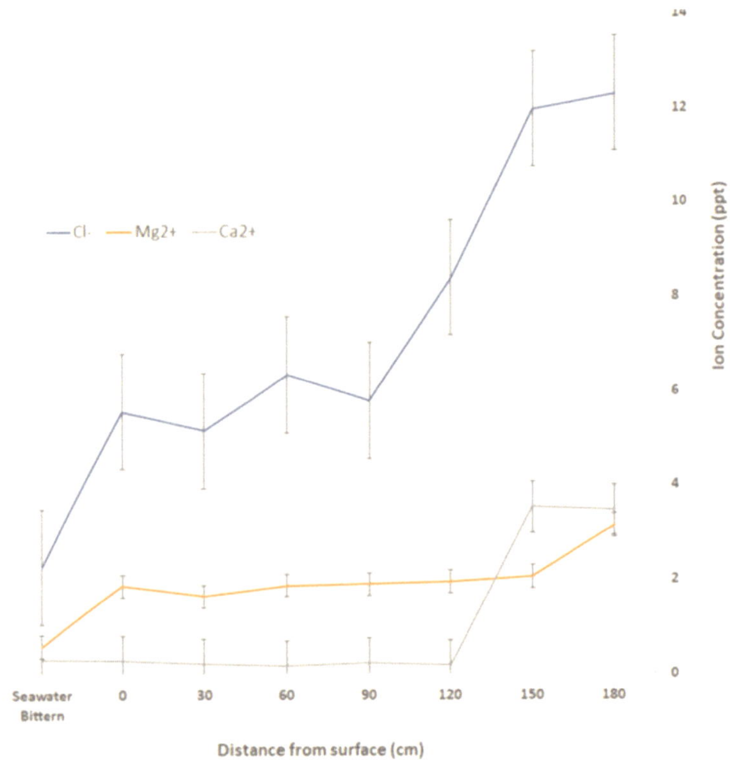

Figure 4
A comparison of the major ion concentration gradient achieved with the constructed solar pond. The samples were obtained using an extraction pump at different depths. The surface data refers to a sample obtained at approximately 5–10 cm below the surface of the pond to allow for the pump to be completely immersed. The bittern sample refers to the ion concentration of the bittern used before the addition of $MgCl_2$.

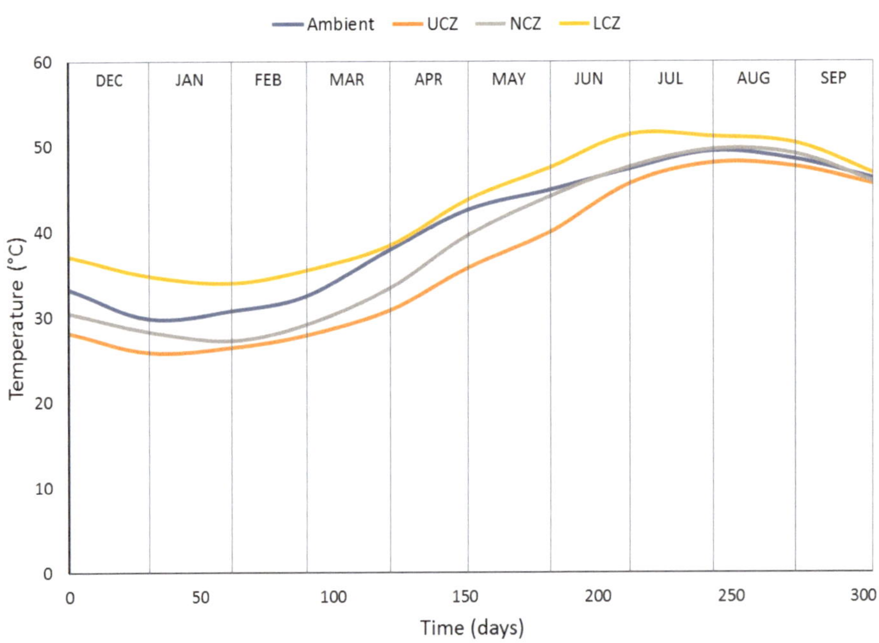

Figure 5
Temperature log of the different zones of the solar pond over the first 300 days of operation. The UCZ, NCZ and LCZ sample were taken at depths of 10, 70 and 150 cm, respectively.

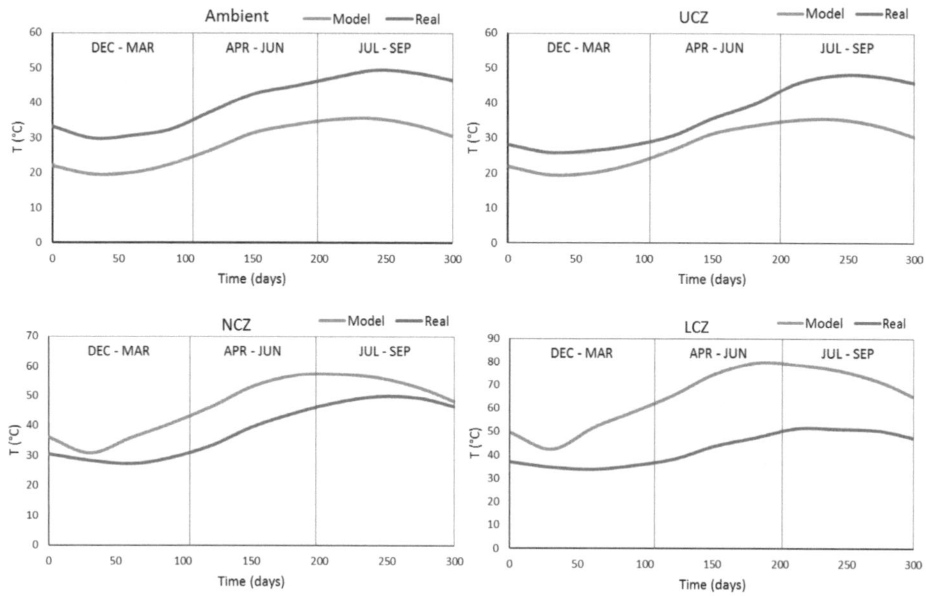

Figure 6
*Comparison of the calculation model predictions vs. the empirical temperature data obtained over the initial 300 days. The NCZ data for the model are the mean values for Ti values reported in **Tables 3** and **4**. The depths of each of the 'real' were obtained at the depths reported in **Figure 5** caption.*

The presence of a well-defined salinity gradient is paramount to the effectiveness of the solar pond's capability for heat storage. Thus, samples were taken at different levels of the pond to observe the concentrations at the expected upper, lower and non-convective zones.

Seawater bittern was the saline material used for the solar pond with additional $MgCl_2$ dissolved in the saline material to ensure a more pronounced salinity gradient was achieved. Future trials were envisioned only to use desalination reject brines as this would not only make use of an industrial waste product, without the conventional energy-intensive processes needed, but it would also improve the potential economic viability of the technology by using an inexpensive resource.

The increase in the salinity of the ponds towards the lower-convective zone can be seen in **Figure 4** and shows promise towards the solar pond's thermal storage potential. An interesting observation of the different ion concentrations were those of calcium and magnesium. The highest concentration cations in the upper and non-convective zones were those of magnesium. It was the case until the deeper depths of the lower-convective zone at which the calcium and magnesium concentrations were much more comparable.

To validate the thermal properties model, the temperatures of the upper-convective, non-convective and lower-convective zones (UCZ, NCZ and LCZ, respectively) were measured periodically as an empirical comparison of the initial 300 days of operation. The temperature measurements were taken at different depths along the centre of the pond; by using RTD with an accuracy of 0.1°C. **Figure 5** shows the temperature measurements obtained from the constructed solar pond while **Figure 6** shows a comparison of the model predictions vs. the actual solar pond temperatures measurements.

The temperature of the lower-convective zone was not expected to reach the heights predicted by the model as the winter month predictions begin with a solar pond with an LCZ already at temperatures of over 45°C stored from the previous summer months. However, the gradual increase in heat storage measured across the early summer months is promising. Except for the lower-convective zone, the thermal behaviour across the initial months of the solar ponds was increased at a similar rate to the predicted data albeit at lower temperatures. The initial increase was hypothesised to be due to the initial heat of dissolution because of adding the salts to the solar pond.

In a nutshell, though the temperature profile of the pond (through the months of operation) follows a similar pattern to the model predictions, the measured temperatures are lower than predicted. The discrepancy can be partly attributed to the fact that the weather data employed in the model is in variance with the actual weather condition experienced during the pond's operation. In addition, the effects of shading of the side walls and turbidity (clarity of the water) were not considered in the model, but in reality, contribute to the reduction of solar radiation received by the pond. The shading due to the side walls tends to reduce the effective surface area of the pond available to receive incident radiation thus consequently resulting in reduced temperature. The pond is prone to dust (due to the geographical location). Increased dust spread on the pond reduces the clarity of the water thereby reducing the penetration of solar radiation into the pond, consequently reducing the pond temperature.

7. Conclusions

Thermal behaviour analysis and a prediction model have been developed which can be effectively used for the construction and operation of a solar salinity pond. For the developed numerical was split into three sections: the UCZ, NCZ and LCZ. However, the NCZ consisted of a much larger depth than the other zones, thereby resulting in much more significant variance in its salinity and density. It was split into 10 sub-sections. By employing the numerical model in a set calculation procedure, the heat transfer coefficient could be first determined followed by the physical parameter of the pond saline material. Following this procedure, the temperature of each layer (and NCZ sub-layer) could be determined for any period in the year.

In this study, the average temperature of each layer was calculated for each month in the year when exposed to the Qatari temperatures. With the high heat climates of Doha, LCZ temperatures were predicted to reach its highest thermal storage potential at temperatures of about 70°C in June with the lowest in January reaching around 40°C.

The developed numerical model is planned for solar desiccant cooling applications in which the salinity gradient solar pond would be used for thermal storage as well as the storage and regeneration of the liquid desiccant used in the proposed air conditioning system.

Acknowledgements

The author acknowledges Qatar National Research Funds (QNRF) for supporting this research through NPRP 7-332-2-138.

Nomenclature

$A_{e(i)}$	effective surface area of the layer that receives the solar radiation
$A_{e(u)}$	effective area that receives the solar radiation

$A_{w(i)}$	corresponding surface area of the wall at the given layer
A_{wU}	surface area of side wall corresponding to the UCZ.
C_w	thermal conductance of the composite wall
C_{P_a}	humid heat capacity of air
h	local time
h_c	convective heat transfer coefficient
I	solar radiation intensity
k_p	thermal conductivity of the insulation material
k_c	thermal conductivity of the concrete wall
N	day of the year
P_{atm}	atmospheric pressure
P_u	vapour pressure of water at the upper layer surface temperature
P_w	partial pressure of the water vapour in the ambient air
$\dot{Q}_{ext}$	heat extraction rate (load) from the LCZ by heat exchanger
$\dot{Q}_G$	heat loss rates through the bottom wall (ground) of the pond
Q_{LN}	heat lost from LCZ to NCZ
Q_{NU}	heat gained from NCZ to UCZ
$\dot{Q}_s$	heat loss rates through the side wall of the pond
Q_{solar}	solar energy absorbed by the layer
Q_U	total heat losses from the UCZ
$\dot{Q}_{Uc}$	convective heat loss rate from the pond's surface
$\dot{Q}_{Ue}$	evaporative of heat loss rate
$\dot{Q}_{Ur}$	heat loss rate due to radiation
$\dot{Q}_{Us}$	heat loss through the side wall
R_H	average monthly relative humidity at location
Rs	specific humidity
R_w	total thermal resistance of the individual resistance of the thermal insulation and concrete
RH	relative humidity
S_c	thickness of concrete
S_p	thickness of insulation material
T_1	temperature in the first layer
T_{amb}	ambient air temperature
T_f	temperature in the last layer
$\overline{T}_g$	annual average earth temperature at the location
$T_g(z,t)$	average soil temperature at depth z (m) and time t (d).
$T_{a(n,t)}$	ambient air temperature at time t of nth day of the year
T_{GU}	ground temperature of the surrounding soil at the layer in consideration
T_{sky}	sky temperature.
z_1	depth of the UCZ
ε	emissivity of water
σ	Stefan Boltzmann's constant
V	average monthly wind speed
β	reflectivity of the radiation
θ_i	angle of incidence
θ_r	angle of refraction
Φ	latitude angle
ω	hour angle
δ	declination angle
λ	latent heat of water evaporation
τ	fraction

Author details

Esam Elsarrag[1]*, Opubo N. Igobo[2] and Philip A. Davies[3]

1 BGG, Manchester, UK

2 Howard Butler, Walsall, UK

3 School of Engineering, University of Birmingham, UK

*Address all correspondence to: elsarrag@hotmail.com

References

[1] Hawlader M. The influence of the extinction coefficient on the effectiveness of solar ponds. Solar Energy. 1980;**25**(5):461-464

[2] Kurt H, Ozkaymak M, Binark AK. Experimental and numerical analysis of sodium-carbonate salt gradient solar-pond performance under simulated solar-radiation. Applied Energy. 2006; **83**(4):324-342

[3] Kurt H, Halici F, Binark AK. Solar pond conception—Experimental and theoretical studies. Energy Conversion and Management. 2000; **41**(9):939-951

[4] Rabl A, Nielsen CE. Solar ponds for space heating. Solar Energy. 1975;**17**(1): 1-12

[5] Tabor H. Solar ponds large-area solar collectors for power production. Solar Energy. 1963;**7**(4):187-188

[6] Tabor H. Non-convecting solar ponds. Philosophical Transactions of the Royal Society of London A: Mathematical, Physical and Engineering Sciences. 1980;**295**(1414):423-433

[7] Bozkurt I, Mantar S, Karakilcik M. A new performance model to determine energy storage efficiencies of a solar pond. Heat and Mass Transfer. 2015; **51**(1):39-48

[8] Bryant H, Colbeck I. A solar pond for London? Solar Energy. 1977;**19**(3): 321-322

[9] Kooi C. The steady state salt gradient solar pond. Solar Energy. 1979;**23**(1): 37-45

[10] Bliss RW. The derivations of several “plate-efficiency factors” useful in the design of flat-plate solar heat collectors. Solar Energy. 1959;**3**(4):55-64

[11] Kooi C. Salt gradient solar pond with reflective bottom: Application to the “saturated” pond. Solar Energy. 1981; **26**(2):113-120

[12] Wang J, Seyed-Yagoobi J. Effect of water turbidity on thermal performance of a salt-gradient solar pond. Solar Energy. 1995;**54**(5):301-308

[13] Karakilcik M, Dincer I, Bozkurt I, Atiz A. Performance assessment of a solar pond with and without shading effect. Energy Conversion and Management. 2013;**65**:98-107

[14] Jaefarzadeh MR, Akbarzadeh A. Towards the design of low maintenance salinity gradient solar ponds. Solar Energy. 2002;**73**(5):375-384

[15] Elsarrag E, Igobo ON, Alhorr Y, Davies PA. Solar pond powered liquid desiccant evaporative cooling. Renewable and Sustainable Energy Reviews. 2016;**58**:124-140

[16] Sayer A, Al-Hussainia H, Campbell A. New comprehensive investigation on the feasibility of the gel solar pond, and a comparison with the salinity gradient solar pond. Applied Thermal Engineering. 2018;**130**(8): 672-683

[17] Amro M, Yusuf B. Integration of pressure retarded osmosis in the solar ponds for desalinationand photo-assisted chloralkali processes: Energy and exergy analysis. Energy Conversion and Management. 2019;**195**:630-640

[18] Kaufmann DW. Sodium Chloride: The Production and Properties of Salt and Brine. New York, USA: Reinhold Publishing; 1960

[19] Ouni M, Guizani A, Lu H, Belghith A. Simulation of the control of a salt gradient solar pond in the south of

Tunisia. Solar Energy. 2003;**75**(2): 95-101

[20] Hillel D. Chapter 12: Soil temperature and heat flow. In: Fundamentals of Soil Physics. New York, USA: Academic Press; 1982. pp. 287-317

Chapter 4

Theoretical and Experimental Analysis of a Thermoelectric Air-conditioning System

Adeyanju Anthony Ademola

Abstract

Thermoelectric devices use the Peltier effect that creates a heat flux between the junctions of two different types of materials. The thermoelectric module also referred to as a heat pump transfers heat from one side to the other when a DC current is applied. This study carried out the theoretical analysis of a thermoelectric air conditioning system. A prototype thermoelectric air conditioner of 286 W cooling capacity was built and a testing enclosure made from plywood and Styrofoam was also constructed in order to validate the theoretical result with an experimentation. It was discovered that thermoelectric air conditioning took 4 minutes to reach its desired temperature of 22°C, whereas the standard air conditioning system (refrigeration cycle) took 20 minutes to cool to a room temperature. Economically, it was also discovered that thermoelectric air conditioning system is 50% cheaper than the refrigeration cycle air conditioning systems. The thermoelectric air conditioner has cheaper maintenance and greater estimated life span of 7 years more than the refrigeration air conditioner. This is because the air conditioner that operates on the refrigeration cycle uses a rotating compressor, while the thermoelectric air conditioner uses thermometric module.

Keywords: thermoelectric, Peltier effect, coefficient of performance, refrigeration cycle

1. Introduction

Air conditioning systems that uses the refrigeration cycle are the most common types of devices that uses air to exchange heat in an occupied space such as a building or a car. These systems consume a lot of electricity, can be bulky and some of the refrigerants used in the system is harmful to the environment. Therefore, there is the need for an alternative air conditioning system, thermoelectric modules are capable of generating both hot and cold temperatures at each side and as such can be used as an alternative to the present system. Although thermoelectric air conditioning systems have many advantages, they are rarely used due to having a lower efficiency as compared to conventional air conditioning systems [1]. It is therefore necessary to design a thermoelectric air conditioning system to achieve similar performance to that of a conventional AC system and also to have less disadvantages by designing a compact and light weigh prototype.

2. Brief history of thermoelectric principles

In the 1820s, it was found that the temperature difference of two dissimilar metals when in contact with each other produced an electromotive force of voltage. This voltage was produced since the temperature difference causes the electrons or other charged carriers to move from the hot side to the cold side of the metal which produced a current as shown in in **Figure 1**. This theory was found by Thomas Seebeck and is known as the Seebeck effect [2].

Approximately 10 years later, a physicist, Jean Peltier found that the reverse of the Seebeck effect is also true. He found that if a current was passed through different metals, the temperature at one side of the metal would increase while at the other side the temperature would decrease. This effect is known as the Peltier effect. The Peltier module work as a heat pump, such that at the cold side of the module, it absorbed the heat to be removed to the other side of the module when a DC voltage is applied [3].

2.1 Operating principle of the refrigeration cycle

The air conditioner consists of two connected coils which contains continuous flowing refrigerant inside of it. The split unit systems are the most common type of air conditioner, in which the coil located inside the room to be cooled is referred to as the evaporator and the coil located outside the room is called the condenser [4].

The operating principle of the refrigeration cycle is to keep the evaporator colder than the temperature of the room and the condenser temperature higher than the surroundings as shown in **Figure 2**. These conditions allow for the continuous flowing fluid to absorb the heat from the room and then eject the heat into the surroundings.

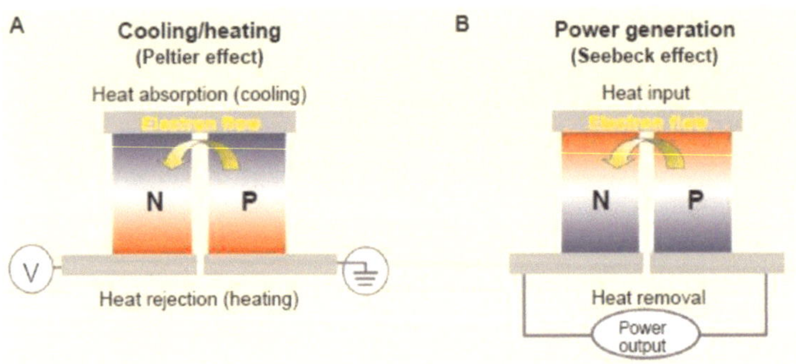

Figure 1.
Thermoelectric principle.

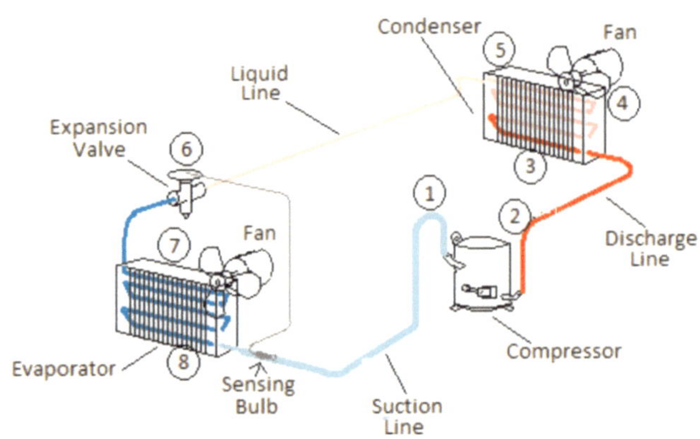

Figure 2.
Refrigeration cycle.

A compressor and an expansion valve are used to achieve these conditions. The compressor, usually a reciprocating compressor, is used to increase the pressure of the refrigerant. The refrigerant in the gaseous state enters the compressor and it is compressed which increases the temperature and pressure of the refrigerant. The temperature at the outlet of the compressor would be much greater than the atmosphere; therefore, when the hot gas passes through the condenser, the heat is easily rejected with the aid of a fan.

During the heat ejection phase, the gas is condensed into a liquid. At the exit of the condenser, an expansion valve is used to reduce the pressure of the fluid and also the temperature drops, which is lower than the room temperature. This is how the cold refrigerant is produced inside an air conditioner.

When the air is passed through the evaporator's coil, the room temperature would drop and the refrigerant is converted to vapor during the heat absorption process.

Therefore, the fundamental rule of the air conditioner is achieved, in which the temperature is lower than room temperature in the coil inside the room and the temperature is more than the atmospheric temperature in the coil outside the room.

2.2 Thermoelectric module

Unlike the conventional air conditioning systems, the Seebeck effect is a reversible process such that heating and cooling can be obtained on both sides depending on the direction of the current applied to the device. In **Figure 3**, when an electric current is supplied to the device, the electrons and holes will move through the P-type and N-type elements thus causing heating and cooling in the respective sides of the module. These elements are an alloy called Bismuth and Tellurium and when exposed to the same temperature, they have different free electron densities. The P-type element has a deficiency of electrons and the N-type element have an excess of electron and when a current is applied, the module tries to establish an equilibrium and as a result, heating and cooling occurs. Alumina ceramic substrates are used on both sides of the module where the heating occurs in one side and cooling

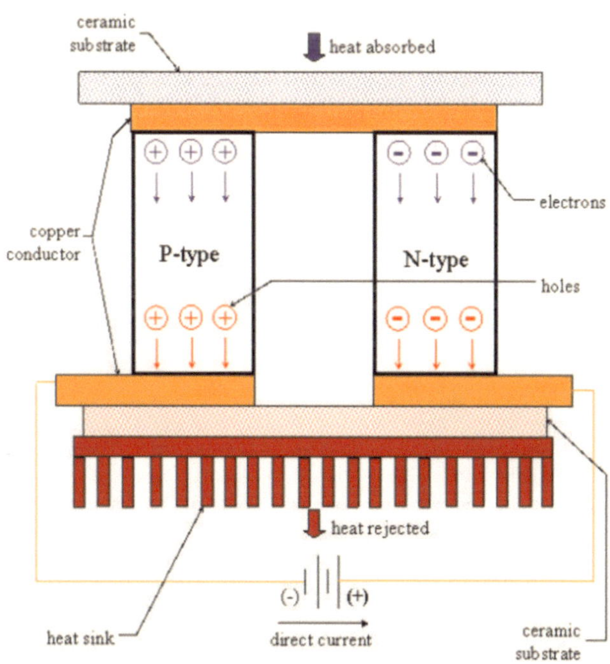

Figure 3.
Principle of thermoelectric module.

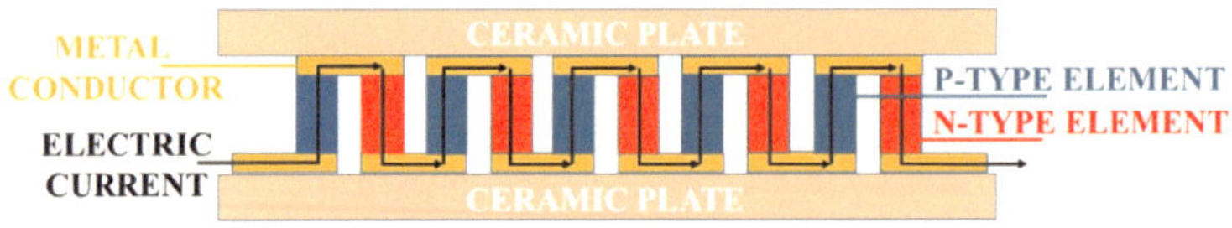

Figure 4.
P-type and N-type elements connected in series.

occurs at the other side. This material is chosen due to being a good insulator of electricity and also being thermally conductive. The coefficient of performance of this device is defined as the ratio of the cooling or heating power to the power supplied to the module.

In order to pump a great amount of heat, the thermoelectric device usually consists of multiple P-type and N-type elements. A typical thermometric device contains around 250 P-type and N-type elements connected in series as shown in in **Figure 4**.

3. Theoretical analysis of thermoelectric air conditioning system

The design of thermoelectric air conditioning system is shown in **Figure 5**, a fan is mounted on top of the cover where air at ambient temperature would be sucked into the device and circulate through the heat sinks and then blow through two rectangular holes such that the direction of the air can be controlled via the flaps. This would allow for the heat from the air to be properly transferred to the heat sinks.

3.1 Determining the cooling load

3.1.1 Cooling and dehumidification of the air

The temperature and relative humidity of the ambient air measured in the laboratory was 31°C and 63% respectively. In order to obtain within thermal comfort range as defined by ASHRAE, it is best to cool and dehumidify the air to 22°C with a relative humidity of 50% as shown in **Figure 6**. The cold side of the thermoelectric module would be at a temperature lower than the dew point temperature and therefore condensation is expected to take place which would decrease the relative humidity of the air [5, 6].

(a) **(b)**

Figure 5.
Thermoelectric air conditioning system.

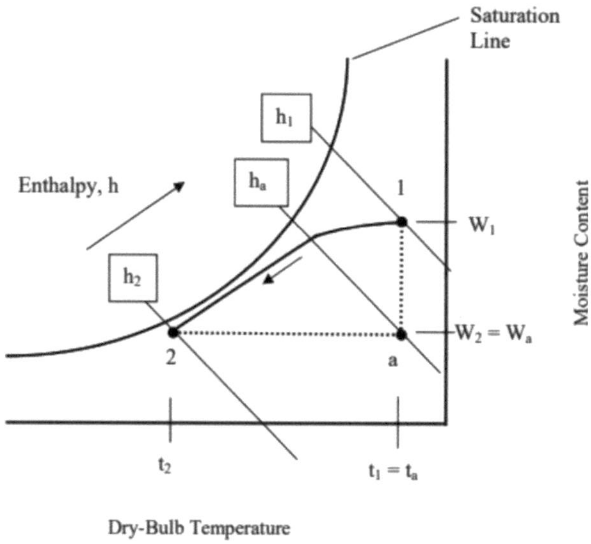

Figure 6.
Psychometric chart: cooling and dehumidifying.

State 1 represents the air properties of the ambient air in the lab while state 2 represents the air properties that we would like to have. The following calculations are used to determine the required cooling capacity for the thermoelectric module.

$$\dot{Q}_T = \dot{Q}_s + \dot{Q}_l \tag{1}$$

$$= \dot{m}_a(h_2 - h_a) + \dot{m}_a(h_a - h_1)h_{f,2} \tag{2}$$

$$= \dot{m}_a\left[(h_1 - h_2) - (W_1 - W_2)h_{f,2}\right] \tag{3}$$

where $\dot{Q}_s$ = sensible heat in kW; $\dot{Q}_l$ = latent heat in kW; h = enthalpy in kJ/kg; $\dot{m}_a$ = mass flow rate in kJ/kg; W = humidity ratio in kg/kg.

At state 1:

31°C, 63% *relative humidity*

$$h_1 = 76.91\ \text{kJ/kg};\ \vartheta_1 = 0.887\text{m}^3/\text{kg};\ W_1 = 0.0179\ \text{kg/kg}$$

At state 2:

22°C, 50% *relative humidity*

$$h_2 = 43.04\ \text{kJ/kg};\ W_2 = 0.0082\ \text{kg/kg};\ h_{\text{f},2} = 83.94\ \text{kJ/kg}$$

A fan of 15 CFM was available at the lab:

15 CFM = 0.00708 m^3/s

$$\dot{m}_a = 0.00708\ \text{m}^3/s \times \frac{1}{0.887\text{m}^3/\text{kg}}$$

$$= 0.00798\ \text{kg/s}$$

$$\dot{Q}_T = \dot{m}_a\left[(h_1 - h_2) - (W_1 - W_2)h_{f,2}\right] \tag{4}$$

$$= 0.00798[(76.91 - 43.04) - (0.0179 - 0.0082)83.94]$$

$$= 0.264\ \text{kW}$$

$$= 264\ \text{W}$$

3.1.2 Sensible cooling load of the enclosure

An enclosure shown in **Figure 7** was designed in order to test the thermoelectric air conditioning system. The enclosure was made of plywood and properly insulated with styrofoam so that the outside temperature would have minimal effect on the testing of the device.

The temperature of the ambient air is 31°C

Thermal comfort temperature is 22°C

Thermal conductivity of wood, $K_{\text{wood}} = 0.151\ \frac{\text{W}}{\text{mK}}$

Thermal conductivity of styrofoam, $K_{\text{styrofoam}} = 0.033\ \frac{\text{W}}{\text{mK}}$

$$R_{\text{Total}} = \frac{L_{\text{wood}}}{K_{\text{wood}}} + \frac{L_{\text{styrofoam}}}{K_{\text{styrofoam}}}$$

$$= \frac{0.01}{0.151} + \frac{0.035}{0.033} \tag{5}$$

$$= 1.13\ \frac{K}{W}$$

Heat transfer through the wall of the box per unit area:

$$q'' = \frac{\Delta T}{R_{total}}$$

$$= \frac{9\,K}{1.13}$$

$$= 7.96\ \frac{\text{W}}{\text{m}^2}$$

$$\text{Area of enclosure} = 2(0.52 \times 0.50) + 2(0.55 \times 0.50) + (0.55 \times 0.50) + (0.57 \times 0.52)\ \text{m}$$

$$= 1.6414\ \text{m}^2$$

Therefore, heat transfer through the wall of the box,

$$\dot{Q} = 7.96 \times 1.6414\ \text{m}^2$$

$$= 13.1\ \text{W}$$

Figure 7.
Enclosure for test of thermoelectric air conditioning system.

3.1.3 Resistor cooling load

A resistor/heating coil was placed in the testing enclosure to observe how long the device takes to remove the heat and maintain a constant temperature. A resistor of 20 W rating was used and the temperature of the enclosure was monitored with time.

3.1.4 Total cooling load required

$$\begin{aligned} \text{Total cooling load} &= \text{cooling/dehumidification} + \text{cooling load of enclosure} + \text{resistor load} \\ &= 264\text{ W} + 13.1\ W + 20\text{ W} \\ &= 297.1\text{ W} \end{aligned} \tag{6}$$

However, the device was sized at a total cooling load of 330 W to take into considerations any additional heat loads that were not accounted for, also the enclosure was not properly sealed.

3.2 Choosing the Peltier module

One of the TEC1-12730 module is capable of producing 250 W of cooling, however in order for one module to produce 250 W of cooling, it needs a dc power supply rated at 30 amps and 18 V. Therefore, by using three modules, a power supply that is available at the lab can be used to supply each module rated at 12 V.

Using three Peltier TEC1-12730 modules:

$$\begin{aligned} \text{Cooling capacity required for each module} &= \frac{330}{3} \\ &= 110\text{ W} \end{aligned}$$

$$\begin{aligned} \text{Temperature difference, } \Delta T &= T_{hot} - T_{cold} \\ &= 50°\text{C} - 18°\text{C} \\ &= 32°\text{C} \end{aligned}$$

The specification graphs for the TEC1-12730 module shown in **Figures 8** and **9** were used in determining the appropriate amperage and voltage needed to supply each thermoelectric module.

At $\dot{Q}_c$ = 110 W and ΔT = 32°C, amperage, I = 21 amps
Using the second graph to determine the voltage to apply:
Using 21 amps and ΔT = 32°C, therefore V = 12 V
Power consumed by the three (3) Peltier device = $3 \times (12 \times 21) = 756$ W

3.3 Sizing the heat sink

Determining the thermal resistance of a required heat sink (**Figure 10**) for the hot side of each thermoelectric module:

Maximum operating temperature of the thermoelectric module: 138°*C* [7].
Thermal resistance of the heat sink needed,

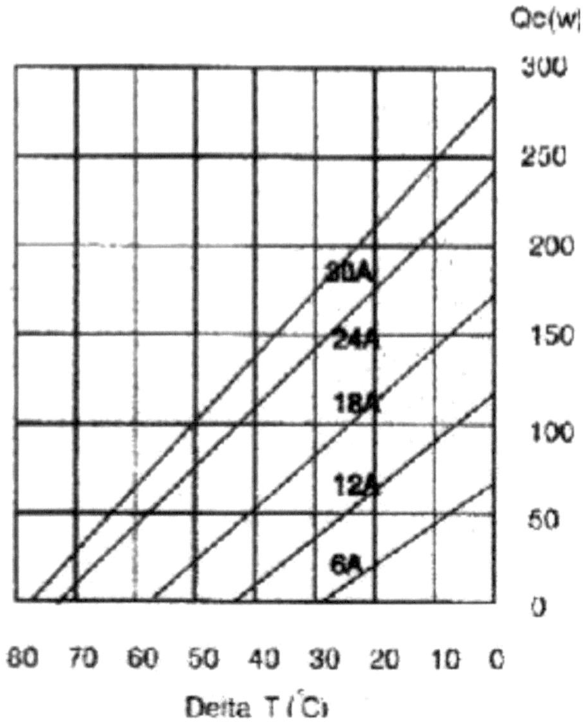

Figure 8.
Specification graph 1 for the TEC1-12730 module.

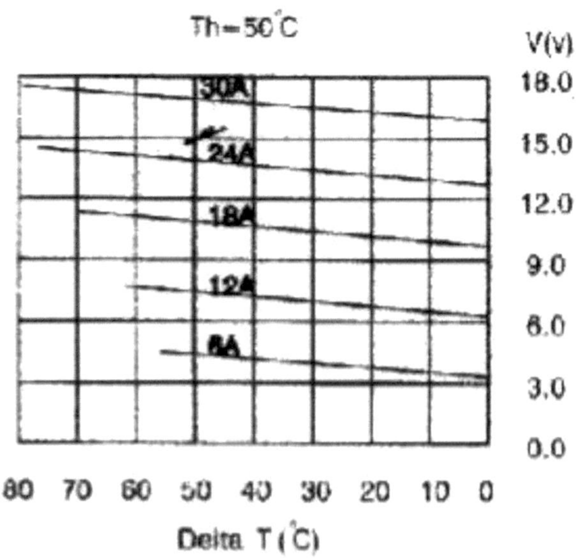

Figure 9.
Specification graph 2 for the TEC1-12730 module.

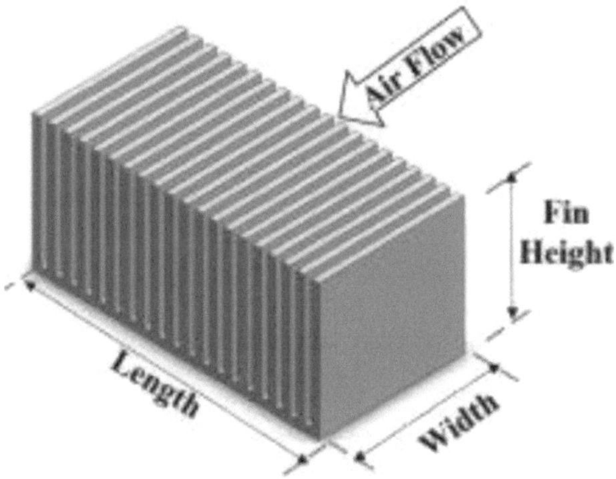

Figure 10.
Sizing the heat sink.

$$R_{hs} = \frac{T_{hot} - T_{amb}}{\dot{Q}_{T,1} + P_{peltier}}$$
$$= \frac{138 - 31}{362}$$
$$= 0.296\ \frac{K}{W}$$

Air flow of fan used for heat sink, ϑ = 0.142 m$^3 s^{-1}$ (CFM of fan available at lab = 300)
Width of heat sink, W = 0.08 m
Width of the fin, L = 0.064 m
Thickness of fin, t_{fin} = 0.0015 m
Number of fins, N_{fin} = 20
Height of fin, H_f = 0.057 m
Height of heat sink, H = 0.069 m
Spacing between the fins [8],

$$b = \frac{W - (N_{fin} \times t_{fin})}{N_{fin} - 1}$$
$$= \frac{0.08 - (20 \times 0.0015)}{20 - 1}$$
$$= 0.003\ \text{m}$$

Velocity of air between the fins,

$$\text{V} = \frac{\vartheta}{N_{fin} \times b \times H_f}$$
$$= \frac{0.142}{20 \times 0.003 \times 0.057}$$
$$= 41.52\ \text{ms}^{-1}$$

The properties of air at one (1) atmosphere and 31°C are as follows:
Density, $\rho = 1.15\,\frac{kg}{m^3}$
Specific heat capacity, $Cp = 1.007\,\frac{kJ}{kg.K}$
Dynamic viscosity, $\mu = 18.718 \times 10^{-6}\,\frac{N.s}{m^2}$
Heat conductance, $K = 26.424 \times 10^{-6}\,\frac{kW}{m.K}$
Prandtl number,

$$\begin{aligned} \text{Pr} &= \frac{\mu \times C_p}{k} \\ &= \frac{18.718 \times 10^{-6} \times 1.007}{26.424 \times 10^{-6}} \\ &= 0.713 \end{aligned}$$

Reynolds number,

$$\begin{aligned} \text{Re} &= \frac{\rho \times V \times b}{\mu} \times \frac{b}{L} \\ &= \frac{1.15 \times 41.52 \times 0.003}{18.718 \times 10^{-6}} \times \frac{0.003}{0.064} \\ &= 358.72 \end{aligned}$$

Nusselt number,

$$\begin{aligned} Nu &= \left[\frac{1}{\left(\frac{\text{Re}\times\text{Pr}}{2}\right)^3} + \frac{1}{\left(0.664\sqrt{\text{Re}}\text{Pr}^{0.33}\sqrt{1+\frac{3.65}{\sqrt{\text{Re}}}}\right)^3} \right]^{\frac{-1}{3}} \\ &= \left[\frac{1}{\left(\frac{358.72\times0.713}{2}\right)^3} + \frac{1}{\left(0.664\sqrt{358.72}0.713^{0.33}\sqrt{1+\frac{3.65}{\sqrt{358.72}}}\right)^3} \right]^{\frac{-1}{3}} \\ &= 12.28 \end{aligned}$$

Heat transfer coefficient,

$$\begin{aligned} \text{h} &= Nu \times \frac{k}{b} \\ &= 12.28 \times \frac{26.424 \times 10^{-3}}{0.003} \\ &= 108.16 \end{aligned}$$

Efficiency of the fin,

$$\begin{aligned} \eta_{fin} &= \frac{\tanh\left(m \times H_f\right)}{m \times H_f} \\ &= \frac{\tanh(29.39 \times 0.057)}{29.39 \times 0.057} \\ &= 0.556 \end{aligned}$$

$$m = \sqrt{\frac{2h}{k_{fin} \times t_{fin}}}$$

$$= \sqrt{\frac{2\,x\,108.16}{167 \times 0.0015}}$$

$$= 29.39$$

Surface area of the exposed base,

$$A_{base} = (N_{fin} - 1)\, b \times L$$

$$= (20 - 1)0.003 \times 0.064$$

$$= 0.00365\ \text{m}^2$$

Area of the fin,

$$A_{fin} = 2 \times H_f \times L$$

$$= 2 \times 0.057 \times 0.064$$

$$= 0.0073\ \text{m}^2$$

Thermal resistance of the heat sink,

$$R_{hs} = \frac{1}{h.\left(A_{base} + \left(N_{fin}\eta_{fin}A_{fin}\right)\right)}$$

$$= \frac{1}{108.16.(0.00365 + (20 \times 0.556 \times 0.0073))}$$

$$= 0.109\ \frac{\text{K}}{\text{W}}$$

$0.109\ \frac{\text{K}}{\text{W}} < 0.296\ \frac{\text{K}}{\text{W}}$, since the calculated thermal resistance of the heat sink available is less than the required thermal resistance of the heat sink then therefore, this heat sink was used.

4. Experimentation of the thermoelectric air conditioning system

4.1 Apparatus

- Thermocouples
- 12 channel thermocouple data recorder
- Power supply
- Multi-meter
- Power strip

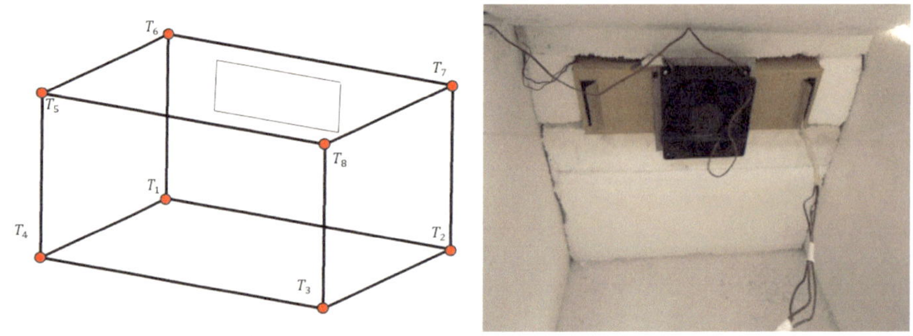

(a) (b)

Figure 11.
Testing enclosure showing points at which temperature were taken.

Figure 12.
Thermoelectric air conditioner placed inside of testing enclosure.

4.2 Method/procedure

i. The thermoelectric air conditioning system was connected to the power supply and connected to the power trip.

ii. The thermocouples were connected to the 12-Channel Thermometer and the location of the thermocouples was noted as shown in **Figure 11**.

iii. The power strip was turned on which simultaneously powered on the thermocouple data recorder and the thermoelectric air-conditioner as shown in **Figure 12** and **Table 1**.

5. Results and discussion

From the graph of temperature (°C) versus time (min) as shown in **Figure 13**, an exponential decay curve was obtained. The graph was linear in the first 3 min and the thermoelectric air conditioner minimum temperature for that time was 22.3°C. For no heat load, the thermoelectric air conditioner was able to cool the enclosure to the desired thermal comfort temperature of 22°C in 6 min. A typical air conditioning system that operates on the refrigeration cycle takes approximately 30 min to cool a room to the thermal comfort temperature of 22°C [9, 10].

Time (min)	Ambient temperate (°C)	Temperature inside enclosure (°C)							
		T1	T2	T3	T4	T5	T6	T7	T8
0	30.5	30.7	30.9	30.8	30.7	30.9	29.7	29.9	31.2
1	31.0	28.2	28.1	28.0	28.1	27.1	27.5	27.1	27.2
2	31.2	26.1	26.1	25.9	26.3	24.8	27.2	26.3	25.5
3	30.0	23.5	23.5	23.4	23.5	22.3	25.1	24.3	23.2
4	31.0	21.7	21.9	21.9	21.9	20.7	24.0	23.5	21.8
5	30.8	20.5	20.7	20.6	20.6	19.6	23.0	22.5	20.6
6	30.0	19.7	19.9	20.0	19.9	18.9	22.4	21.6	20.0
7	31.0	18.4	18.7	18.8	18.6	17.8	21.3	20.4	18.9
8	31.0	18.0	18.2	18.3	18.1	17.3	20.9	20.0	18.5
9	31.0	17.8	18.1	18.2	18.0	17.1	20.1	19.3	18.3
10	31.0	17.6	17.9	18.0	17.7	16.9	20.0	18.8	18.1
11	30.6	17.4	17.7	17.9	17.5	16.7	18.9	18.1	18.0
12	30.6	17.2	17.4	17.5	17.2	16.4	18.7	18.1	17.7
13	30.8	17.4	17.7	17.8	17.4	16.6	18.9	18.2	17.9
14	30.5	17.2	17.6	17.6	17.4	16.5	18.8	18.0	17.7
15	31.0	17.0	17.2	17.3	17.2	16.5	18.6	17.8	17.6
16	31.1	16.7	16.9	17.3	16.7	16.6	18.3	17.5	17.1
17	30.9	16.4	16.8	16.6	16.5	16.5	17.7	17.2	17.1
18	30.9	16.4	16.9	16.7	16.5	16.4	17.7	16.8	16.9
19	30.5	16.5	16.6	16.6	16.5	16.3	17.5	16.7	17.0
20	30.6	16.4	16.5	16.6	16.7	16.3	17.2	16.6	16.9
21	31.0	16.5	16.4	16.6	16.5	16.3	17.1	16.7	16.8
22	30.5	16.5	16.4	16.6	16.4	16.2	16.8	16.6	16.8
23	29.8	16.4	16.5	16.7	16.4	16.3	16.6	16.5	16.7

Table 1.
Temperature readings obtained from testing enclosure of thermoelectric air conditioner.

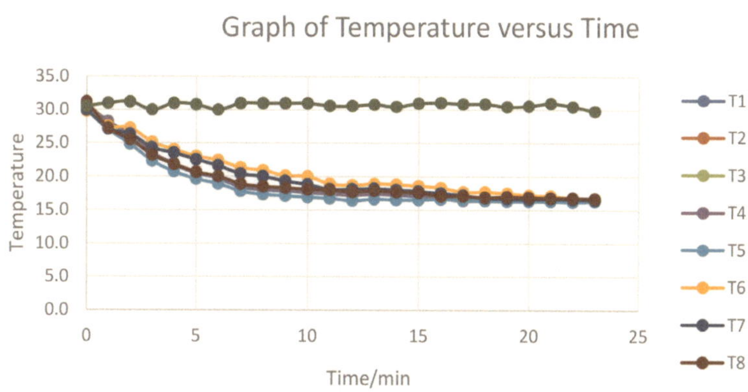

Figure 13.
Graph of temperature/°C versus time/min.

Therefore, the thermoelectric air conditioner has a much faster cooling rate as compared to the conventional air conditioning system. The thermometric device was able to cool the enclosure and maintain a minimum temperature of 26.5°C, however it was expected that the device would cool the enclosure much less than this temperature. This was limited to the high temperature on the hot side of the module. The temperature on the hot side of the thermometric module was stable at 39.6°C when the minimum temperature of the enclosure was 26.5°C. From the specification graph of the module, the temperature difference chosen was 30°C. Therefore, if the hot side of the module was maintained at a much lower temperature then the cold side of module would result in a lower temperature reading (**Tables 2** and **3**).

Specific volume at inlet, $\vartheta_1 = 0.882\,\frac{m^3}{kg}$

15 CFM fan used = $0.00708\,\frac{m^3}{s}$

Mass flow rate, $m_a = 0.00708\,\frac{m^3}{s} \times \frac{1}{0.882\frac{m^3}{kg}}$

Mass flow rate, $m_a = 0.008027\,\frac{kg}{s}$

Using a psychometric chart, the enthalpy was determined for both the inlet and outlet of the thermoelectric air conditioning unit.

Enthalpy of the air at the inlet, $h_i = 71.2\,\frac{kJ}{kg}$

Enthalpy of the air at the outlet, $h_o = 35.6\,\frac{kJ}{kg}$

Difference in enthalpy,

$$h_d = h_i - h_o$$

$$= 71.2\text{–}35.6$$

$$= 35.6\,\frac{kJ}{kg}$$

Thermoelectric module	Voltage (V)	Current (A)	Power (W)
Module 1	11.85	18.11	214.60
Module 2	11.79	17.61	207.62
Module 3	11.96	17.32	207.15

Table 2.
Voltage and current readings obtained from the multi-meter.

No.	Inlet air temperature (°C)		Outlet air temperature (°C)		Power (W)
	Dry bulb	Relative humidity %	Dry bulb	Relative humidity %	
1	30.7	60.6	17.1	52.1	614.25
2	28.1	60.9	18.8	54.2	617.10
3	31.5	59.7	16.9	54.8	613.98
Average	30.1	60.4	17.6	53.7	615.11

Table 3.
Experimental results of thermal properties of thermoelectric air-conditioner.

Cooling capacity of the unit,

$$\begin{aligned}\dot{Q}_{cc} &= \dot{m}_a(h_i - h_o)\\ &= 0.008027\ \frac{\text{kg}}{\text{s}} \times 35.6\ \frac{\text{kJ}}{\text{kg}}\\ &= 0.286\ \text{kW}\\ &= 286\ \text{W}\end{aligned}$$

Coefficient of performance,

$$\begin{aligned}\text{C.O.P.} &= \frac{\dot{Q}_{cc}}{P_{in}}\\ &= \frac{0.286}{0.61511}\\ &= 0.465\end{aligned}$$

In comparing the owning and operating costs of the thermoelectric air conditioner to an air conditioner that operates using the refrigeration cycle, the thermoelectric air conditioner was the better choice of approximately 47.5% cheaper in the overall costs. Although the overall cost of the thermoelectric air conditioner was cheaper, it consumes a considerable large amount of power of 695 W more. This resulted in a higher cost of electricity per year of $156.12 more than the refrigeration air conditioner.

However, the major factors which influenced the thermoelectric air conditioner in being the overall cheaper choice are the; life span of the device and the operating costs. The thermoelectric air conditioner has a greater estimated life span of 7 years more than the refrigeration air conditioner. This is because the life span of the air conditioner that operates on the refrigeration cycle uses a compressor which is the "heart" of that air conditioning system, contains a lot of moving parts and is therefore more prone to failure. On the other hand, the thermoelectric air conditioner uses thermometric module which is the main component of the system, contains no moving parts. Maintenance of the thermometric air conditioner is also cheaper in which it does not need re-gassing or regular inspection check for gas leaks as compared to the refrigeration air conditioning system.

The coefficient of performance was calculated for the thermoelectric air conditioning system, which was found to be 0.465. This value is small in comparison with the average coefficient of performance for the vapor refrigeration cycle of 3.0 [11, 12]. The main reason for this vast difference is the power consumption of the thermoelectric air conditioner. The experimental cooling capacity was found to be 286 W while the system was sized for 330 W. In comparison, this value showed a 13.1% reduction of cooling capacity which may be due to the power supply used to power the thermoelectric air conditioner, since the system was only consuming 615.11 W while it was calculated that the system needed 756 W in order to produce a cooling capacity of 330 W. Additional causes may be inaccuracies in the experiment or heat leaks between the cold and hot side of the thermoelectric module.

6. Conclusion

A prototype of a thermoelectric air conditioner was designed. Designed calculations were produced. The hot side of the module did not exceed the maximum operating temperature of 138°C, which means that the design of the heat sink for the hot side was sufficient. In additionally, CFM rating for the exhaust fans were calculated which led to the proper selection of the fans to be used. The prototype of the thermoelectric air conditioner was successfully built and tested. From the design calculations, the appropriate materials for the device were selected which enabled the device to cool and dehumidify to the thermal comfort zone. The performance of the prototype was determined where:

i. A cost analysis was done in which the thermoelectric air conditioner was compared to a conventional air conditioner that operates based on the refrigeration cycle and the thermoelectric air conditioner was the better choice with an overall 47.5% cheaper in overall cost.

ii. The coefficient of performance of the system was calculated and found to be 0.465.

iii. The device was able to cool and dehumidify the air within the thermal comfort zone of 22°C and 53.7% relative humidity.

Author details

Adeyanju Anthony Ademola
Mechanical and Manufacturing Engineering Department, University of West Indies, St. Augustine, Trinidad and Tobago

*Address all correspondence to: anthony.adeyanju@sta.uwi.edu

References

[1] Patyk A. Thermoelectrics: impacts on the environment and sustainability. Journal of Electronic Materials. 2010

[2] Wan Q, Deng Y, Su C, Wang Y. Optimization of a localized air conditioning system using thermoelectric coolers for commercial vehicles. Journal of Electronic Materials. 2017

[3] Nikam AN, Hole JA. A review on use of Peltier effects. International Journal of Science, Spirituality, Business and Technology. 2018:2277-7261

[4] Attar A, Lee H, Weera S. Optimal design of automotive thermoelectric air conditioner (TEAC). Journal of Electronic Materials. 2014

[5] Pache GS, Tamse AB. HVAC (heat ventilation and air conditioning system) using TEC (thermoelectric couple). IOSR Journal of Mechanical and Civil Engineering. 2014:28-33

[6] Irshad K, Habib K, Basrawi F, Thirumalaiswamy N, Saidur R, Saha BB. Thermal comfort study of a building equipped with thermoelectric air duct system for tropical climate. Applied Thermal Engineering. 2015

[7] Hebei IT. (Shanghai) Co., Ltd. Thermoelectric Cooler TEC1-12730. 2019 Available from: http://peltiermodules.com/peltier.datasheet/TEC1-12730.pdf [Accessed: 24 January 2019]

[8] Adeyanju AA, Ekwue E, Compton W. Experimental and theoretical analysis of a beverage chiller. Research Journal of Applied Sciences. 2010;5(3):195-203

[9] Adeyanju AA. Experimental comparison of thermoelectric refrigeration and vapor power compression refrigeration. Journal of Engineering and Applied Science. 2010; 5(3):221-225

[10] Zuazua-Ros A, Martín-Gómez C, Ibañez-Puy E, Vidaurre-Arbizu M, Gelbstein Y. Investigation of the thermoelectric potential for heating, cooling and ventilation in buildings: Characterization options and applications. Renewable Energy. 2019: 229-239

[11] Tipsaenporm W, Rungsiyopas M, Lertsatitthanakorn C. Thermodynamic analysis of a compact thermoelectric air conditioner. Journal of Electronic Materials. 2014

[12] Ibañez-Puy M, Bermejo-Busto J, Martín-Gómez C, Vidaurre-Arbizu M, Antonio Sacristán-Fernández J. Thermoelectric cooling heating unit performance under real conditions. Applied Energy. 2017:303-314

Chapter 5

An Experimental Study of Synthetics and Natural Refrigerants Gases

Eraldo Cruz dos Santos, Carlisson Azevedo, Caio Macêdo, Carla Azevedo, Raíssa Araújo, Sílvia Soares Gomes, Lana Baia and Yves Alexandrino

Abstract

In Brazil, there has been a significant change in the types of refrigerants used in air conditioning and refrigeration systems. This chapter seeks to compare, based on legislation established by the European Parliament and of the Council, which has been understood and obeyed all over the world, the use of F-gas and different types of fluids intended for use in such systems and also discuss the use of natural fluids such as hydrocarbons, ammonia, and carbon dioxide (CO_2), in refrigeration or air conditioning systems, considering the characteristics and properties of these types of fluids, such as thermodynamic and psychometric parameters, the global warming potential (GWP), application limits, flammability factor, and performance as refrigerant. Some specific examples of the use of fluids such as R-R290, R-410a, and R-600a, instead of R-22 fluid, will also be discussed. For this purpose, a test bench was developed with an equipment chiller in the Climatization and Thermal Comfort Laboratory—ClimaTConT—of the Federal University of Pará, in order to compare the types of refrigerant fluids, usually used in air conditioning systems, considering the conditions, evaluating the possibilities of reuse of these fluids, without performance losses, in significant values, always seeking the reduction of the use of synthetic fluids, which are more aggressive to nature, generating pollution and degradation to local environments.

Keywords: F-gas, hydrocarbons, natural refrigerants, synthetic fluids, chiller bench

1. Introduction

When it comes to environmental pollution, the first idea is the one caused by the operation of a motor vehicle, consuming a fossil fuel, nonrenewable, or the irregular disposal of garbage in areas near urban centers. However, refrigeration and air conditioning systems are highly polluting the local environment, because these systems use refrigerants with specific characteristics, and some of their components can react with the ozone layer or be released into the atmosphere, which is equivalent to a few hundred or thousands of times the emission of CO_2.

A great example is a cooling system that works with R-404A refrigerant, which is a synthetic fluid. Under normal operating conditions, this system has no ozone-depleting chemical component. But, if an approximately 10 kg of this refrigerant leaks into the atmosphere, the contribution to the greenhouse effect/ global warming will be equivalent to the emission of 40 tons of CO_2 [1, 2].

In Brazil, all refrigerants, currently available on the market, are the basis of HCFC and HFC, but these fluids have their days counted for their discontinuation given the contribution to environmental pollution and the action of the Montreal and Kyoto Protocols that made societies aware of the harmful action of these products when released into the local environment [1–3].

Due to these prohibitions imposed by world laws and protocols, the air conditioning and refrigeration market in Brazil aims solutions for the use of refrigerants, which vary between synthetic fluids that has a high value of global warming potential (GWP) and natural fluids, which has a lower value of global warming potential (GWP) than the synthetic ones. However, natural fluids have some impasses that are directly related to their composition, such as ammonia, which has excellent thermodynamic qualities, but it is a highly toxic fluid. The CO_2 itself is another fluid that can be used in refrigeration systems, but as a downside, it works under high pressures, close to 100 bar. The hydrocarbon refrigerants, which are better known as R-290 and R-600a, are also excellent thermodynamic fluids, but these fluids are highly flammable and have their maximum amount to be applied to limited climate systems for security reasons [4–10].

In order to encourage the use of natural fluids in air conditioning and refrigeration systems, some laws and protocols have been created in the world to regulate, standardize, and inform mainly about safety issues related to the use of these types of refrigerants [4–6, 11]. From the point of view of use, there are scholarly works in air conditioning and refrigeration systems produced by companies that already work with some natural fluids and have great yields, both on the issue of capacity and cooling efficiency, as the issue of energy consumption [12, 13].

The present work intends to develop a test bench so that reliable data on synthetic and natural fluids can be obtained. In the experimental stage, the refrigerants R-22 and R-290 were used. Nevertheless, in the simulation stage, some data were taken of Danfoss software in order to evaluate the thermodynamic, psychometric, and electric parameters, the ozone depletion potential (ODP), the GWP, the working pressure limits, and the local conditions. Hence, it has become possible to indicate the ideal fluid that should operate in a safe condition and with less aggressive potential to the environment.

2. Refrigerant legislation

With the discoveries of the harmful effects of synthetic refrigerants, various organizations and institutions around the world have begun to work together to develop methods for control and to standardize and also to develop some kind of awareness about the use of these refrigerants as a working fluid in refrigeration and air conditioning systems.

Protocols were created first to fight the fluids that negatively affect the ozone layer [1] and later to fight against the use of synthetic fluids, which are massive contributors to the greenhouse effect. Since the 1970s, some scientists have proven that there is a direct relationship between ozone depletion and the use of CFC (fluorine and chlorine-based hydrocarbons) compounds by industry, not only as refrigerants fluids but as aerosol propellants and foam blowing agents. According to photochemical reaction models involving ultraviolet irradiation, the depletion of the ozone layer is the result of a chain effect promoted by chlorine (or bromine) atoms [2].

2.1 Montreal Protocol

The Montreal Protocol is an international agreement developed by the United Nations, focusing on the environment. This protocol aims at the gradual reduction to the full transition of synthetic fluids with high ozone layer depletion potential by other types of environmentally friendly refrigerants. Since its inception in 1987, data collection through 2014 indicates that, with the Montreal Protocol, it has been possible to reduce the supply of ozone-depleting fluids by 98%, as most of this amount was used in cooling and air conditioning systems [1]. In European countries, in October 2016, HFC-lowering steps were agreed and became part of the Montreal Protocol, which entered into force on January 2019, pending ratification by 20 states, which appears to be a formality at the moment.

In the Montreal Protocol, in addition to the mechanisms for the gradual reduction and elimination of refrigerant use discussed above, many governments are applying tax measures to reduce the consumption of high potential global warming refrigerants, such as the creation of weighted taxes according to potential for the use of these products.

The production and use of HFCs such as R-32, R-125, R-134a, and R-143A and their mixtures (R-404, R-407C and R-410A) are not regulated by the Montreal Protocol but must be country-specific regulations made individually, so countries such as Spain, Denmark, Norway, and Sweden have imposed taxes on the use of HFCs in their air conditioning and refrigeration systems to reduce their use and control.

2.2 Kyoto Protocol

Like the Montreal Protocol, the Kyoto Protocol is an international agreement, managed by the United Nations, that seeks to reduce pollutant emissions into the atmosphere. It is known that the last 150 years of industrial activities related to developed countries generated an increase in emissions and strengthening the greenhouse effect [2].

This protocol was prepared in Kyoto, Japan, in 1997 and, after adjustments, began to be implemented in 2005 and had its first stage, known as the first period, accounted for between 2001 and 2012. During that period 37 countries industrialized, and the European community have if pledged to reduce pollutant emissions by 5% compared to the 1990s. In the second period, in 2012 there was a meeting between the leaders of the countries that signed the Kyoto Protocol and the United Nations. At this meeting it was agreed that between 2013 and 2020 countries would continue to work on reducing emissions by 18% of the values stipulated in the 1990s [2].

2.3 F-gas regulation

The 517/2014 Law of the European Parliament and of the Council became known as regulating the use of fluorine-containing gases, known as F-gas, with the main objective of reducing emissions of gases with high potential for the destruction of the ozone layer, because they contribute to the greenhouse effect. Brazil is a signatory to this Law, which defines the rules on storage, recovery of synthetic fluids, as well as conditioning the trade of equipment and the refrigerants themselves in use in the Brazilian territory and also sets the practical limits on the use of F-gas in Brazil [3–5].

The first highlight of 517/2014 Law is the prohibition of the intentional release of any amount of F-gas into the atmosphere; however, it does not cite any criminal implication on anyone caught doing the act of this way.

Equipment		Prohibition date
Household freezers and refrigerators with HFC refrigerant and greater than 150 GWP		January 1, 2015
Cold rooms for commercial use	HFC with GWP 2500+	January 1, 2020
	HFC with GWP 150+	January 1, 2022
Fixed refrigeration equipment with HFC refrigerant and with GWP greater than 2500		January 1, 2020
Commercial central cooling system with a capacity of 40 kW or higher that works with GWP 150+		January 1, 2022
Mobile residential air conditioning equipment using HFC refrigerant with GWP 150+		January 1, 2020
Fixed residential air conditioners with load less than 3 kg and GWP 750+		January 1, 2025

Source: Regulation (EU) N°. 517/2014 [5].

Table 1.
Equipment with dates for prohibiting the use of refrigerants.

The Article 4 of this Law establishes that stationary refrigeration equipment, air conditioners, and heat pumps must have an operator that checks for leakage points on the equipment according to the equivalent amount of CO_2 in refrigerant charge where 5 or more equivalent tons of CO_2 should be checked at least once a year. For equipment with an equivalent load above 50 tons of CO_2, this check should be performed every 6 months. Systems with loads over 500 tons equivalent must be checked every 3 months.

Article 13 cites restrictions on the use of F-gas and imposes that from January 2020, the use of F-gas with GWP equal or greater than 2500 shall be prohibited in refrigeration systems with an equivalent load of 40 tons of CO_2 or more. Nevertheless, if the fluid used in the system can be recovered, reprocessed, and reused, it will have its utility extended until January 1, 2030.

Table 1 shows some air conditioning and refrigeration equipment with their respective prohibition dates for their refrigerants.

The questions of the regulation of refrigerants is the application of control mechanisms over the use and disposal of F-gas in each country in order to comply with the dates set out in **Table 1**.

For the purpose of facilitating the understanding about this chapter and the application in HVAC and refrigeration's systems, some definitions used by law and by some educational and research entities and institutions are presented below:

- **Global warming potential (GWP):** This is the direct mass comparison of CO_2 that a fluid causes when released into nature over a 100-year period [5].

- **Tons of CO2 equivalent:** Quantity of mass with a global warming potential multiplied by a global warming potential coefficient.

- **Recovery or regeneration:** Process of gathering and direct F-gas storage equipment, i.e., the removal of a refrigerant from system to store it in an external tank [14]. In this process some contaminants found in the systems are removed, like oil, water, oleic acid, and hydrochloric acids. In this sequence, the refrigerant is distilled and then filtered, condensed, and analyzed, and if it complies with ARI 700 Standard (purity, humidity, acidity, non-condensable gases, and maximum allowable residue levels), then it will be refilled.

- **Recycling:** Reuse of F-gas after a treatment process, that is, is the process involving the cleaning of refrigerant by an oil separator and a filter drier for reuse. It is to remove some contaminants like oil, water, and oleic and hydrochloric acids that can be found in system. In this process, the cooling fluid is distilled, filtered, then condensed, and finally, bottled [5, 14].

- **Reuse:** Reprocessing of the refrigerant, involving the manufacturing processes of new products from the refrigerant recovered.

- **Repair:** Restoration of equipment, for some reason, has lost its tightness, and this has resulted in the release of the working fluid to the atmosphere.

- **Leak detection system:** Mechanical or electronic equipment that allows to identify possible leakage points in the system components [14].

- **Destination:** Currently in Brazil there are regeneration plants capable of receiving refrigerants to be reused within the standards required by law.

In underdeveloped countries, such as Brazil, despite the knowledge of the mechanism of ozone layer depletion and the creation of specific legislations to address the issue, there is still no concern with the use and disposal of F-gases, which makes it difficult to meet the dates stipulated by 517/2014 Law [5].

3. Refrigerants

Refrigerants can be defined as substances that serve as a means of transport and heat transfer, absorbing heat at low temperatures and rejecting it at higher temperatures [15]. From that definition, in principle, any substance that changes phase in the refrigeration or air conditioning cycle, from liquid to vapor during heat absorption and from vapor to liquid during heat rejection, may function as a refrigerant. Their commercial usefulness depends on the temperature ranges and pressure at which the system will operate and, more recently, the environmental impacts that eventually this refrigerant can cause when released into the environment.

Water and air can also be considered as fluid refrigerants, although they do not show phase change during the process as well as other synthetic fluids or not form aqueous solutions that are characterized by operating at temperatures below 273.16 K (0°C) without crystallizing.

Among the various properties that a refrigerant should include for use in air conditioning and refrigeration systems, the most important are [15]:

i. Favorable thermodynamic properties.

ii. High chemical stability when operating within the system and low chemical stability outside the system.

iii. Nontoxic.

iv. Non-flammable.

v. Having small specific volume (causing less compressor work).

vi. Having high latent heat of vaporization.

vii. Evaporate at pressures above atmospheric (in case leakage air will not enter the system avoiding the risk of explosions).

viii. Having compatibility with the lubricating oil of the compressor.

ix. Having adequate compatibility with the materials of the refrigeration system.

x. Be easy to detect.

xi. In case of leakage, it should not attack or spoil food, should not contribute to global warming, and should not attack the ozone layer.

xii. Not posing a danger to the environment.

xiii. Be commercially available at a reasonable cost.

The refrigerant groups available in the Brazilian market are divided into [15, 16]:

- **CFCs (chlorofluorocarbons):** They are refrigerant fluids whose molecule is formed by the element's chlorine, fluorine, and carbon, for example, the fluids R-11, R-12, R-113, R-114, R-115, and R-502, among others. The lifetime of CFCs in the atmosphere ranges from 60 to 540 years, causing high depletion of the ozone layer. Most CFCs have an ozone layer depletion potential (ODP) index ranging from 0.6 to 1.

CFCs were no longer manufactured by industrialized countries in January 1996 and, with some exceptions, by developing countries. These fluids feature high ODP and high GWP. According to the Montreal Protocol, in 1996 these gases were banned from developed countries, and in 2010 they should be banned from developing countries such as Brazil, but these gases are still found in old refrigeration and air conditioning systems. These fluids are used in automotive air conditioners, commercial refrigeration, and home refrigeration (refrigerators and freezers).

- **Hydrochlorofluorocarbons (HCFCs):** In these refrigerants some chlorine atoms are replaced by hydrogen, as an example of these fluids, have it R-22, R-123, R-401A, and R-4012A. These fluids are used in climate control systems, window air conditioners, split air conditioners, self-containers, cold rooms, etc.

These gases represent the second generation of fluorinated gases and were the main alternative to CFCs. They have lower ODP, but the GWP still between 1000 and 5000, so they are a little less harmful to the environment than their CFC predecessors, considering the ozone depletion. HCFCs are not fully halogenated like CFCs. The atmospheric lifetime of HCFCs ranges from 2 to 22 years, and therefore their ODP ranges from 0.02 to 0.1 [15, 17].

Currently, in Brazil, HCFC-22 or (R-22) has been the most commonly used refrigerant in air conditioning systems since the early 1990s. HCFCs have been used as transition fluids, and their restriction on use and manufacturing started in 2004. According to the Montreal Protocol, the elimination of HCFCs in developed countries is expected by 2030 and in Brazil by 2040. Nevertheless, some countries are already well advanced on their elimination. In the European Union the use of HCFCs was terminated in January 2010 and may still be used until January 2015 in some specific cases.

- **Hydrofluorocarbons (HFCs):** In these fluids all chlorine atoms are replaced by hydrogen, for example, R-134a, R-404A, R-407C, R-410A, etc. These fluids are used in automotive air conditioning systems, commercial refrigeration and domestic refrigeration (refrigerators and freezers), etc.

As there are no chlorine atoms in HFCs refrigerants, they do not cause ozone layer depletion. HFCs are expected to become the most widely used in air conditioning systems in the coming decades. These fluids can be said to represent the third generation of fluorinated gases that have zero ODP and median and high GWP values. Currently, HFCs are largely used in commercial refrigeration and residential air conditioning.

- **Hydrofluoroolefins (HFOs):** The HFOs represent the fourth generation of fluorinated gases, with zero ODP and very low GWP values. It is already being used in automotive air conditioning systems in developed countries (the United States and Europe). They are fluids used in place of R-134a as they are suitable for high temperature applications and operate at similar pressures.

- **Natural refrigerants:** These fluids are generated through natural biochemical processes, so they pose no risk to the ozone layer and have very low or zero GWP indices. The most common natural fluids are R717 (ammonia), R744 (CO_2), hydrocarbons, R718 (water), and R729 (air).

Although the application of natural fluids in refrigeration systems is a worldwide trend, each fluid has its own characteristics and requires special care to be implanted.

- **Hydrocarbons:** They consist of a group of nontoxic gases with zero ODP and low GWP. They are environmentally friendly and have excellent thermodynamic properties, allowing good efficiency in similar or even better cooling systems than HCFC and HFC fluids.

Because these refrigerant fluids are considered as flammable gas, hydrocarbon refrigeration systems must meet a number of safety guidelines. Typically, these gases are applied in small systems such as refrigerators or indirect systems and cooling secondary systems with other fluids (such as a CO2 cascade system). This ensures a low risk of fire if a leak occurs.

The most commonly used hydrocarbons in refrigeration are propane (R-290) and isobutane (R-600a). In addition to the GWP, the ozone depletion potential (ODP) is another parameter that compares the ozone layer depletion potential regarding the R-11 refrigerant, which is assigned the value 1. This value is compared to other types of fluids.

Refrigerants can be classified as simple, which are not mixed with other fluids such as R-22, R-134a, and R-32 and natural or blend fluids, which are mixtures of fluids, such as R-410A. The blend fluids may also be azeotropic or non-azeotropic. And this is another classification of refrigerants, which divides them into two broad categories, which are:

- **Azeotropic mixtures:** In these types of mixtures, the components cannot be distilled off. The mixture evaporates and condenses as if it were a single substance. Its properties are totally different from those of its components. As an example, we have CFC/HFC-500, HCFC/CFC-501, and HCFC/CFC-502. For example, HCFC/CFC-501 is a mixture of 75% HCFC-22 with 25% CFC-12 on a mass basis. Azeotropic mixtures causing ozone layer depletion were no longer manufactured in 1996 in developed countries.

- **Zeotropic or non-azeotropic mixtures:** In a zeotropic mixture, its components are separated by distillation process. Thus, the mixture evaporates and condenses at different temperatures. Examples of such fluid types are R-400 and R-401A/B/C. Currently, these are the most promising alternative refrigerants for retrofit in air conditioning and refrigeration systems.

The CFCs and HCFCs are used worldwide in a variety of applications. By the middle of 1980, about 1/3 of CFC consumption occurred in the United States (USA). In 1985, the consumption of these fluids was around 278 tons.

These fluids served industrial sectors of insulating foams, automotive air conditioning, refrigeration and residential air conditioning, and commercial and industrial and other products. The CFC-expanded insulating foams have been widely used in industrial processes. Of this total, 19% was used in automotive air conditioning and 5% in new refrigeration and air conditioning systems. According to ABRAVA the percentage of halogenated refrigerant consumption is as follows: 77% HCFC-22, 10% CFC-11, and 10% CFC-12 [14].

Synthetic refrigerants (CFC, HCFC, HFC, and HFO) are considered safer, simpler to handle, and cheaper; however, most of them present some types of environmental risks. The presence of chlorine in the composition of most synthetic fluids is what makes them harmful to the ozone layer. Fluids that do not have chlorine in their composition are considered ecological but still favor the greenhouse effect, so the group of HFOs represents the promising generation of synthetic fluids.

The natural fluids such as CO_2, ammonia, and hydrocarbons are considered more complex and have a higher cost. In addition, they are considered highly flammable, and care should be taken to install the systems that will receive these fluids. They are recommended for low load operations such as vending machines and have good applicability at any temperature.

Ammonia, used in ammonia fluid R-717, is considered toxic and slightly flammable, has good thermodynamic characteristics, and should be used in systems combined with glycol (chiller system), or CO_2 (cascade system), for example [18].

The fluids with behavior of one substance, known as simple fluids, are easier to work because they have well defined properties, can be carried gaseous or liquid loads and in case of leaks, composition still the same in the installation, the only change will be in the volume. Azeotropic blend fluids such as R-22, despite being mixtures, have simple fluid characteristics. Non-azeotropic substances (such as R-404A, which has R-125, 44%, R-143 to 52%, R134 to 4%) have fluids with different boiling points and do not mix perfectly, so do not mix well. Thus, do not behave like simple fluid.

The use of these types of refrigerant implies extra care, as in the case of leaks, the composition of the leaked fluid and the composition of the remaining fluid may be unknown requiring a new charge. There is also the characteristic of glide temperature, where the evaporation temperature is not constant.

All these substances used in air conditioning and refrigeration systems are classified by safety groups as defined by AHSRAE. The classification considers the flammability and toxicity of each substance, as shown in **Table 2**, which presents the safety classification according to ASHAE [16]. Each refrigerant group requires a set of safety procedures for their use, installation, and disposal. The refrigerants found in the CFC, HCFC, and HFC categories are mostly classified in group A1 as non-flammable and of low toxicity. They are safe but environmentally aggressive fluids.

HFO refrigerants are less environmentally friendly but are classified in group A2L, i.e., there is a low risk of spreading flames in the event of a leak, and therefore require some caution and safety devices in equipment using these types of refrigerants.

High flammability	A3	B3
Low flammability	A2	B2
	A2L	B2L
Non-flammable	A1	B1
	Low toxicity	High toxicity

Source: ASHRAE Fundamentals [16].

Table 2.
ASHAE safety classification.

Natural fluids, while an alternative more environmentally friendly, require a lot of care, and their facilities must follow the safety guidelines defined for each application. Propane and isobutane hydrocarbons are classified as highly flammable (group A3) while ammonia as highly toxic and low flammability (group B2L). Although carbon dioxide falls within group A1, it is a choking gas, so your systems must follow a series of precautions to ensure the safety of the installation.

In a Refrigeration system, the refrigerant is contained within the equipment, there is no direct contact between users and the refrigerant.

When a coolant leak in a refrigeration system and air conditioning, depending on the design and the refrigerant charge amount used in the system, choking problems may occur in people or product contamination.

Most leaks in the air conditioning and refrigeration systems are related to maintenance aspects, from the use of poorly maintained, deteriorated or even inadequate components and equipment as well as precarious labor that only acts when problems occur.

In the refrigeration, the word "retrofit" has been used to designate the adaptations that are made to equipment that works with CFCs so that they can work with alternative fluids (HCFCs, HFCs), making them efficient, modern, and economical. Another solution is the alternative fluid line, also called "blends," which is a good alternative for converting equipment that is operating in the field as it requires minimal changes to the original system, and in most cases no compressor replacement is required [19].

Table 3 shows a comparison of the ODP, GWP, and ASHRAE safety ratings of some refrigerants.

In the Brazilian market, as shown in **Table 3**, it is observed that the replacement of refrigerant fluid R-22 (HCFC) with R-410A (HFC) is occurring; however, it is worth noting that this substitution does not alter the potential for global warming.

In a direct comparison, R-134a refrigerant (tetrafluoroethane) has similar physical and thermodynamic properties to R-12, but it has lower ozone depletion potential (ODP) due to the absence of chlorine and the shorter lifetime in the atmosphere (16 years versus 120 years for R012). This fluid has a 90% reduction in greenhouse potential when compared to R-12. In addition, it is non-flammable and nontoxic, has high thermal and chemical stability, has compatibility with the materials used, and has adequate physical and thermodynamic properties. In addition, the R-134a refrigerant is free of chlorine and therefore has good compatibility with elastomers. One of the consequences of this change is that during the maintenance process of the air conditioning and refrigeration equipment, the R-22 fluid is released to nature, without any kind of control.

According to the Brazilian Ministry of Environment (MMA), the most commonly used refrigerant in cooling and air conditioning is HCFC-22, accounting for 82% of consumption [15, 19]. Considering the indicators shown in **Table 3** and the

Number/name	Formula	ODP	GWP	ASHAE safety classification
CFCs				
R-11/trichlorofluoromethane	CCl_3F	1	4000	A1
R-12/dichlorodifluoromethane	CCl_2F_2	1	8500	A1
R-502/CFC blend	R-115 (51%), R-22 (49%)	0.23	5590	A1
HCFCs				
R-22/chlorodifluoromethane	$CHClF_2$	0.055	1700	A1
R-123/dichlorotrifluoroethane	$C_2HCl_2F_3$	0.02	93	A1
HFCs				
R-32/difluoromethane	CH_2F_2	0.0	650	A1
R-134a/tetrafluoroethane	$C_2H_2F_4$	0.0	1300	A1
R-410a/AZ-20	R-32 (50%), R-125 (50%)	0.0	1730	A1
R-507/AZ-50	R-125 (50%), R-143a (50%)	0.0	3300	A1
HFOs				
R-1234yf	$C_3H_2F_4$	0.0	4	A2L
R-1234ze	$C_3H_2F_4$	0.0	6	A2L
R-218/perfluoropropane	C_3F_8	0.0	7000	A2L
Natural refrigerants (NRs)				
R-290/propane, Care 40	C_3H_8	0.0	~5	A3
R-600a/isobutane, Care 10	C_4H_{10}	0.0	~5	A3
R-717/ammonia	NH_3	0.0	<1	B2L
R-718/water	H_2O	0.0	<1	A1

Source: ASHRAE Fundamentals [16].

Table 3.
Comparison of available refrigerants for air conditioning and refrigeration systems.

properties listed above, it is important to realize that most refrigerants do not meet the requirements for commercial use in order to comply with Brazilian and worldwide legislation.

The scenario of replacing the R-22 in the refrigeration and air conditioning industry points to a number of alternative fluids. In this context, manufacturers in the refrigeration and air conditioning industry are looking for solutions that meet their goals without neglecting good performance.

According to ABRAVA [14], **Table 4** presents the suggestions of alternative fluids to the use of R-22 according to each application:

In a survey by the members of the UFPA Resfriar Project, in the metropolitan region of Belém do Pará, along with companies that work with the sale, installation, and maintenance of HVAC systems, more than 95% of people involved with the services did not have either training courses or were qualified to develop the services.

In Brazil, it is possible to observe that, in the practical activities, during the maintenance processes, the people qualified for the development of installation,

Equipment types	Usual application	Main refrigerant substitutes
Window air conditioners (WAC)	Residential	R-410A
Air conditioners and heat pumps	Residential/small commercial air conditioning	R-410A
Hot and cold air conditioners (multi-split systems)	Commercial	R-410A
Multi-split systems	Residential/commercial	R-407C or R-410A
Large air conditioning systems	Commercial	R-134a or R-410A
Air condensing chillers	Central systems	R-134a or R-410A
Water condensing chillers	Central systems	R-134a
Commercial refrigeration	Commercial	R-134a, R-404A, R-410A, or R-507A
Industrial and transport refrigeration	Industrial/transport	R -134a or ammonia

Source: ABRAVA [14].

Table 4.
Alternative refrigerants for R-22 refrigerant according to your applications.

operation, and maintenance activities of air conditioning and refrigeration systems, called in Brazil as "refrigeration professional," are the people who, not necessarily, are technicians of maintenance of air conditioning and refrigeration systems and, in many cases, are called "curious".

It was found that during the development of the analysis and maintenance services, these people either lack adequate equipment, such as a vacuum pump or a refrigerant recovery unit, or lack of knowledge, or because they could not afford to purchase such basic equipment and instruments required to perform the service and release the refrigerants from the equipment to the local environment. Besides that, they always have a recharge for normal system operation and, in many cases, do not have a basic analysis or concern about the consequences of these acts.

This common and uncontrolled practice releases thousands of kilograms of refrigerant fluids per year, to the environment, contributing significantly to global warming.

4. Test bench type chiller

In this topic, the theoretical foundations about vapor compression type cooling systems, with synthetic and natural fluids, necessary for the understanding about the experimental apparatus used are initially approached. After that, the experimental apparatus developed in the ClimatCont laboratory to simulate a chiller system has been described.

4.1 Theoretical fundamentals

Some scientific studies specifically dedicated to the comparison of natural fluids with the most common synthetic ones, such as R-22 fluid, have been carried out and published. The research conducted by Park and Jung's [20] related the replacement of R-22 with R-290 with the need of electrical adaptations to ensure the safety of the installation, which resulted in performance coefficients (COP) up to 11.5% when compared to the R-22, i.e., the system with R-290 required 11.5% less electrical power to generate the same refrigeration capacity. Another important point was the reduction of compressor discharge temperatures, which were reduced from 358.16 K (85°C) to 338.16 K (65°C), indicating better use of the lubricant life [20].

Sharmas and Babu comparatively analyzed mixtures between HC and HFC in relation to their characteristics as refrigerants, observing that mixtures containing HCs provided the system with a COP up to 2% higher than a system with HFC, also reducing the compressor discharge temperature, up to 295.16 K (22°C), and for the HCs mixture, a reduction of up to 1 kW per cooling tons in electric power consumption had been achieved [21]. In a computational analysis, these authors assessed that the HC mixture provided a COP of 5.35% over R-22, plus 286.16 K (13°C) lower discharge temperature for the same fluid and with reduced electrical power consumption by 5% [22].

Agrawal et al. compared a mixture of HC (R-290 and R-600a) with HFC (R-134a) and found that the optimum charge for the mixture was 60 g, less than half of the original charge 140 g of HFC, and the nearly 40% improvement in COP. With the optimum loads, HFC consumed 0.5 kW/h, and the HC mixture consumed 0.4 kW/h, providing the cooling capacities of 70 W and 76 W, respectively [23].

4.2 Chiller test bench

To carry out a practical analysis of the use of types of refrigerants, a test bench with a chiller was developed at the Federal University of Pará (UFPA) in partnership with the Refrigeration and Climatization and Thermal Comfort Laboratories (ClimaTConT) along with the research group "Resfriar Project," from the Energy, Biomass and Environment Group (EBMA). The main purpose of the bench is to create a demonstration of the operation of the refrigeration cycle, considering the local psychometric conditions, with a variety of refrigeration fluids.

In this test bench, all parameters are controlled and components have been assembled to facilitate understanding of fluid behavior and performance. After doing a research in the technical literature and articles on refrigerants and their properties and characteristics as environmental contaminants, the conclusion was that the bench will be a good system for a comparison between the original working fluid, HCFC R-22, and natural fluids, hydrocarbon base, which after investigation of the negative points of natural fluids, were the best suited to the project conditions [17, 24].

The test bench is composed of a condenser unit Elgin TUM-2053E 220 V, 60 Hz, single phase, with 1.6 kW of cooling capacity and power consumption of 880 W, with R-22 fluid and with a 1.5 l capacity liquid tank. The expansion device is a Thermostatic Expansion Valve from Danfoss, model TX2, for R-22, with number 01 orifice, with a maximum capacity of 2.5 kW and a single evaporator made of cooper tube with 3/8" of diameter and 1/16" of thickness, contained in an insulated box that can hold until 45L of water as shown in **Figure 1**.

Systems that use water as a refrigerant can both remove heat and add heat of an ambient making the environment conditioned. Refrigerant circulates inside pipes between the heater and the cooler. These systems can be classified according to operating temperature, flow type, and degree of pressurization [25].

The chilled water system is a type of cooling system that operates with water being a secondary fluid in the temperature range of 277.16 K (4°C) to 286.16 K (13°C), usually between 279.16 K (6°C) and 280.16 K (7°C), with working

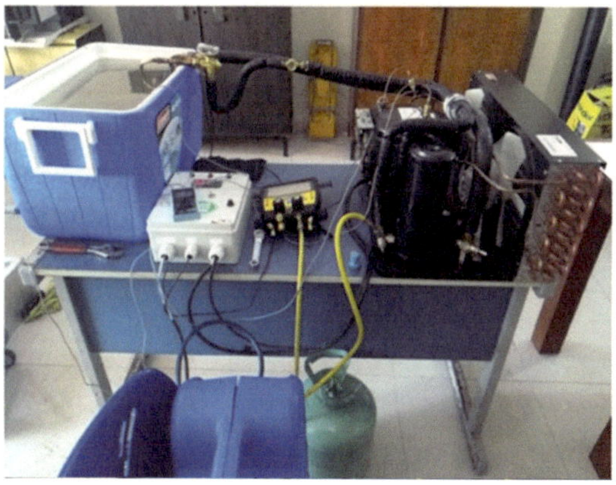

Figure 1.
Bench tests with chiller mounted by ClimaTConT.

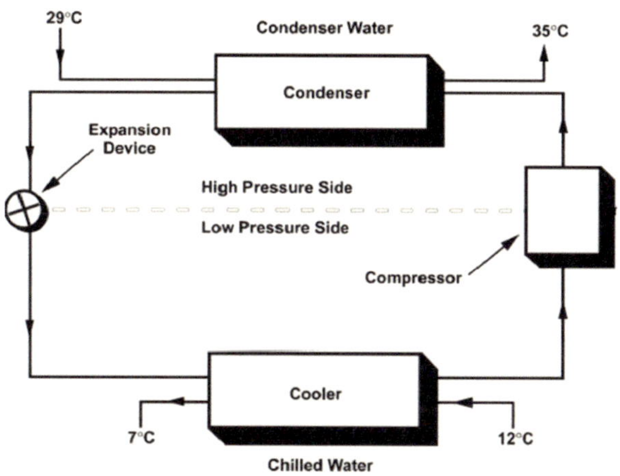

Figure 2.
Schematic representation of the chiller cooling cycle. Source: ASHRAE Fundamentals (adapted) [16].

pressure close to 800 kPa as presented in **Figure 2**, which shows a schematic of the test bench stand with a chiller, which must be used for the cooling of environments [26].

Preliminary data observed from the test bench with chiller shows that its working pressure is 310.26 kPa (45 psig), which indicates an evaporation temperature of 267.16 K (6°C); considering the ΔT of 283.16 K (10°C), the fluid is cooled to 279.16 K (6°C). In this way, the bench tests consisted of evaluating the energy consumption of the R-22 fluid operation, in addition to the working temperatures at specific points, with the values of the refrigerant R-290, which is the fluid with characteristics similar to HCFC.

To measure the temperature parameters, five-point thermometers were installed that acquired the following temperatures: T_1, for room temperature; T_2, for the temperature in the suction pipe; T_3, for the temperature in the discharge pipe; T_4, for condenser air outlet temperature; and T_5, for water temperature. For the measurement of pressure values in the suction (p_{Suc}) and discharge (p_{Desc}) lines of the compressor, a digital pressure gauge was installed in the respective test bench pipes. The energy consumed (E_{Cons}) during the tests was acquired by a power meter installed on the chiller test bench electrical system, as shown in **Figure 1**.

4.3 Weather conditions

In Belém do Pará, which is a city located in the northern region of Brazil, the season with precipitation is overcast, and the dry season is partly cloudy. The city is surrounded by the Guamá River, which is responsible for the high rainfall index of the city.

All year round, the climate of Belém is hot and with high thermal sensation. Throughout the year, the average temperature generally ranges from 297.16 K (24°C) to 309.16 K (36°C), with a relative air unit that is approximately 90%. This city has a thermal sensation ranging from 307.16 K (34°C) to 319.16 K (42°C).

The average altitude of Belém do Pará ranges from 0 to 20 m above sea level, with average barometric pressure of 1010.28 kPa. In summary, Belém do Pará is a city of tropical climate.

5. Results and discussions

By using the chiller test bench, the following results were obtained for R-22 and R-290 refrigerants.

5.1 Evaluation of R-22 refrigerant in bench tests with chiller

To perform this test, the bench was initially charged with a load of 910 g of R-22 refrigerant. **Table 5** shows the results obtained with R-22 refrigerant on the chiller test bench.

When using the R-22 refrigerant, the test bench responded by reducing the water temperature by approximately 286.16 K (13°C) within 30 min of operation, where 0.5 kW of electricity was consumed to perform cooling.

For refrigerant R-22, which is known by high discharge temperatures that occur during its use in operation, values around 340.16 K (70°C) were measured, as shown in **Figure 3** that presents the variation in discharge temperature. Usually, these high values are disadvantageous given that they influence lubrication and hence compressor life.

Figure 3 above shows a large variation in the discharge temperature of the chiller test bench, as a function of operating time. In parallel to the measurement of the discharge temperature variation, the values of the system discharge pressure during operation were also measured, because the higher the temperature of the water to be cooled, the higher the condensing pressure, as can be seen in **Figure 4**, which shows the variation of discharge pressure.

The suction pressure determines at which temperature the refrigerant will evaporate, i.e., the vaporization pressure of the fluid. At the test bench, pressures ranged from 489.52 kPa (71 psig) to 335.08 kPa (48.7 psig), which correspond to the

Time (min)	T1 (K)	T2 (K)	T3 (K)	T4 (K)	T5 (K)	p_{Suc} (kPa)	p_{Desc} (kPa)	E_{Cons} (kW)
10	303.56	292.76	340.26	309.56	289.66	489.52	1613.37	—
15	304.16	280.36	340.26	309.66	280.76	380.59	1668.53	12,019
30	304.26	275.96	341.96	307.96	277.66	351.63	1560.97	12,019
45	304.36	273.96	331.96	307.56	276.76	335.08	1548.56	12,019
60	304.66	273.96	327.16	308.46	276.36	344.73	1489.95	12,020

Table 5.
Test results for R-22 refrigerant on the chiller test bench.

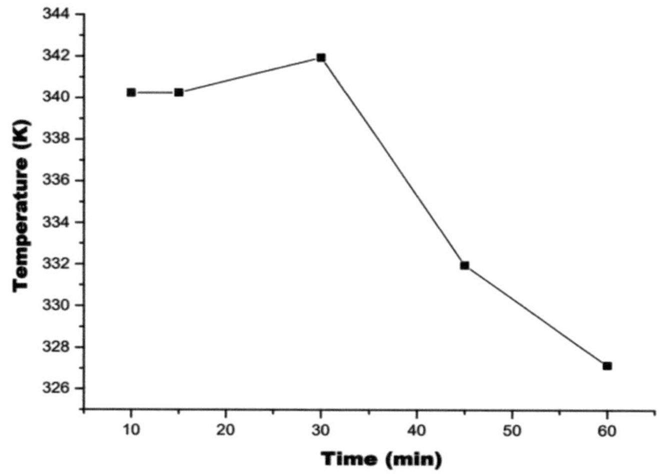

Figure 3.
Variation in discharge temperature on test bench using R-22 refrigerant.

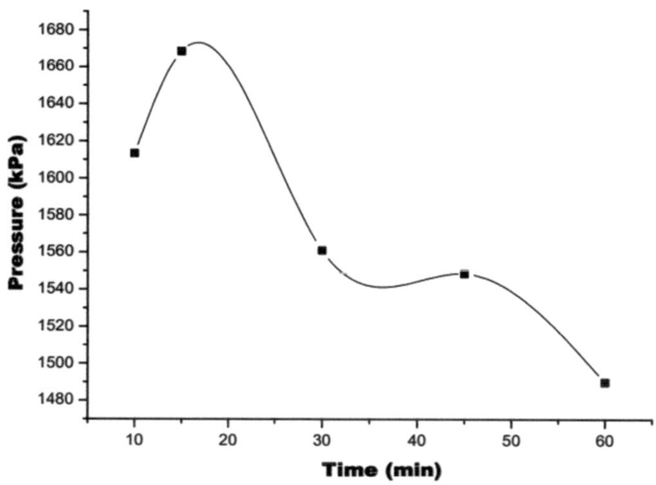

Figure 4.
Variation of discharge pressure on test bench using R-22 refrigerant.

evaporation temperatures of 278.46 K (5.4°C) and 269.16 K (−4.0°C), respectively. During 1 h of operation, the system was able to reduce the temperature and keep the water between 276.16 K (3°C) and 277.16 K (4°C).

5.2 Evaluation of R-290 refrigerant in bench tests with chiller

To perform this, the test bench with chiller was charged with a 370 g load of refrigerant R-290 for 90 min. **Table 6** shows the results obtained with R-290 refrigerant on the chiller test bench.

For the operation of the test bench, the load of R-22 refrigerant was removed and recovered, and then leakage was verified in the components of the test bench with nitrogen. Posteriorly, the bench was charged with 370 g of R-290 refrigerant.

During the tests, it was possible to keep the water temperature close to 274.56 K (1.56°C) and consume 1.0 kW in 70 min of operation. The temperature differences measured at discharge were significantly smaller than those for R-22 refrigerant, as

Time (min)	T_1 (K)	T_2 (K)	T_3 (K)	T_4 (K)	T_5 (K)	P_{suc} (kPa)	P_{Desc} (kPa)	E_{cons} (kW)
0	305.66	294.26	324.96	306.76	275.66	317.15	1206.58	12,022
10	305.36	292.16	331.36	309.06	275.36	320.60	1292.76	12,022
20	306.56	287.96	325.16	307.46	275.36	344.75	1268.63	12,023
30	306.26	275.96	325.26	307.56	274.86	343.35	1268.63	12,023
40	306.46	275.26	325.26	307.46	274.86	341.29	1268.63	12,023
50	306.16	274.96	324.46	307.06	274.36	334.39	1247.95	12,023
60	305.46	274.66	323.46	306.16	274.66	327.50	1206.58	12,023
70	305.36	274.16	323.16	306.16	274.66	327.50	1206.53	12,023
80	304.66	273.76	322.76	305.76	274.76	327.50	1206.53	12,023
90	305.26	273.76	322.66	305.86	274.66	324.05	1213.48	12,024

Table 6.
Test results with R-290 refrigerant on the chiller test bench.

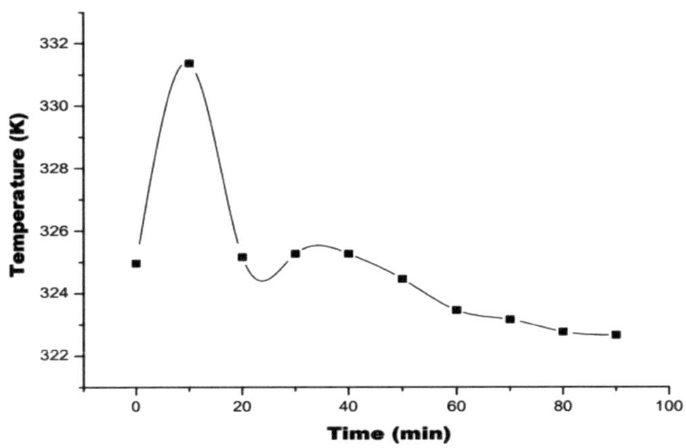

Figure 5.
Variation in discharge temperature on test bench using R-290 fluid.

shown in **Figure 5**, which highlights the variation in bench discharge temperature with R-290 refrigerant.

It is observed in **Figure 5** that the maximum discharge temperature measured on the chiller test bench, with R-290 refrigerant, was less than 331.36 K (59°C). The discharge pressure compared to that of R-22 refrigerant showed values up to 29% lower, with little fluctuation, as shown in **Figure 6** that presents the pressure variation in the discharge plumbing.

It was observed that the suction pressure ranged from 317.16 to 344.75 kPa (46 and 50 psig), corresponding to the respective evaporation temperatures of 269.16 K (−4°C) and 271.16 K (−2°C).

When comparing energy consumption, the chiller test bench with R-22 refrigerant consumed 1.0 kW of electricity to cool 40 l of water in 50 min of operation, while with R-290 refrigerant consumed 1.0 kW in 70 min of operation, i.e., with R-22 the bench consumed 1.2 kW/h, while with the R-290 the consumption was 0.857 kW/h.

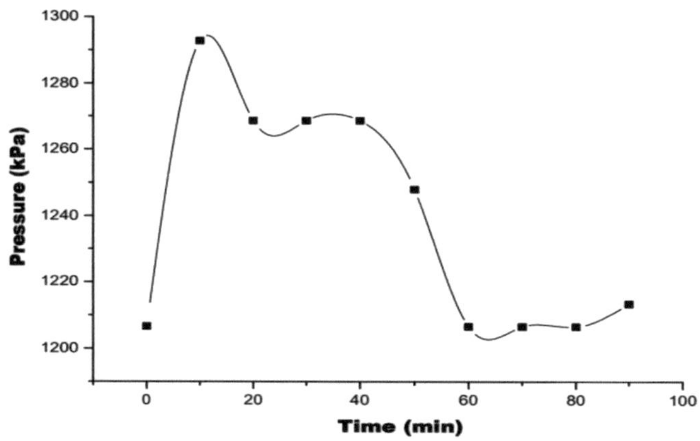

Figure 6.
Variation of discharge pressure on test bench using R-290 refrigerant.

The cooling capacity of system $\dot{Q}_{Ref}$ was 2326 W, so the performance coefficient (COP) can be calculated as shown in Eq. (1) and Eq. (2):

$$COP_{R-22} = \frac{\dot{Q}_{Ref}}{E_{Cons}} = \frac{2326}{1200} = 1.9334 \tag{1}$$

$$COP_{R-290} = \frac{\dot{Q}_{Ref}}{E_{Cons}} = \frac{2326}{857} = 2.714 \tag{2}$$

As calculated above, the R-290 refrigerant COP was 40% higher than of the R-22 refrigerant.

5.3 Comparison between types of refrigerants

The graphs shown below are comparing pressure and temperature of the most common synthetic and natural fluids taken from the Danfoss application [27] and based on the National Institute of Standards and Technology (NIST) [28].

As presented in **Tables 4** and **5**, the R-22 substitute refrigerant is R-410A refrigerant. The R-410A is an almost azeotropic mixture composed of R-125 and R-32. It is a chemically stable product with low temperature glide and low toxicity. According to ASHRAE this fluid is classified as A1 in group L1.

Figure 7 shows a comparative graph between these two fluids for temperature and discharge pressure.

When compared to other refrigerants, R-410A has the highest working pressure, as shown in **Figure 8**, which evidences a comparative graph between dew temperature and operating pressure. Both refrigerants have 0 (zero) ODP, but the GWP is virtually unchanged around 1700, which implies in no advantage in this replacement. Besides that, R-410A has its days counted for discontinuation given the action of laws and protocols, in order to reduce the environmental pollution [1–3].

It can be seen from **Figure 8** that R-410A refrigerant has higher cooling capacity and works at higher pressures than R-22 refrigerant. Because this product is not azeotropic, it should always be charged in air conditioning and liquid refrigeration systems. This gas is also not miscible with mineral oils, and therefore polyol-esters (POE) should be used.

Some comparative studies [20–23] have concluded that R-290 fluid is the natural one with thermodynamic properties more similar to R-22, as shown in **Figure 9**.

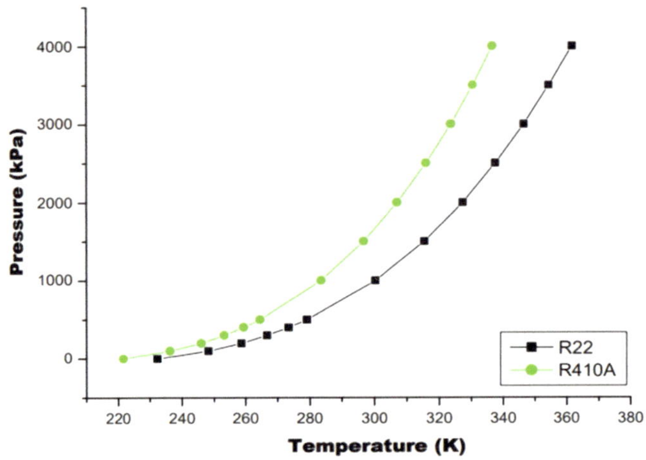

Figure 7.
Comparative graph between R-22 and R-410A refrigerants. Source: Danfoss [27].

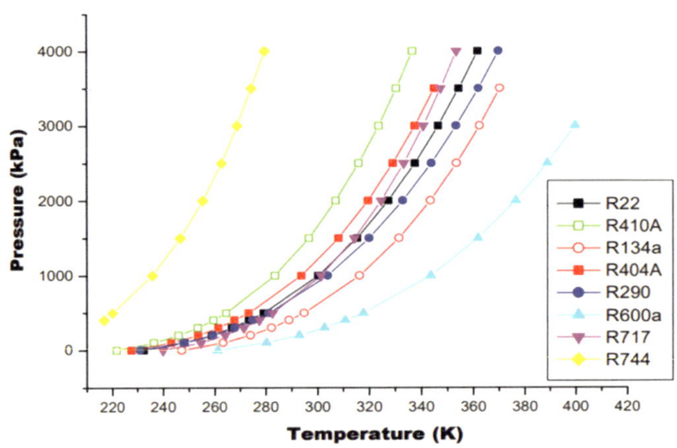

Figure 8.
Comparative graph between the most commonly used refrigerants in Brazil. Source: Danfoss [27].

Furthermore, it would make it possible to simply replace one fluid to another with less but rigorous modifications to the equipment's electrical system in order to prevent possible fire in case of leakage.

Notwithstanding the similarity with R-22, the R-290 fluid also works at temperatures and pressures close to those of the R-404A, indicating that it can be applied smoothly in these systems, as shown in **Figure 10**.

Between the most commonly used synthetic fluids, it can be seen from **Figure 11** that there is almost a pattern in temperature and pressure variation according to fluid composition.

By observing **Figure 12**, there is remarkable divergence, especially in the case of CO_2 due to it is low critical temperature, which does not allow its direct use in air/water condensation systems such as synthetic fluids.

It is important to note that each type of refrigerant has its applications for each type of air conditioning and refrigeration system and that they should not be mixed, requiring specific procedures for replacing the refrigerant charge.

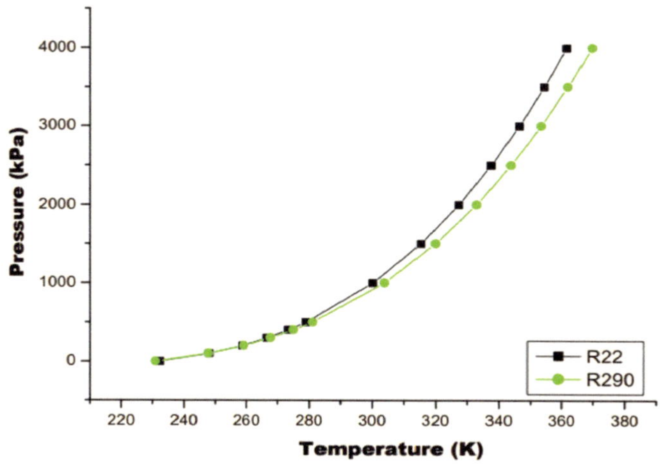

Figure 9.
Pressure–temperature relation between R-22 and R-290 refrigerants. Source: Danfoss [27].

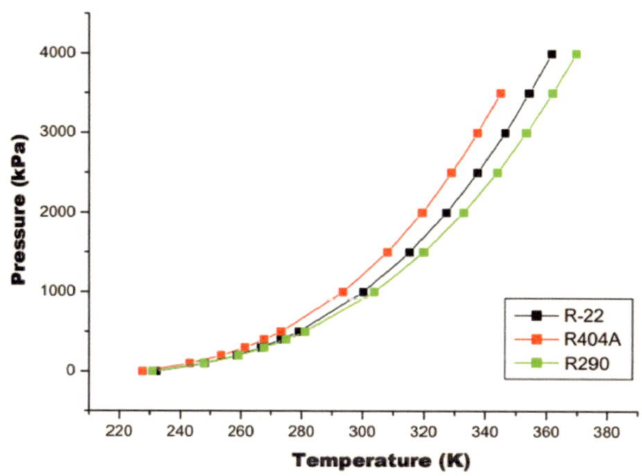

Figure 10.
Parameters of synthetic and natural fluids. Source: Danfoss [27].

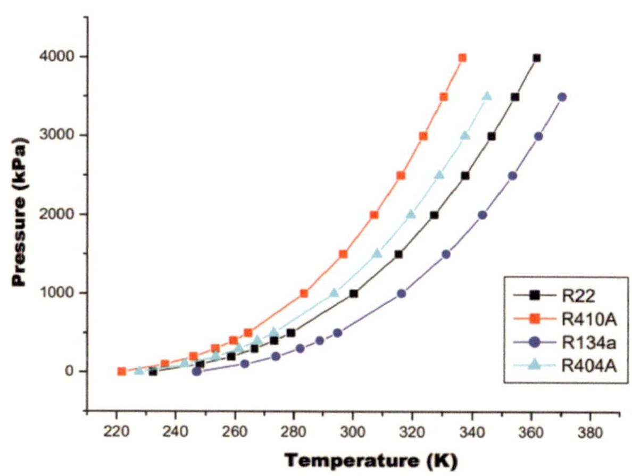

Figure 11.
Pressure-temperature relation between azeotropic and zeotropic synthetic fluids. Source: Danfoss [27].

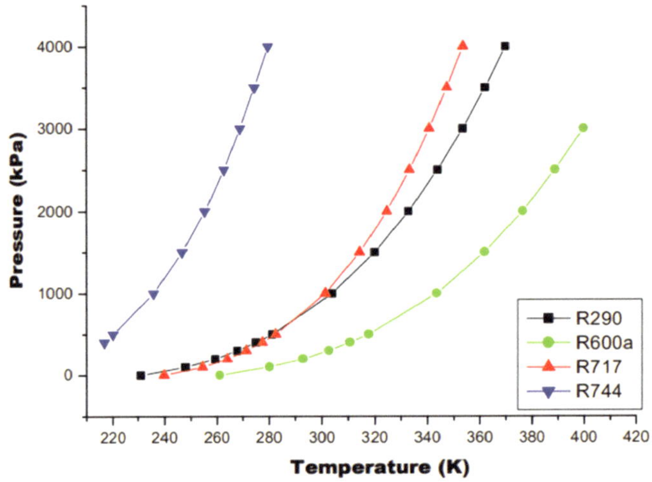

Figure 12.
Pressure-temperature relation between natural fluids. Source: Danfoss [27].

6. Conclusions

What is expected of this work is the development of discussions regarding the use and disposal of F-gases, so that they can create more effective control mechanisms, not only to the process of replacement of refrigerants CFCs by HFCs but also to controlling the use and especially the disposal of these refrigerants in a controlled manner, preventing the reuse and release of these types of refrigerants in nature.

The tests performed on ClimaTConT of the UFPA were extremely satisfactory, and with the information obtained from the test bench with chiller, it is possible to realize that the current replacement of R-22 with R-410A is not necessarily the best choice for air conditioning systems, because the various indicators, such as ODP and GWP, do not vary widely, which does not contribute to the reduction of ozone depletion.

In the face of all the information listed, some immediate actions should be taken:

- Refrigerant replacement or retrofit:
    - The impact of CFCs on the ozone layer is a serious risk to human survival. Therefore, some short-term actions should be taken, such as:
        1. Effecting the transition phase where R-12 should be replaced by R-134a, and R-11 by R-123. Other substitutes with ODP less than 0.05 should be used to replace CFCs. It is important to note that HCFCs themselves will begin to be restricted from the year 2004.
        2. HFCs and their zeotropic mixtures can be used without restrictions: The R-404A replacing R-502 and R-407C replacing R-22. There are refrigeration systems where these replacements can be made with very small cooling capacity losses and efficiency.
        3. Developing control mechanisms for the recovery of refrigerants used in residential, commercial, and industrial air conditioning and refrigeration equipment, such as recovery of fluids for recycling and/or disposal, preventing these fluids from being released into the local environment.

4. Development of studies, individualized, aimed at reducing the refrigerant charge used in air conditioning and refrigeration systems.

    - It is important to be careful with the safety equipment required when working with natural refrigerants such as R-290 and R-600a as they are flammable and must have their concentrations controlled.

    - Responsible companies must be required to use basic equipment such as a vacuum pump, refrigerant recovery unit, and precision scale for the installation and maintenance of HVAC systems.

- Popular programs must be created with training and recycling courses for those involved in the installation, maintenance, and operation of HVAC systems.

- During the manufacture, installation, operation, and maintenance of air conditioning and refrigeration systems using CFCs and HCFCs, deliberate leakage of these products into the environment should be avoided.

- Avoid emissions of CFCs and HCFCs by recovery, recycling, and reprocessing of refrigerants.

It is important that users are aware of the origin of the refrigerant to be used in air conditioning and refrigeration systems or in industrial processes, obeying the standards of purity and quality, within international standards. One must also pay attention to the classification of flammability and toxicity and know the rules and regulations governing this area (ASHRAE 34 and ARI 700).

According to the experience of the members of the UFPA "Resfriar Project," coupled with the knowledge of ABRAVA [14], **Table 7** points out the alternatives and trends for the replacement of the most commonly used refrigerant in cooling and air conditioning systems according to their applications, in the transition or retrofitting phase.

Considering the actual scenario in Brazil, the lack of reliable information about leaks in air conditioning and refrigeration systems precludes further analysis. In any case, informal data obtained from companies operating in the maintenance

Application	Refrigerant used	Alternative	Trend
Domestic refrigeration	R-11 to R-123 R-12 to R-134a	Recycling	R-600a, HFO
Commercial refrigeration	HCFC-22	HFC-404A, CO_2, ISCEON, recycling, HFO, HC, R-134a, glycol	HFC-404A, CO_2, ISCEON, recycling, HFO, HC, R-134a, glycol
Industrial refrigeration	Ammonia, HCFC-22, HFC-134a	R-134a, ammonia, recycling	Ammonia, R-134a, recycling, HFO
Automotive refrigeration	R-22 to R-134a	R-134a, recycling	R-123yf, CO_2
Climatization systems	HCFC-22	HFCs (R-134a, R-290, R-407C, R-410A, and R-600a)	R-290, R-410A, R-32, R-600a, recycling, HFO

Source: ABRAVA [14].

Table 7.
Alternatives and substitution trends of refrigerants according to their applications.

segment indicate that annual leakage rates can be between 40 and 100%, depending on the frequency of maintenance performed [15].

This work proved that even with the reduction in the refrigerant charge used on the chiller test bench, it was possible to improve the system efficiency by replacing an HCFC refrigerant with a natural refrigerant, following a worldwide trend of reduction of the refrigerant charge used in air conditioning and cooling systems.

Acknowledgements

The authors would like to thank the EBMA, ITEC, and ClimaTConT, for their technical and financial support.

Author details

Eraldo Cruz dos Santos*, Carlisson Azevedo, Caio Macêdo, Carla Azevedo, Raíssa Araújo, Sílvia Soares Gomes, Lana Baia and Yves Alexandrino
Federal University of Pará (UFPA), Belém, Pará, Brazil

*Address all correspondence to: eraldocs@ufpa.br

References

[1] United Nations Development Programme. Montreal Protocol; 2019

[2] United Nations Framework Convention on Climate Change. Kyoto Protocol; 2019

[3] Europa Policies. F-Gas; 2019

[4] The Institute of Refrigeration. F Gas Handbook 2019. ACR News; 2019

[5] Regulation (EU) N°. 517. 2014

[6] Refrigerants, Naturally! Natural Refrigerants; 2019

[7] Bitzer. Refrigerant Report 18. 2014

[8] AREA. Low GWP refrigerants—Guidance on minimum requirements for contractors' training and certification. 2012

[9] Hydrocarbons 21. What every Technician Should Know; 2019

[10] Hasse V, Ederberg L, Colbourne D. Natural Refrigerants—Sustainable Ozone and Climate-Friendly Alternatives to HCFCs. 1st ed. Eschborn; 2008

[11] British Refrigeration Association. Guide to Flammable Refrigerants. 2012

[12] AREA. Equipment for Refrigerants with Lower (A2L) and Higher (A3) Flammability. 2016

[13] Choudhari C, Sapali S. Performance investigation of natural refrigerant R290 as a substitute to R22 in refrigeration systems. RAAR. 2017. DOI: 10.1016/j.egypro.2017.03.084

[14] ABRAVA. Associação Brasileira de Refrigeração, Ar Condicionado, Ventilação e Aquecimento. Conheça sobre a Aplicação e Destinação de Fluídos Refrigerantes: Série Você Sabia, cartilha 2ª. Edição; 2014

[15] Marcagnan MH. Princípios Básicos de Refrigeração. São Leopoldo, RS: Universidade do Vale do Rio dos Sinos, Unisinos; 2015

[16] ASHRAE. Fundamentals. Atlanta, GA: American Society of Heating, Refrigerating and Air Conditioning Engineers; 2001

[17] Calm JM, Domanski PA. R-22 replacement status. ASHRAE Journal. 2004;**46**(8):29-39

[18] Emerson. Guide for subcritical and transcritical CO_2 applications. 2014

[19] Radermacher R, Hwang Y. Vapor Compression Heat Pumps with Refrigerant Mixtures. Boca Raton: Taylor&Francis; 2005

[20] Park K, Jung D. Performance of R290 and R1270 for R22 applications with evaporator and condenser temperature variation. Journal of Mechanical Science and Technology. 2008;**22**:532-537. DOI: 10.1007/s12206-007-1028-3

[21] Sharmas S, Babu A. Theoretical performance investigation of vapour compression refrigeration system using HFC and HC refrigerants mixtures as alternatives to replace R22. Energy Procedia. 2017;**109**:235-242. DOI: 10.1016/j.egypro.2017.03.053

[22] Theoretical computation of performance of sustainable energy efficient R-22 alternatives for residential air conditioners. Energy Procedia. 2017;**138**:710-716. DOI: 10.1016/j.egypro.2017.10.205

[23] Agrawal N, Patil S, Nanda P. Experimental studies of a domestic refrigerator using R290/R600a zeotropic blends. Energy Procedia. 2017;**109**:425-430. DOI: 10.1016/j.egypro.2017.03.051

[24] Palm B. Refrigeration systems with minimum charge of refrigerant. Applied Thermal Engineering. 2007;**17**:1693-1701

[25] ASHRAE. Handbook HVAC, Systems and Equipment's. Atlanta, GA; 2016

[26] ASHRAE. Fundamentals of Water System Design (SI edition). Atlanta, GA; 2006

[27] Danfoss. Refrigerant Slider; 2019

[28] NIST. Reference Fluid Thermodynamic and Transport Properties Database. 2019

Chapter 6

Effect of Nanoparticles on Performance Characteristics of Refrigeration Cycle

Ravinder Kumar

Abstract

Since few years, researchers are introduced nanorefrigerant in the recent development of refrigeration systems. This chapter briefly summarizes the behavior of different nanoparticles in vapor compression refrigeration cycle. The nanoparticles' infusion in refrigeration cycle affects the viscous behavior of the refrigerant. It is found that very limited studies have been done on viscosity characteristics of nanolubricants, but few reports states that viscosity increases with nanoparticles additive which improves the tribology of compressor. But inversely, the increment of viscosity promotes to the possible pressure drop in the system which eventually drops the cycle performance. The optimum contribution of nanorefrigerant with high thermal conductivity and low viscosity is the success key of system performance with nanoparticles. However, it is found that the contribution of nanoparticles on the basis of physical phenomena that are affecting the vapor compression cycle is limited in the literature. This chapter aims to make a review on the mechanism of improving vapor compression cycle by using nanorefrigerants.

Keywords: nanorefrigerants, nanolubricants, compressor work, coefficient of performance, heat transition

1. Introduction

Nanorefrigerant was defined to complete the objective of improving thermal performance using little possible fraction of nanoparticles in the base refrigerant. An improved term of refrigerant known as "nanorefrigerant" firstly introduced and experimentally implemented by Wang et al. [1]. The addition of nanoparticles in refrigerant improves the system performance in terms of improvement in flow and pool boiling heat transfer characteristics as well as flowing pool condensation heat transfer [2, 3]. Nowadays, nanorefrigerant and nanolubricant became a great alternate to enhance the performance of cooling cycles in terms of tribology performance, heat mass transfer properties and refrigerant/oil mixture relations [4–7]. Specifically, nanorefrigerant improves the heat transfer coefficient at evaporation and condensation whereas a nanolubricant improves the tribology characteristics which ultimately increase the compressor performance. The expected volume concentration of nanoparticles is calculated using Eq. (1).

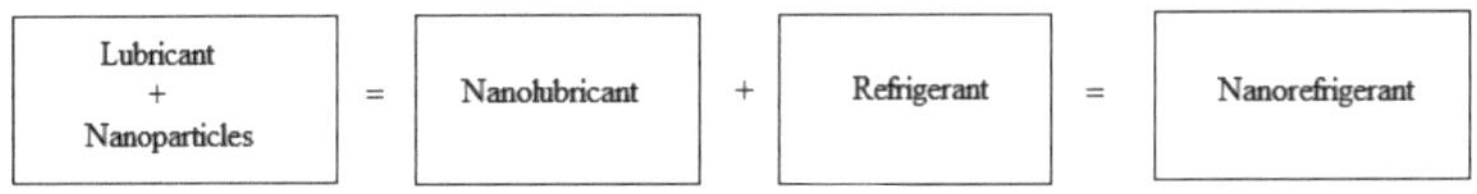

Figure 1.
Formation of nanorefrigerant in VCR cycle.

$$ø = \frac{\frac{m_p}{\rho_p}}{\frac{m_p}{\rho_p} + \frac{m_L}{\rho_L}} \times 100 \quad (1)$$

where, ø is the volume fraction in percentage, ρ_p and ρ_L are the density of nanoparticles and density of the lubricant respectively; and m_p and m_L are the masses of nanoparticles and lubricants respectively. The formation of nanorefrigerant is possible as shown in **Figure 1**.

The present chapter aims to define the mechanism that steers towards improvement in overall VCR cycle performance using nanolubricants and nanorefrigerants. The authors' hope that this paper will be useful to define the research gaps and understands the contribution of nanorefrigerants and nanolubricants in refrigeration cycle.

2. Improvement in VCR system performance

The section consists of four sub-sections. First part concerns research findings related to Al_2O_3 nanoparticles and the other parts deals with review studies related to CuO, TiO_2 and CNT nanoparticles. The schematic diagram of refrigeration cycle is shown in **Figure 2**.

2.1 Studies related to Al_2O_3

Till date, researchers developed various methods to improve heat transfer characteristics of refrigerants inside the vertical and horizontal tubes. Adding nanoparticles to base refrigerant is one of the most efficient methods to enhance the thermal characteristics of refrigerant.

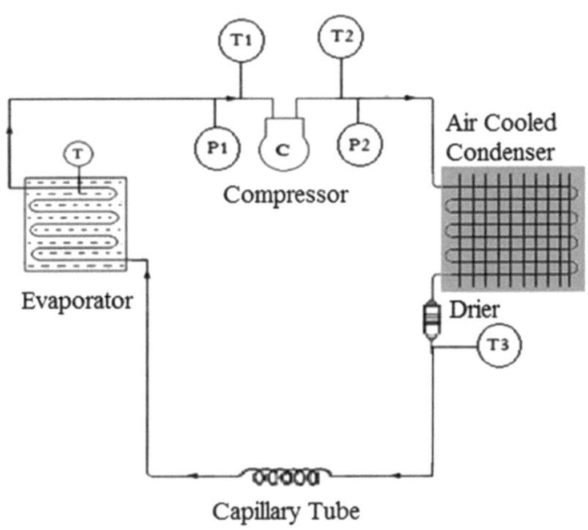

Figure 2.
Schematic diagram of vapor compression refrigeration cycle.

Jwo et al. [8] dispersed 0.1% mass fraction of Al_2O_3 particles in polyester oil and reported 2.4% reduction in compressor work. Mahbubul et al. [9, 10] reported significant increment in thermal conductivity and viscosity with Al_2O_3 nanoparticles dispersed in R141b. Later, the author observed increment in heat transfer coefficient and pressure drop up to 383 and 181%, respectively using particle volume fraction between 1 and 5% with fixed mass flux of 100 kg/m^2 s. Mahbubul et al. [11] reported large frictional pressure drop of R134a/Al_2O_3 nanorefrigerant as compared to R113/CuO nanorefrigerant flows inside horizontal smooth tube due to higher vapor quality [12].

Sun and Yang [13] studied the effect of Alumina-R141b, Cu-R141b, Al-R141b and CuO-R141b with mass fractions 0.1–0.3 wt% in a computer aided test rig on flow boiling heat transfer in horizontal tube and reported that Cu-R141b nanorefrigerant had the highest heat transfer coefficient as compared to other mixtures. Kedzierski [14] dispersed 1.6% volume fraction of Al_2O_3 in R134a/POE mixture flows on a horizontal and rough flat surface and found that higher volume fraction of nanoparticles with low average diameter has greater positive effect on heat transfer characteristics of base refrigerant. Further, Kedzierski [15] investigated the effect of Al_2O_3 nanoparticles on the pool boiling characteristics of R134a/POE mixture inside the rectangular finned surface and reported that 3.6% nanoparticle volume fraction enhanced the boiling heat transfer performance up to 113%. Tang et al. [16] reported that using δ-Al_2O_3 with R141b significantly improved pool boiling heat transfer coefficient in the system but adding surfactant at higher concentration corrupted the heat transfer process.

Mahbubul et al. [17] dispersed Al_2O_3 in R141b refrigerant for thermal conductivity and viscosity investigation. The author reported that viscosity and thermal conductivity of R141b/Al_2O_3 nanorefrigerant at 2% volume fraction are 179 and 1.626 times greater than pure refrigerant. Jwo et al. [8] used 0.05–2% weight fraction of Al_2O_3 particles with R134a and R12, respectively and reported that R134a refrigerant replaces R12, as polyester oil replaces mineral oil. Further, 0.1 wt% fraction of nanoparticles in R134a refrigerant reduced the energy consumption by 2.4% which significantly improved the COP of refrigerator. Kumar and Elansezhian [18] experimentally investigated the effect of R134a/Al_2O_3/PAG blend on the overall performance of VCR cycle and observed lower energy consumption by 10.32%. Author stated that using nanoparticle in refrigeration system is a cost effective method which improve its COP and length of capillary tube is reduced.

Subramani and Prakash [19] observed 25% less energy consumption and 33% overall COP enhancement in VCR cycle using Al_2O_3 nanorefrigerant. The freezing capacity of the cycle was also improved. Yusof et al. [20] dispersed 0.2% Al_2O_3 particles in polyester (POE) lubricant and reported 7% improvement in system COP and 2.1% reduction at compressor energy consumption. Cremaschi et al. [21] studies that alumina nanoparticles did not improve the solubility between refrigerant and lubricant, while addition of nanoparticles had slightly lowered the solubility of R410a/POE.

2.2 Studies related to CuO

Kedzierski and Gong [22, 23] observed heat transfer improvement between 50 and 275% using 0.5% mass fraction of CuO particles with R134a/RL68H and R134a/POE blend. Moreover, R134a/RL68H blend shows higher heat transfer enhancement as compared to the R134a/POE blend. In Later study, the author used 2 Vol% fraction of CuO particles in R134a refrigerant and reported nanorefrigerant has higher heat flux. Bartelt et al. [24] dispersed 0.5–1% mass fraction of CuO nanoparticles in R134a/polyester blend in horizontal flow boiling conditions and found 42–82% and 50–101% heat transfer enhancement for 1 and 2% mass fraction respectively. Peng et al. [25] dispersed 0.1 and 0.5 wt% CuO nanoparticles in R113 refrigerant to study heat transfer performance inside a horizontal rough pipe and reported 29.7% HTC using nanoparticles in base refrigerant. Henderson et al. [26] reported lower heat transfer performance with 0.5 and 0.05 vol% of CuO and SiO_2 nanoparticles dispersed in R134a and R134a/POE blend during boiling flow conditions in horizontal tube. Further, the author used 0.02, 0.04 and 0.08 vol% of CuO nanoparticles in R134a/POE blend and observed that nanoparticle with 0.04 and 0.08 vol% improved heat transfer performance up to 52 and 76%, respectively. Kedzierski and Gong [27] dispersed 0.5% mass fraction of CuO nanoparticles in polyester oil and observed 275% improvement in heat transfer with base refrigerant R134a.

Later, Bartelt et al. [28] extended the experiment of Kedzierski and Gong [27] and observed that 2% concentration of CuO nanoparticles gives the highest improvement up to 101% (**Tables 1** and **2**) [29, 46].

Abdel-Hadi et al. [30] experimentally found that CuO nanoparticles with average size 25 nm and concentration 0.55% is an optimum value which significantly enhanced the evaporative heat transfer coefficient. Kumar et al. [29] observed 7% reduction in compressor energy consumption and 46% enhancement in COP with dispersion of 0.2–1 wt% fraction of CuO nanoparticles in compressor lubricant. Moreover, the author reported reduction in friction and wear in compressor using nanoparticles in base lubricant. Peng et al. [31] used Cu nanoparticle in R113/VG68 (ester oil) mixture. It was observed that using Cu nanoparticles with average size of 20 nm strongly improved the heat transfer performance up to 23.8% as compared to other particles sizes of 50 and 80 nm. Akhavan-Behabadi et al. [32] found 83% increment in heat transfer rate with 1.5% mass fraction of CuO nanoparticles dispersed in R600a/polyester oil condensed inside the smooth horizontal tube (**Figures 3** and **4**) [29, 41, 46].

2.3 Studies related to TiO_2

Wang et al. [1] performed among the first experimental study using nanorefrigerant which proves that cooling speed and COP of domestic refrigerator significantly enhanced by utilizing TiO_2 nanoparticles in R134a based system. Further, Jiang et al. [33] studied the thermal conductivity using new theory and compared with the experimental data using different R22 nanorefrigerant fractions. The investigation proved that particle aggregation theory and the resistance network is a useful method

Nanoparticles concentration	0%	0.2%	0.4%	0.6%	0.8%	1.0%
Energy consumption (kW)	0.113	0.112	0.10	0.107	0.105	0.105
Energy saving (%)	—	0.79	3.37	5.55	7.31	7.31

Table 1.
Compressor energy consumption using distinct nanoparticle fractions [29].

Nanoparticles concentration	0%	0.2%	0.4%	0.6%	0.8%	1.0%
Energy consumption (kW)	0.1323	0.1278	0.1250	0.1236	0.1224	0.1224
Energy saving (%)	—	3.40	5.51	6.57	7.48	7.48

Table 2.
Compressor energy consumption of LPG/MO mixture using distinct nanoparticle fractions [46].

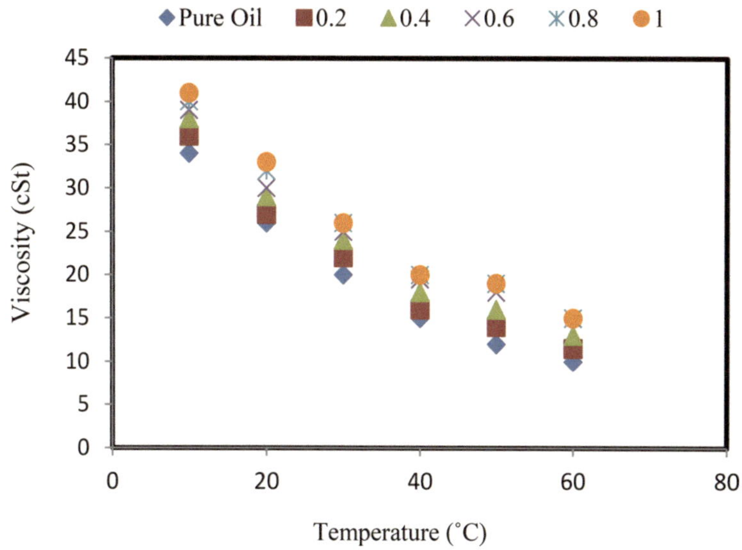

Figure 3.
Viscous behavior of mineral oil appended with CuO nanoparticles [29].

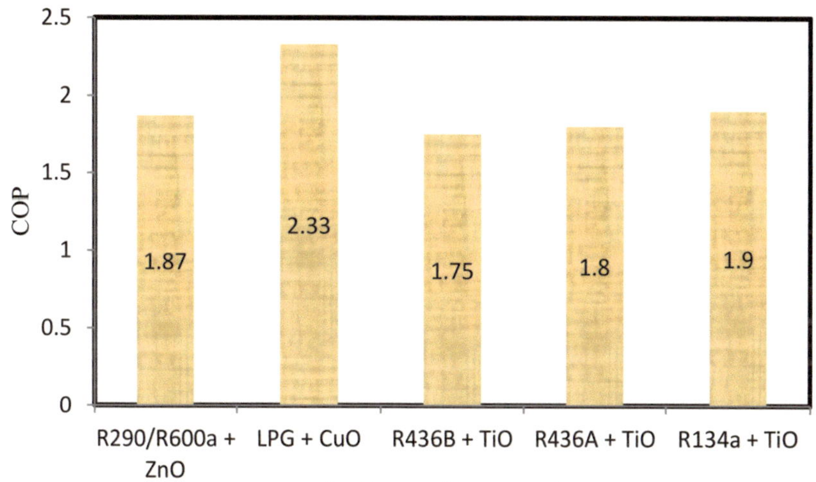

Figure 4.
COP comparison of the refrigerant cycle appended with different refrigerants/nanoparticles.

to calculate the thermal conductivity properties of nanorefrigerant. Li et al. [34] observed that significant improvement at heat transfer coefficient of R11 refrigerant using TiO_2 particles. Trisaksri and Wongwises [35] experimentally observed that nucleate pool boiling heat transfer coefficient (HTC) of using R141b refrigerant decreases with increment of TiO_2 nanoparticle fractions at high heat fluxes.

Mahbubul et al. [36] found that pressure drop increases with the addition of TiO_2 gradually in base refrigerant. Trisaksri and Wongwises [37] reported that increasing the particle volume fraction decreased the boiling heat transition rate and adding TiO_2 particles in base refrigerant R141b deteriorated the heat flow. Bobbo et al. [38] experimentally investigated a study in order to define the effect of TiO_2/carbon nanohorns (SWCNH) dispersion on the tribology of compressor polyester oil (SW32). The author reported better performance of TiO_2/SW32 blend compared to pure SW32 and SW32/SWCNH mixtures. Bi et al. [39] dispersed TiO_2 nanoparticles in R600a refrigerant through compressor lubricant and observed significant increment in freezing capacity about 9.60% by reducing energy consumption up to 5.94% using 0.1 and 0.5 g/L TiO_2 nanoparticle. Sabareesh et al. [40] used 0.05–0.015 vol% TiO_2 nanoparticles in compressor lubricant and observed that 0.01 vol% is optimum fraction value for better tribological properties. The author reported 17% improvement in COP and 11% reduction compressor energy consumption using nanoparticles with R12. Adelekan et al. [41] observed better COP in refrigerant cycle using TiO_2/MO nanolubricant instead of R134a/MO.

2.4 Studies related to CNTs and other nanoparticles

Jiang et al. [42] reported 50–104% increment in thermal conductivity with distinct particle fractions and diameters of CNTs dispersed in R113 refrigerant. The author declared 1.0 vol% fraction of CNT as optimum value. Park and Jung [43] performed nucleate boiling heat transfer analysis for CNTs with 1 vol% fraction using R123 and R134a refrigerants. The results found that heat transfer rate was improved (up to 36.6%) at low heat fluxes while it starts decreases at large heat fluxes. Peng et al. [44] reported 61% improvement in heat transfer coefficient with CNTs dispersed in R113/oil blend. Moreover, the author found that higher length and smaller outer diameter increased the heat transfer coefficient. Jiang et al. [42] conducted an experimental study on carbon nanotubes (CNTs) based nanofluids and proposed a modified Yu-Choi model which defined a decent deviation about 5.5% from the experimental result. Henderson et al. [45] used SiO_2 particles in polyester oil and reported 55% reduction in flow boiling performance due to the difficulties in nanolubricant dispersion and stability. Whereas, using Al_2O_3/POE nanolubricant results the great heat transfer improvement.

Kumar and Singh [46] reported 7.48% less energy consumption and 48% higher COP with 1.0 wt% of ZnO nanoparticles dispersed in R290/R600a/MO blend. Peng et al. [47] dispersed diamond nanoparticles in R113/VG68 blend to study nucleate boiling heat transfer coefficient. The author reported 63.4% enhancement in HTC using 0.05–0.5 wt% of nanoparticles fractions. Moreover, the study was compared with CuO/oil blend and found that diamond nanoparticles have higher impact on heat transfer characteristics. Kedzierski [48] reported 98% improvement in boiling heat transfer with 0.5%, 1% and 2% mass fractions of diamond nanoparticles dispersed in R134a. Naphon et al. [49] concentrated on the effect of Ti nanoparticles on the efficiency of copper heat pipe using R11 as base refrigerant. The study reported 0.01% nanoparticle fraction give highest efficiency ratio. Wang et al. [50] used a new category of nano oil which is created by mixing $NiFe_2O_4$ nanoparticles into naphthene-based oil B32 as an alternate to polyester VG32 and observed 6% improvement in overall COP.

3. Conclusions and future work

In present chapter, the studies of the previous findings on the heat transition characteristics, solubility and system performance of vapor compression refrigeration cycle has been presented. The literature related to pool boiling heat transfer enhancement of refrigerants and rheological behavior of lubricants using nanoparticles critically reviewed. The use of nanorefrigerants were improved the heat transfer performance of VCR system especially in nucleate and pool boiling heat transfer. CNTs can be considered as a best candidate for heat transfer enhancement of base refrigerants in comparison with the other nanoparticles.

The heat transfer rate increases with decreases nanoparticle dimension while the pressure drop decreases with decreases nanoparticles dimension.

The nanolubricants have great tribology characteristics in terms of wear rate and friction reduction in comparison with base lubricants. TiO_2 and CuO are best nanoparticles especially in improvement in tribology characteristics of compressor. It is reported that COP and freezing speed in cooling cycles is increased through application of nanoparticles in base refrigerants. It is found that improvement in VCR system parameters strongly depends on nanoparticle concentration. Thus, optimum value of particle fraction must be defined for better and stable results.

Furthermore, the below mentioned suggestions can be considering in future works:

1. As the nanoparticles concentration has great effect on thermophysical and tribology properties of refrigerants/lubricants, the study is needed to find the optimum nanorefrigerants and nanolubricants from the perspective of particle size, shapes and flow conditions.

2. The study on analytical models for the prediction of physical properties is so limited. Future works are needed to viral this study.

3. The study on condensation and evaporation flow of nanorefrigerants is very limited. It needs to elaborate.

4. There are no studies available related to effect of nanoparticles on new blend refrigerants such as R1234f.

5. The studies on the use of nanoparticles with natural refrigerants such as NH_3 and CO_2 not performed.

6. As per literature survey, several studies have been done on positive behavior of nanoparticles on heat transfer enhancement of refrigerants. The studies related to effect of nanoparticles on basic physical phenomena are somewhere missing.

7. There are no studies available on the flow of nanorefrigerants inside the microchannels and corrugated tubes.

Author details

Ravinder Kumar
Department of Mechanical Engineering, Dr. B.R. Ambedkar National Institute of Technology, Jalandhar, India

*Address all correspondence to: gsp.ravinder@gmail.com

References

[1] Wang RX, Hao B, Xie GZ. A refrigerating system using HFC134a and mineral lubricant appended with n TiO_2 as working fluids. In: Proceedings of the 4th International Symposium on HAVC; Beijing, China: Tsinghua University Press; 2003. pp. 888-892

[2] Azmi WH, Sharma KV, Sarma PK, Mamat R, Najafi G. Heat transfer and friction factor of water based TiO_2 and SiO_2 nanofluids under turbulent flow in a tube. International Communications in Heat and Mass Transfer. 2014;**59**: 30-38

[3] Hamid KA, Azmi WH, Mamat R, Usri NA, Najafi G. Effect of temperature on heat transfer coefficient of titanium dioxide in ethylene glycol-based nanofluid. Journal of Mechanical Engineering Science. 2015;**8**:1367-1375

[4] Alawi OA, Sidik NAC. Applications of nanorefrigerant and nanolubricants in refrigeration, air-conditioning and heat pump systems: A review. International Communication of Heat Mass Transfer. 2015;**68**:91-97

[5] Alawi OA, Sidik NAC, Mohammed H. A comprehensive review of fundamentals, preparation and applications of nanorefrigerants. International Communications in Heat and Mass Transfer. 2014;**54**:81-95

[6] Azmi WH, Sharif MZ, Yusof TM, Mamat R, Redhwan AAM. Potential of nanorefrigerant and nanolubricant on energy saving in refrigeration system—A review. Renewable and Sustainable Energy Reviews. 2017;**69**:415-428

[7] Redhwan AAM, Azmi WH, Sharif MZ, Zawawi NNM. Thermal conductivity enhancement of Al_2O_3 and SiO_2 nanolubricants for application in automotive air conditioning (AAC) system. In: MATEC Web of Conferences; Vol. 90. 2016. p. 01051

[8] Jwo CS, Jeng LY, Teng TP, Chang H. Effects of nanolubricant on performance of hydrocarbon refrigerant system. Journal of Vacuum Science and Technology B: Microelectronics and Nanometer Structures Processing. 2009; **27**(3):1473-1477

[9] Mahbubul IM, Saidur R, Amalina MA. Thermal conductivity, viscosity and density of R141b refrigerant based nanofluid. Procedia Engineering. 2013;**56**:310-315

[10] Mahbubul IM, Saidur R, Amalina MA. Heat transfer and pressure drop characteristics of Al_2O_3-R141b nanorefrigerant in horizontal smooth circular tube. Procedia Engineering. 2013;**56**:323-329

[11] Mahbubul IM, Fadhilah SA, Saidur R, Leong KY, Amalina MA. Thermophysical properties and heat transfer performance of Al_2O_3/R-134a nanorefrigerants. International Journal of Heat and Mass Transfer. 2013;**57**:100-108

[12] Peng H, Ding G, Jiang W, Hu H, Gao Y. Measurement and correlation of frictional pressure drop of refrigerant-based nanofluid flow boiling inside a horizontal smooth tube. International Journal of Refrigeration. 2009;**32**:1756-1764

[13] Sun B, Yang D. Experimental study on the heat transfer characteristics of nanorefrigerants in an internal thread copper tube. International Journal of Heat and Mass Transfer. 2013;**64**:559-566

[14] Kedzierski MA. Viscosity and density of aluminum oxide nanolubricant. International Journal of Refrigeration. 2013;**36**:1333-1340

[15] Kedzierski MA. R134a/Al_2O_3 nanolubricant mixture pool boiling on a rectangular finned surface. Journal of Heat Transfer. 2012;**134**:12-15

[16] Tang X, Zhao YH, Diao YH. Experimental investigation of the nucleate pool boiling heat transfercharacteristics of δ-Al_2O_3-R141b nanofluids on a horizontal plate. Experimental Thermal and Fluid Science. 2014;**52**:88-96

[17] Mahbubul IM, Saidur R, Amalina MA. Influence of particle concentration and temperature on thermal conductivity and viscosity of Al_2O_3/R141b nanorefrigerant. International Communications in Heat and Mass Transfer. 2013;(43): 100-104

[18] Kumar DS, Elansezhian RD. Experimental study on Al_2O_3-R134a nanorefrigerant in refrigeration system. International Journal of Modern Engineering Research. 2012;**2**(5): 3927-3929

[19] Subramani N, Prakash MJ. Experimental studies on a vapour compression system using. International Journal of Engineering, Science and Technology. 2011;**3**(9):95-102

[20] Yusof TM, Arshad AM, Suziyana MD, Chui LG, Basrawi MF. Experimental study of a domestic refrigerator with POE-Al_2O_3 nanolubricant. International Journal of Mechanical Engineering and Technology. 2015;**11**:2243-2252

[21] Cremaschi L, Wong T, Bigi AAM. Thermodynamic and heat transfer properties of Al_2O_3 nanolubricants. In: International Refrigeration and Air Conditioning Conference; Purdue: Purdue E-Pubs; 2014. pp. 1-10

[22] Kedzierski MA, Gong M. Effect of CuO nanolubricant on R134a pool boiling heat transfer. International Journal of Refrigeration. 2009;**32**: 791-799

[23] Kedzierski MA. Effect of CuO nanoparticle concentration on R134a/ lubricant pool boiling heat transfer. Journal of Heat Transfer. 2009;**131**: 043205

[24] Bartelt K, Park Y, Liu L, Jacobi A. Flow-boiling of R-134a/POE/CuO nanofluids in a horizontal tube. In: International Refrigeration and Air Conditioning Conference; Indiana: West Lafayette; 2008

[25] Peng H, Ding G, Jiang W, Hu H, Gao Y. Heat transfer characteristics of refrigerant-based nanofluid flow boiling inside a horizontal smooth tube. International Journal of Refrigeration. 2009;**32**:1259-1270

[26] Henderson K, Park Y, Liu L, Jacobi AM. Flow-boiling heat transfer of R-134abased nanofluids in a horizontal tube. International Journal of Heat and Mass Transfer. 2010;**53**:944-951

[27] Kedzierski MA, Gong M. Effect of CuO Nanolubricant on R134a Pool Boiling Heat Transfer with Extensive Measurement and Analysis Details. USA: US Department of Commerce, National Institute of Standards and Technology; 2007

[28] Bartelt K, Park Y, Liu L, Jacobi A. Flow-boiling of R-134a/POE/CuO nanofluids in a horizontal tube. In: International Refrigeration and Air Conditioning Conference; Purdue: Purdue e-Pubs; 2008. pp. 1-8

[29] Kumar R, Singh J, Kundal P. Effect of CuO nanolubricant on compressor characteristics and performance of LPG based refrigeration cycle: Experimental investigation. Heat and Mass Transfer. 2018;**54**(5):1405-1413

[30] Abdel-Hadi EA, Taher SH, Torki AHM, Hamad SS. Heat transfer analysis of vapor compression system using nano CuO-R134a. In: International Conference on Advanced Materials Engineering IPCSIT; Vol. 15. 2011. pp. 80-84

[31] Peng H, Ding G, Hu H, Jiang W. Effect of nanoparticle size on nucleate pool boiling heat transfer of refrigerant/oil mixture with nanoparticles. International Journal of Heat and Mass Transfer. 2011;**54**:1839-1850

[32] Akhavan-Behabadi MA, Sadoughi MK, Darzi M, Fakoor-Pakdaman M. Experimental study on heat transfer characteristics of R600a/POE/CuO nano-refrigerant flow condensation. Experimental Thermal and Fluid Science. 2015;**66**: 46-52

[33] Jiang W-T, Ding G-L, Wang K-J. Calculation of the conductivity of nanorefrigerant based on particles aggregation theory. Journal of Shanghai Jiaotong University. 2006;**8**:1-3

[34] Li P, Wu X, Li H, Wang W. Investigation of pool boiling heat transfer of R11 with TiO_2 nano-particles. Journal of Engineering Thermophysics. 2008;**29**(1):124

[35] Trisaksri V, Wongwises S. Nucleate pool boiling heat transfer of TiO_2-R141b. International Communications in Heat and Mass Transfer. 2018;**92**: 56-63

[36] Mahbubul IM, Saidur R, Amalina MA. Pressure drop characteristics of TiO_2-R123 nanorefrigerant in a circular tube. Engineering Transactions. 2011;**6**: 124-130

[37] Trisaksri V, Wongwises S. Nucleate pool boiling heat transfer of TiO_2-R141b nanofluids. International Journal of Heat and Mass Transfer. 2009;**52**: 1582-1588

[38] Bobbo S, Fedele L, Fabrizio M, Barison S, Battiston S, Pagura C. Influence of nanoparticles dispersion in POE oils on lubricity and R134a solubility. International Journal of Refrigeration. 2010;**33**:1180-1186

[39] Bi S, Guo K, Liu Z, Wu J. Performance of a domestic refrigerator using TiO_2-R600a nano-refrigerant as working fluid. Energy Conversion and Management. 2011;**52**:733-737

[40] Sabareesh RK, Gobinath N, Sajith V, Das S, Sobhan CB. Application of TiO_2 nanoparticles as a lubricant-additive for vapor compression refrigeration systems—An experimental investigation. International Journal of Refrigeration. 2012;**35**:1989-1996

[41] Adelekan DS, Ohunakin OS, Babarinde TO, Odunfa MK, Leramo RO, Oyedepo SO, et al. Experimental performance of LPG refrigerant charges with varied concentration of TiO_2 nano-lubricants in a domestic refrigerator. Case Studies in Thermal Engineering. 2017;**9**:55-61

[42] Jiang W, Ding G, Peng H. Measurement and model on thermal conductivities of carbon nanotube nanorefrigerants. International Journal of Thermal Sciences. 2009;**48**:1108-1115

[43] Park KJ, Jung D. Boiling heat transfer enhancement with carbon nanotubes for refrigerants used in building air-conditioning. Energy and Buildings. 2007;**38**:1061-1064

[44] Peng H, Ding G, Hu H, Jiang W. Influence of carbon nanotubes on nucleate pool boiling heat transfer characteristics of refrigerant-oil mixture. International Journal of Thermal Sciences. 2010;**49**:2428-2438

[45] Henderson K, Park YG, Liu L, Jacobi AM. Flow-boiling heat transfer of R-134a-based nanofluids in a horizontal tube. International Journal of Heat and Mass Transfer. 2010;**53**(5-6):944-951

[46] Kumar R, Singh J. Effect of ZnO nanoparticles in R290/R600a (50/50) based vapour compression refrigeration system added via lubricant oil on compressor suction and discharge

characteristics. Heat and Mass Transfer. 2017;**53**(5):1579-1587

[47] Peng H, Ding G, Hu H, Jiang W, Zhuang D, Wang K. Nucleate pool boiling heat transfer characteristics of refrigerant/oil mixture with diamond nanoparticles. International Journal of Refrigeration. 2010;**33**:347-358

[48] Kedzierski MA. Effect of diamond nanolubricant on R134a pool boiling heat transfer. Journal of Heat Transfer. 2012;**134**:051001

[49] Naphon P, Thongkum D, Assadamongkol P. Heat pipe efficiency enhancement with refrigerant–nanoparticles mixtures. Energy Conversion and Management. 2009;**50**:772-776

[50] Wang R, Wu Q, Wu Y. Use of nanoparticles to make mineral oil lubricants feasible for use in a residential air conditioner employing hydro-fluorocarbons refrigerants. Energy and Buildings. 2010;**42**:2111-2117

Chapter 7

Modeling of the Energy for Bound Water Freezing in Logs Subjected to Refrigeration

Nencho Deliiski and Natalia Tumbarkova

Abstract

This study suggests an approach for modeling of the total thermal energy needed for freezing the bound water in logs subjected to refrigeration. The approach maximally considers the physics of the freezing process of the bound water in wood. It considers the nonstationary change in the icing degree of logs formed by the crystallization of the bound water in them, as well as the influence of the fiber saturation point of each wood species on its current amount of nonfrozen bound water depending on temperatures below −1°C. Mathematical descriptions of the thermal energy of the phase transition of bound water in logs and also of the latent thermal energy of the bound water released in logs during their freezing have been executed. These descriptions were introduced in our own 2D nonlinear mathematical model of the freezing process of logs. The model was transformed in a form suitable for programming with the help of explicit schemes of the finite difference method. For the solution of the model and energy simulations with it, a software program was prepared, which was input into the calculation environment of Visual Fortran Professional.

Keywords: logs, refrigeration, bound water, freezing, latent heat, thermal energy

1. Introduction

It is known that the duration and the energy consumption of the thermal treatment of frozen logs in winter, aimed at their plasticizing for the production of veneer, depend on the degree of the logs' icing [1–10].

In the accessible specialized literature, there are limited reports about the temperature distribution in frozen logs subjected to defrosting [8, 11–21], and there is very little information about the research of the temperature distribution in logs during their freezing given by the authors only [22–24]. That is why the modeling and the multiparameter study of the freezing process of logs are of considerable scientific and practical interest.

For different engineering, technological, and energy calculations, it is necessary to be able to determine the nonstationary temperature field in logs depending on the temperature of the gaseous or liquid medium influencing them and on the duration of their staying in this medium. Such calculations are carried out using mathematical models, which describe adequately the complex processes of freezing both the free and bound water in the wood. In the specialized literature, information about 2D temperature distribution in logs subjected to freezing was given by the authors [22–24]. In the available literature for hydrothermal treatment of frozen wood materials, there is no information at all about the quantitative determination of the energy characteristics of both the free and bound water during their freezing in the wood or in other capillary porous materials.

The aim of the present work is to suggest a methodology for mathematical modeling and research of two mutually connected problems: 2D nonstationary temperature distribution in logs subjected to refrigeration and change in two important energy characteristics of the bound water in logs during its freezing—thermal energy of the gradual phase transition of bound water from liquid into solid state and latent thermal energy of the bound water released in the logs at temperatures below −1°C.

2. Mathematical model of the 2D temperature distribution in logs subjected to refrigeration

2.1 Mechanism of the temperature distribution in logs during their refrigeration

The mechanism of the temperature distribution in logs during their refrigeration can be described by the equation heat conduction [2, 5, 8–10].

When the length of the logs is less than their diameter by at least 3–4 times, for the calculation of the change in the temperature in the longitudinal sections of the logs (i.e., along the coordinates r and z of these sections) during their refrigeration in the air medium, the following 2D mathematical model can be used [23]:

$$c_{\mathrm{we}}\rho_{\mathrm{w}}\frac{\partial T(r,z,\tau)}{\partial\tau}=\lambda_{\mathrm{wr}}\left[\frac{\partial^2 T(r,z,\tau)}{\partial r^2}+\frac{1}{r}\cdot\frac{\partial T(r,z,\tau)}{\partial r}\right]+\frac{\partial\lambda_{\mathrm{wr}}}{\partial T}\left[\frac{\partial T(r,z,\tau)}{\partial r}\right]^2 + \lambda_{\mathrm{wp}}\frac{\partial^2 T(r,z,\tau)}{\partial z^2}+\frac{\partial\lambda_{\mathrm{wp}}}{\partial T}\left[\frac{\partial T(r,z,\tau)}{\partial z}\right]^2+q_{\mathrm{v}} \tag{1}$$

with an initial condition

$$T(r,z,0)=T_{\mathrm{w0}} \tag{2}$$

and boundary conditions for convective heat transfer:

- along the radial coordinate r on the logs' frontal surface during the cooling process

$$\frac{\partial T(r,0,\tau)}{\partial r}=-\frac{\alpha_{\mathrm{wp\text{-}fr}}(r,0,\tau)}{\lambda_{\mathrm{wp}}(r,0,\tau)}\left[T(r,0,\tau)-T_{\mathrm{m\text{-}fr}}(\tau)\right], \tag{3}$$

- along the longitudinal coordinate z on the logs' cylindrical surface during the cooling

$$\frac{\partial T(0,z,\tau)}{\partial z}=-\frac{\alpha_{\mathrm{wr\text{-}fr}}(0,z,\tau)}{\lambda_{\mathrm{wr}}(0,z,\tau)}\left[T(0,z,\tau)-T_{\mathrm{m\text{-}fr}}(\tau)\right]. \tag{4}$$

Equations (1)–(4) represent a common form of a mathematical model of the logs' freezing process.

2.2 Mathematical description of the heat sources in logs during their freezing

The volume internal heat source in the logs, q_v, reflects in Eq. (1) the influence of the latent heat of water in the wood on the logs' freezing process. In the available literature for hydrothermal treatment of frozen wood materials, no information can be found on the approaches for quantitative determination of the heat source q_v.

That is why as a methodology for the determination of q_v during the freezing of logs, in [23], a perspective is used, which is already applied for determination of the volume heat source, q_{vM}, during the process of solidification of melted metal [25–28]. According to this methodology, the following equations for determination of the volume internal sources of latent heat separately of the free and the bound water in wood during the logs' freezing process have been obtained [23]:

$$q_{\text{v-fw}} = K_{\psi\text{-fw}} \rho_{\text{w}} L_{\text{cr-ice}} \frac{\partial \psi_{\text{ice-fw}}}{\partial \tau}, \tag{5}$$

$$q_{\text{v-bw}} = K_{\psi\text{-bw}} \rho_{\text{w}} L_{\text{cr-ice}} \frac{\partial \psi_{\text{ice-bw}}}{\partial \tau}, \tag{6}$$

where $L_{\text{cr-ice}} = 3.34 \cdot 10^5\ \text{J}\cdot\text{kg}^{-1}$ [2, 5, 22–24] and

$$K_{\psi\text{-fw}} = \frac{\rho_{\text{w}} - (1 - \Psi_{\text{ice-fw}}) \cdot \rho_{\text{wUfsp}}}{\Psi_{\text{ice-fw}} \cdot \rho_{\text{w}}}, \tag{7}$$

$$K_{\psi\text{-bw}} = \frac{\rho_{\text{wUfsp}} - (1 - \Psi_{\text{ice-bw}}) \cdot \rho_{\text{wUnfw}}}{\Psi_{\text{ice-bw}} \cdot \rho_{\text{w}}}. \tag{8}$$

A numerical approach and an algorithm for the computation of $\Psi_{\text{ice-fw}}$ and $\Psi_{\text{ice-bw}}$ are given in [23]. The difference $\rho_{\text{w}} - \rho_{\text{wUfsp}}$ in the right-hand part of Eq. (7) reflects the entire mass of free water (in kg), which is contained in 1 m^3 of the logs. The wood densities ρ_{w} and ρ_{wUfsp}, which participate in Eqs. (7) and (8), are determined above the hygroscopic range according to the equations below [1–19, 29]:

$$\rho_{\text{w}} = \rho_{\text{b}} \cdot (1 + u), \tag{9}$$

$$\rho_{\text{wUfsp}} = \rho_{\text{b}} \cdot (1 + u_{\text{fsp}}). \tag{10}$$

The density of the wood ρ_{wUnfw} in Eq. (8) is determined according to the following equation in relation to the present entirely liquid quantity of nonfrozen water in the wood, u_{nfw}, corresponding to the current wood temperature $T < 272.15$ K [2, 10]:

$$\rho_{\text{wUnfw}} = \rho_{\text{b}} \cdot \frac{1 + u_{\text{nfw}}}{1 - \frac{S_{\text{v}}}{100}\left(u_{\text{fsp}}^{272.15} - u_{\text{nfw}}\right)}, \tag{11}$$

where

$$u_{\text{fsp}}^{272.15} = u_{\text{fsp}}^{293.15} + 0.021, \tag{12}$$

$$u_{\text{nfw}} = 0.12 + \left(u_{\text{fsp}}^{272.15} - 0.12\right) \cdot \exp\left[0.0567(T - 272.15)\right] @ 213.15\text{K} \le T \le 272.15\text{K}. \tag{13}$$

2.3 Introducing the mathematical description of heat sources into the mathematical model

After substituting q_v in Eq. (1) by the expression for $q_{v\text{-fw}}$ from Eq. (5) and of $\frac{\partial \psi_{\text{ice-fw}}}{\partial \tau}$ in this expression by $\frac{\partial \psi_{\text{ice-fw}}}{\partial \tau} \cdot \frac{\partial T}{\partial T}$, it is obtained that

$$\rho_w \cdot c_{\text{we-fr-fw}} \frac{\partial T(r,z,\tau)}{\partial \tau} = \lambda_{\text{wr}} \left[\frac{\partial^2 T(r,z,\tau)}{\partial r^2} + \frac{1}{r} \cdot \frac{\partial T(r,z,\tau)}{\partial r} \right] + \frac{\partial \lambda_{\text{wr}}}{\partial T} \left[\frac{\partial T(r,z,\tau)}{\partial r} \right]^2 + \lambda_{\text{wz}} \frac{\partial^2 T(r,z,\tau)}{\partial z^2} + \frac{\partial \lambda_{\text{wz}}}{\partial T} \left[\frac{\partial T(r,z,\tau)}{\partial z} \right]^2 + K_{\psi-\text{fw}} \rho_w L_{\text{cr-ice}} \frac{\partial \psi_{\text{ice-fw}}}{\partial \tau} \cdot \frac{\partial T(r,z,\tau)}{\partial T}. \tag{14}$$

For the freezing of the free water in the wood, Eq. (14) is transformed in the following form:

$$\rho_w \left[c_{\text{we-fr-fw}} - K_{\psi-\text{fw}} L_{\text{cr-ice}} \frac{\partial \psi_{\text{ice-fw}}}{\partial T} \right] \cdot \frac{\partial T(r,z,\tau)}{\partial \tau} = \lambda_{\text{wr}} \left[\frac{\partial^2 T(r,z,\tau)}{\partial r^2} + \frac{1}{r} \cdot \frac{\partial T(r,z,\tau)}{\partial r} \right] + \frac{\partial \lambda_{\text{wr}}}{\partial T} \left[\frac{\partial T(r,z,\tau)}{\partial r} \right]^2 + \lambda_{\text{wz}} \frac{\partial^2 T(r,z,\tau)}{\partial z^2} + \frac{\partial \lambda_{\text{wz}}}{\partial T} \left[\frac{\partial T(r,z,\tau)}{\partial z} \right]^2. \tag{15}$$

Equation (15) can be represented, as follows:

$$\rho_w \cdot c_{\text{we-fr-fw}^*} \cdot \frac{\partial T(r,z,\tau)}{\partial \tau} = \lambda_{\text{wr}} \left[\frac{\partial^2 T(r,z,\tau)}{\partial r^2} + \frac{1}{r} \cdot \frac{\partial T(r,z,\tau)}{\partial r} \right] + \frac{\partial \lambda_{\text{wr}}}{\partial T} \left[\frac{\partial T(r,z,\tau)}{\partial r} \right]^2 + \lambda_{\text{wz}} \frac{\partial^2 T(r,z,\tau)}{\partial z^2} + \frac{\partial \lambda_{\text{wz}}}{\partial T} \left[\frac{\partial T(r,z,\tau)}{\partial z} \right]^2. \tag{16}$$

The effective specific heat capacity $c_{\text{we-fr-fw}^*}$ in Eq. (16) is equal to

$$c_{\text{we-fr-fw}^*} = c_{\text{we-fr-fw}} - K_{\psi-\text{fw}} L_{\text{cr-ice}} \frac{\partial \psi_{\text{ice-fw}}}{\partial T} = c_{\text{we-fr-fw}} - c_{\text{Lat-fw}}, \tag{17}$$

where $c_{\text{we-fr-fw}}$ is the effective specific heat capacity of the wood during freezing only of the free water in it, J·kg^{-1}·K^{-1}, and $c_{\text{Lat-fw}}$ is the specific heat capacity, which is formed by the release of the latent heat of the free water during its crystallization in the wood, J·kg^{-1}·K^{-1}.

After substituting q_v in Eq. (1) by the expression for $q_{v\text{-bw}}$ from Eq. (6) and of $\frac{\partial \psi_{\text{ice-bw}}}{\partial \tau}$ in this expression by $\frac{\partial \psi_{\text{ice-bw}}}{\partial \tau} \cdot \frac{\partial T}{\partial T}$, it is obtained that

$$\rho_w c_{\text{we-fr-bw}} \frac{\partial T(r,z,\tau)}{\partial \tau} = \lambda_{\text{wr}} \left[\frac{\partial^2 T(r,z,\tau)}{\partial r^2} + \frac{1}{r} \cdot \frac{\partial T(r,z,\tau)}{\partial r} \right] + \frac{\partial \lambda_{\text{wr}}}{\partial T} \left[\frac{\partial T(r,z,\tau)}{\partial r} \right]^2 + \lambda_{\text{wp}} \frac{\partial^2 T(r,z,\tau)}{\partial z^2} + \frac{\partial \lambda_{\text{wp}}}{\partial T} \left[\frac{\partial T(r,z,\tau)}{\partial z} \right]^2 + K_{\psi-\text{bw}} \rho_w L_{\text{cr-ice}} \frac{\partial \psi_{\text{ice-bw}}}{\partial \tau} \cdot \frac{T(r,z,\tau)}{\partial T}. \tag{18}$$

For the case of freezing the bound water in the wood, Eq. (18) is transformed in the following form:

$$\rho_{\mathrm{w}}\left[c_{\text{we-fr-bw}} - K_{\psi\text{-bw}} L_{\text{cr-ice}} \frac{\partial \psi_{\text{ice-bw}}}{\partial T}\right] \cdot \frac{\partial T(r,z,\tau)}{\partial \tau} = \lambda_{\mathrm{wr}}\left[\frac{\partial^2 T(r,z,\tau)}{\partial r^2} + \frac{1}{r}\cdot\frac{\partial T(r,z,\tau)}{\partial r}\right] + \frac{\partial \lambda_{\mathrm{wr}}}{\partial T}\left[\frac{\partial T(r,z,\tau)}{\partial r}\right]^2 + \lambda_{\mathrm{wp}}\frac{\partial^2 T(r,z,\tau)}{\partial z^2} + \frac{\partial \lambda_{\mathrm{wp}}}{\partial T}\left[\frac{\partial T(r,z,\tau)}{\partial z}\right]^2. \quad (19)$$

Equation (19) can be represented, as follows:

$$\rho_{\mathrm{w}} \cdot c_{\text{we-fr-bw}^*} \cdot \frac{\partial T(r,z,\tau)}{\partial \tau} = \lambda_{\mathrm{wr}}\left[\frac{\partial^2 T(r,z,\tau)}{\partial r^2} + \frac{1}{r}\cdot\frac{\partial T(r,z,\tau)}{\partial r}\right] + \frac{\partial \lambda_{\mathrm{wr}}}{\partial T}\left[\frac{\partial T(r,z,\tau)}{\partial r}\right]^2 + \lambda_{\mathrm{wp}}\frac{\partial^2 T(r,z,\tau)}{\partial z^2} + \frac{\partial \lambda_{\mathrm{wp}}}{\partial T}\left[\frac{\partial T(r,z,\tau)}{\partial z}\right]^2. \quad (20)$$

The effective specific heat capacity $c_{\text{we-fr-bw}^*}$ in Eq. (20) is equal to

$$c_{\text{we-fr-bw}^*} = c_{\text{we-fr-bw}} - K_{\psi\text{-bw}} L_{\text{cr-ice}} \frac{\partial \psi_{\text{ice-bw}}}{\partial T} = c_{\text{we-fr-bw}} - c_{\text{Lat-bwm}}, \quad (21)$$

where $c_{\text{we-fr-bw}}$ is the effective specific heat capacity of the wood during freezing only of the bound water in it, J·kg^{-1}·K^{-1}, and $c_{\text{Lat-bwm}}$ is the specific heat capacity, which is formed by the release of the latent heat of the maximum possible amount of bound water during its crystallization in the wood, J·kg^{-1}·K^{-1}.

2.4 Mathematical description of the relative icing degree of logs caused from the freezing of the bound water

The relative icing degree of the logs, which is caused from the freezing of only the bound water in them, $\Psi_{\text{ice-bw}}$, can be determined as a relationship of the mass of the ice formed by the bound water in 1 kg wood, $m_{\text{ice-bw}}$, to the sum of the mass of this ice and of the mass of the nonfrozen water in 1 kg wood at $u = u_{\text{nfw}}$ and at $T < 272.15$ K, m_{nfw}, according to the following equation:

$$\Psi_{\text{ice-bw}} = \frac{m_{\text{ice-bw}}}{m_{\text{ice-bw}} + m_{\text{nfw}}} = \frac{u_{\text{fsp}} - u_{\text{nfw}}}{u_{\text{fsp}} - u_{\text{nfw}} + u_{\text{nfw}}} = 1 - \frac{u_{\text{nfw}}}{u_{\text{fsp}}}. \quad (22)$$

Using experimental data of Stamm [30, 31], in [10] the following equation was suggested for the calculation of the fiber saturation point of the wood species, depending on T:

$$u_{\text{fsp}} = u_{\text{fsp}}^{293.15} - 0.001(T - 293.15), \quad (23)$$

where $u_{\text{fsp}}^{293.15}$ is the standardized fiber saturation point at T = 293.15 K, i.e., at t = 20°C.

After substituting u_{nfw} and u_{fsp} in Eq. (22) by the expressions for u_{nfw} from Eq. (13) and for u_{fsp} from Eq. (23), it is obtained that

$$\Psi_{\text{ice-bw}} = 1 - \frac{0.12 + \left(u_{\text{fsp}}^{272.15} - 0.12\right) \cdot \exp\left[0.0567(T - 272.15)\right]}{u_{\text{fsp}}^{293.15} - 0.001(T - 293.15)}. \quad (24)$$

and

$$\frac{\partial \Psi_{\text{ice-bw}}}{\partial T} = -\frac{\left\{0.0567\left(u_{\text{fsp}}^{272.15} - 0.12\right) \cdot \exp\left[0.0567(T - 272.15)\right]\right\} \cdot \left[u_{\text{fsp}}^{293.15} - 0.001(T - 293.15)\right]}{\left[u_{\text{fsp}}^{293.15} - 0.001(T - 293.15)\right]^2} + \frac{0.001 \cdot \left\{0.12 + \left(u_{\text{fsp}}^{272.15} - 0.12\right) \cdot \exp\left[0.0567(T - 272.15)\right]\right\}}{\left[u_{\text{fsp}}^{293.15} - 0.001(T - 293.15)\right]^2}. \tag{25}$$

The sign "minus" in the right parts of Eq. (25) reflects mathematically the circumstances that the icing degree $\Psi_{\text{ice-bw}}$ decreases with an increasing temperature T.

2.5 Mathematical description of the thermophysical characteristics of logs

In **Figure 1**, the temperature ranges are presented, at which the refrigeration process of logs is carried out when $u > u_{\text{fsp}}$. The thermophysical characteristics of the logs and of both the frozen free and bound water in them have also been shown. The information on these characteristics is very important for solving the mathematical model given above.

During the first range from T_{w0} to $T_{\text{fr-fw}}$, only a cooling of the logs with full liquid water in them occurs. During the second range from $T_{\text{fr-fw}}$ to $T_{\text{fr-bwm}}$, a further cooling of the logs occurs until reaching the state needed for starting the crystallization of the free water. During this range also, the phase transition of this water into ice is carried out. The second range is absent when the wood moisture content is less than u_{fsp}.

During the third range from $T_{\text{fr-bwm}}$ to $T_{\text{w-fre-avg}}$, a further cooling of the logs is carried out until reaching the state needed for starting the crystallization of the bound water. During this range also, the phase transition of the bound water into ice is gradually performed.

The generalized effective specific heat capacities of the logs, $c_{\text{we-fr-gen}}$, during the above pointed three ranges of the freezing process above the hygroscopic range, are equal to the following:

$$\text{I.Range}: c_{\text{we-fr-gen1}} = c_{\text{w-nfr}}, \tag{26}$$

$$\text{II.Range}: c_{\text{we-fr-gen2}} = c_{\text{w-nfr}} + c_{\text{fw}} - c_{\text{Lat-fw}}, \tag{27}$$

$$\text{III.Range}: c_{we\text{-fr-gen3}} = c_{\text{w-fr}} + c_{\text{bwm}} - c_{\text{Lat-bwm}}. \tag{28}$$

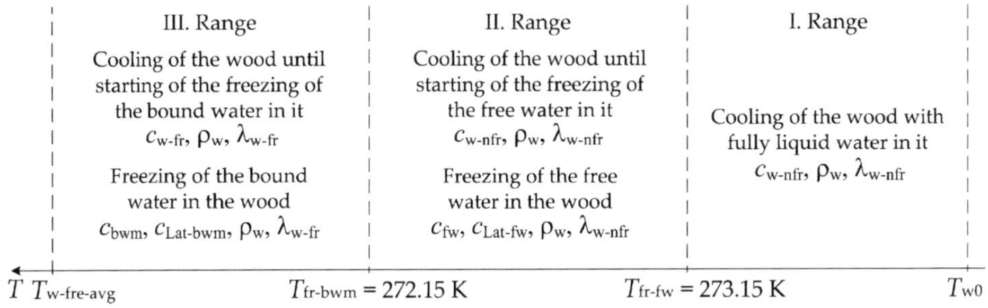

Figure 1.
Temperature ranges of the logs' refrigeration process above the hygroscopic range and thermophysical characteristics of the wood and the frozen water in it.

Mathematical descriptions of the specific heat capacities c_{w-nfr}, c_{w-fr}, c_{fw}, and c_{bwm} and also of the thermal conductivity of nonfrozen and frozen wood, λ_w, have been suggested by the first co-author earlier [9, 10] using the experimentally determined data in the dissertations by Kanter [32] and Chudinov [2] for their change as a function of t and u.

These relations are used in both the European [5–10] and the American specialized literature [11–16] when calculating various processes of the wood thermal treatment.

Using Eqs. (26)–(28), Eqs. (16) and (20) can be united in the following equation, with the help of which, together with the initial condition (2) and the boundary conditions (3), and (4) the 2D change of the temperature in subjected to freezing logs can be calculated:

$$\rho_w \cdot c_{\text{we-fr-gen1, 2, 3}} \frac{\partial T(r,z,\tau)}{\partial \tau} = \lambda_{wr} \left[\frac{\partial^2 T(r,z,\tau)}{\partial r^2} + \frac{1}{r} \cdot \frac{\partial T(r,z,\tau)}{\partial r} \right] + \frac{\partial \lambda_{wr}}{\partial T} \left[\frac{\partial T(r,z,\tau)}{\partial r} \right]^2 + \lambda_{wp} \frac{\partial^2 T(r,z,\tau)}{\partial z^2} + \frac{\partial \lambda_{wp}}{\partial T} \left[\frac{\partial T(r,z,\tau)}{\partial z} \right]^2. \tag{29}$$

For the convective boundary conditions (3) and (4) of the logs' freezing process, the following relations for the heat transfer coefficients are most suitable [33]:

- at cylindrical surface of the horizontally situated logs

$$\alpha_{\text{wr-fr}} = 2.56[T(0,z,\tau) - T_{m-fr}(\tau)]^{E_{fr}}, \tag{30}$$

- at frontal surface of the logs

$$\alpha_{\text{wp-fr}} = 1.123[T(r,0,\tau) - T_{m-fr}(\tau)]^{E_{fr}}, \tag{31}$$

where E_{fr} is determined during the validation of the nonlinear mathematical model of the freezing process through minimization of the root square mean error (*RSME*) between the computed by the model and experimentally obtained transient temperature fields in logs subjected to freezing.

2.6 An approach for modeling of the energy needed for freezing the bound water in logs

The total energy consumption, which is needed for freezing the bound water in 1 m^3 of logs, $Q_{\text{fr-bw-total}}$, can be calculated according to the following equation:

$$Q_{\text{fr-bw-total}} = Q_{\text{fr-bw}} - Q_{\text{Lat-bw}}, \tag{32}$$

where $Q_{\text{fr-bw}}$ is the energy needed for the phase transition of the bound water from liquid to solid state in 1 m^3 of logs subjected to freezing, kWh·m^{-3}, and $Q_{\text{Lat-bw}}$ is the latent thermal energy of the bound water in 1 m^3 of logs released in them during its crystallization, kWh·m^{-3}.

It is known that the energy consumption for heating 1 m^3 of wood materials, Q_w, with an initial mass temperature T_{w0} to a given average mass temperature T_{w-avg} is determined using the following equation [2, 5, 10]:

$$Q_w = \frac{c_w \cdot \rho_w}{3.6 \cdot 10^6} \cdot \left(T_{w-avg} - T_{w0} \right). \tag{33}$$

The multiplier $3.6{\cdot}10^6$ in the denominator of Eq. (37) ensures that the values of Q_w are obtained in $kWh{\cdot}m^{-3}$, instead of in $J{\cdot}m^{-3}$.

Based on Eq. (33), it is possible to calculate the energy $Q_{fr\text{-}bw}$ according to the equation

$$Q_{fr\text{-}bw} = \frac{c_{bw\text{-}avg} \cdot \rho_w}{3.6 \cdot 10^6} \cdot \left(272.15 - T_{w-fr-avg}\right), \tag{34}$$

where according to [8, 9]

$$c_{bw-avg} = 1.8938 \cdot 10^4 \left(u_{fsp}^{272.15} - 0.12\right) \cdot \frac{\exp\left[0.0567\left(T_{w-fr-avg} - 272.15\right)\right]}{1+u} \quad @T_{w-fre-avg} \le T \le 272.15\text{K}, \tag{35}$$

$$T_{w-fr-avg} = \frac{1}{S_w} \iint_{S_w} T(r,z,\tau) dS_w @T_{w-fre-avg} \le T(r,z,\tau) \le 272.15\text{K}, \tag{36}$$

and the area of ¼ of the longitudinal section of the log subjected to freezing, S_w, is equal to

$$S_w = \frac{D \cdot L}{4}. \tag{37}$$

Based on Eq. (33), it is possible to calculate also the energy $Q_{Lat\text{-}bw}$ according to the equation

$$Q_{Lat\text{-}bw} = \frac{c_{Lat\text{-}bw} \cdot \rho_w}{3.6 \cdot 10^6} \cdot \left(272.15 - T_{w-fr-avg}\right), \tag{38}$$

where the specific heat capacity, which is formed by the release of the latent heat of the bound water, c_{Lat-bw}, is calculated according to Eq. (21) and the density ρ_w is calculated according to Eq. (9).

3. Experimental research of 2D temperature distribution in logs subjected to freezing

For the validation of the suggested above mathematical model, it is necessary to have experimentally obtained data about the temperature distribution in logs during their freezing. The logs subjected to freezing in our experimental research were with a diameter of 240 mm, length of 480 mm, and $u > u_{fsp}$. This means that the logs contained the maximum possible amount of bound water for the separate wood species. They were produced from the sapwood of a freshly felled pine (*Pinus sylvestris* L.) and spruce (*Picea abies* L.) trunks. Before the experiments, four holes with diameters of 6 mm and different lengths were drilled in each log parallel to its axis until reaching the characteristic points of the log [22].

The coordinates of the characteristic points of the logs are given in **Figure 2**. These coordinates of the points allow the determination of the 2D temperature distribution in logs during their freezing. For refrigeration of the logs according to the suggested methodology by the authors [22], a horizontal freezer was used with adjustable temperature range from −1 to −30°C.

Sensors Pt100 with long metal casings were positioned in the four drilled holes of the logs. The automatic measurement and record of t_m, ϕ_m, and t in the

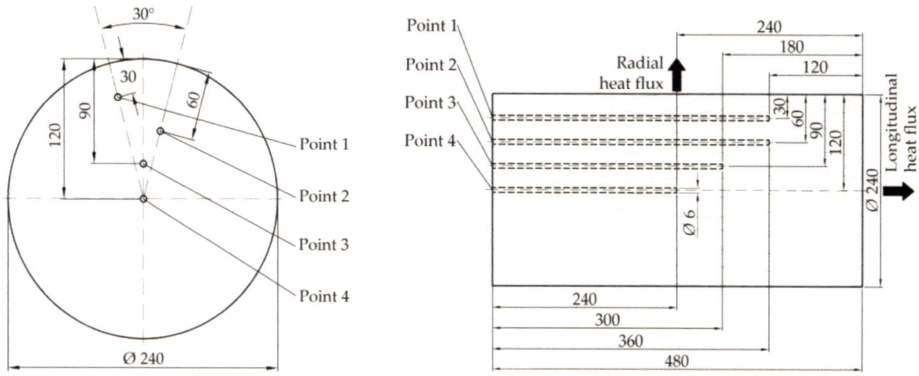

Figure 2.
Radial (left) and longitudinal (right) coordinates of four characteristic points for measurement of the temperature in logs subjected to refrigeration.

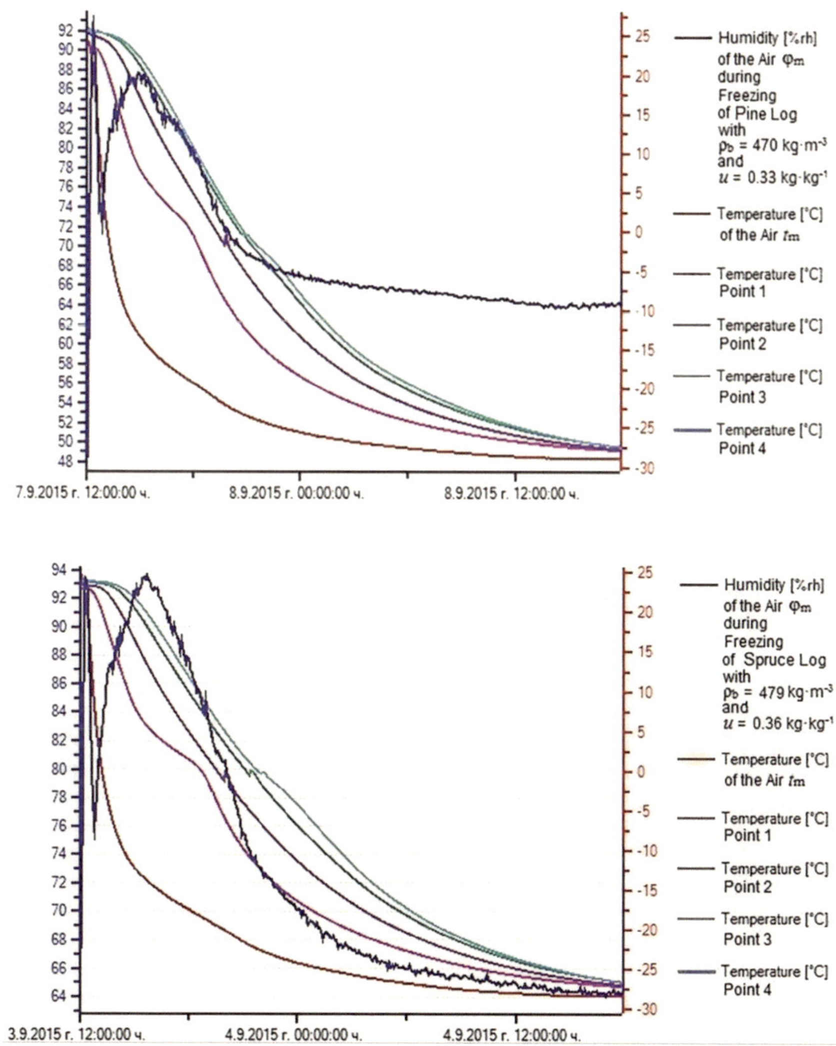

Figure 3.
Experimentally determined change in t_m, ϕ_m, and t in four points of the studied logs P1 (above) and S1 (below) with D = 0.24 m and L = 0.48 m during their 30 h refrigeration.

characteristic points of the logs during the experiments were realized by data logger-type HygroLog NT3 produced by ROTRONIC AG (http:/www.rotronic.com). The data logger has software HW4 for graphical presentation of the data.

In **Figure 3**, the change in the temperature of the processing air medium, t_m, and in its humidity, ϕ_m, and also in the temperature in 4 characteristic points of pine log named below as P1 with u = 0.33 kg·kg^{-1} and ρ_b = 470 kg·m^{-3} and spruce log named as S1 with u = 0.36 kg·kg^{-1} and ρ_b = 479 kg·m^{-3} during their separate 30 h refrigeration is presented. The record of all data was made automatically by the data logger with intervals of 5 min.

4. Numerical solution of the mathematical model of the logs' freezing process

For numerical solution of the mathematical model, a software package was prepared in Visual Fortran Professional developed by Microsoft. Using the package, computations were carried out for the calculation of the 2D nonstationary change of t in the characteristic points of ¼ of the longitudinal sections of the studied logs, whose experimentally determined temperature fields are presented in **Figure 3**.

The model has been solved with the help of explicit schemes of the finite difference method in a way analogous to the one used and described in [9, 10, 33]. For the computation of the temperature distribution in ¼ of the longitudinal section of the logs, which is symmetrical towards the remaining ¾ of the same section, the model was solved with step Δr = Δz = 0.006 m along the coordinates r and z and with the same initial and boundary conditions, as they were during the experimental research.

The interval between the time levels, $\Delta\tau$ (i.e., the step along the time coordinate), has been determined by the software package according to the condition of stability for explicit schemes of the finite difference method [10], and in our case it was equal to 6 s.

During the solving of the model, the mathematical descriptions of the thermophysical characteristics of pine wood with $u_{\text{fsp}}^{293.15} = 0.30$ kg·kg^{-1} and volume shrinkage S_v = 11.8% and also of spruce sapwood with $u_{\text{fsp}}^{293.15} = 0.32$ kg·kg^{-1} and S_v = 11.4% were used [8, 10].

4.1 Mathematical description of T in the freezer during logs' refrigeration

The curvilinear change in the freezing air medium temperature, $T_{\text{m-fr}}$, which is shown in **Figure 1**, with high accuracy (correlation 0.98 for the both studied logs and root square mean error (RSME) σ = 1.28°C for P1 and σ = 1.22°C for S1) has been approximated with the help of the software package TableCurve 2D [34] by the following equation:

$$T_{\text{m-fr}} = \frac{a_{\text{fr}} + c_{\text{fr}}\tau^{0.5}}{1 + b_{\text{fr}}\tau^{0.5} + d_{\text{fr}}\tau}, \tag{39}$$

whose coefficients are a_{fr} = 309.7863391, b_{fr} = 0.007125039, c_{fr} = 1.321533597, and $d_{\text{fr}} = -2.769 \cdot 10^{-6}$ for log P1 and to a_{fr} = 305.6335660, b_{fr} = 0.005833651, c_{fr} = 1.061216339, and $d_{\text{fr}} = -2.275 \cdot 10^{-6}$ for log S1. Equation. (39) and its coefficients were introduced in the software for solving Eqs. (3) and (4) of the model.

4.2 Computation of 2D temperature distribution in logs during their refrigeration

The mathematical model of the logs' freezing process has been solved with different values of the exponent E_{fr} in Eqs. (30) and (31). The calculated by the model change of the temperature in four characteristic points of the longitudinal logs' sections with each of the used values of E_{fr} during the freezing has been compared mathematically with the corresponding experimentally determined change of t in the same points with an interval of 5 min. The aim of this comparison was to find the value of E_{fr}, which ensures the best qualitative and quantitative compliance between the calculated and experimentally determined temperature fields in the logs' longitudinal sections.

As a criterion of the best compliance between the compared values of the temperature total for the four characteristic points, the minimum value of RSME, σ_{avg}, has been used. For the determination of RSME, a software program in the calculation environment of MS Excel has been prepared. With the help of the program, RSME simultaneously for a total of 1440 temperature-time points during a separate 30 h refrigeration of the logs has been calculated. During the simulations the same initial and boundary conditions have been used as during the experiments. It was determined that the minimum values of RSME overall for the studied four characteristic points are σ_{avg} = 1.67°C for P1 and σ_{avg} = 1.54°C for S1. The minimum values of σ_{avg} were obtained with the values of E_{fr} = 0.52 for P1 and E_{fr} = 0.48 for S1 in Eqs. (30) and (31).

Figure 4 presents, as an example, the calculated change in t_{m-fr}, log's surface temperature t_s, and t of four characteristic points of the studied pine log P1.

The comparison to each other of the analogical curves in **Figure 3**—above and **Figure 4** shows good conformity between the calculated and experimentally determined changes in the very complicated temperature field of the pine log during its refrigeration.

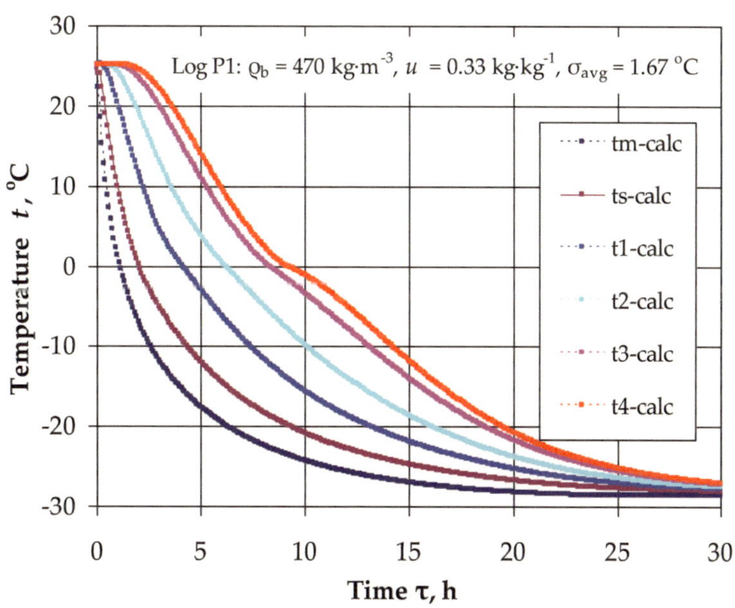

Figure 4.
Experimentally determined and calculated change in t_m, t_s, and t in four points of the studied pine log P1 during its 30 h refrigeration.

During our wide simulations with the mathematical model, we observed good compliance between computed and experimentally established temperature fields during refrigeration of logs from different wood species with different moisture content. The overall RSME for the studied four characteristic points in the logs does not exceed 5% of the temperature ranging between the initial and the end temperatures of the logs subjected to refrigeration.

4.3 Change of the relative icing degree of logs $\Psi_{ice\text{-}bw\text{-}avg}$ and its derivative

The calculation of the average mass icing degree of the logs caused from the freezing the bound water, $\Psi_{ice-bw-avg}$, is carried out according to the following equation, which was obtained in [23] using Eq. (24):

$$\Psi_{ice-bw-avg} = \frac{1}{S_w} \cdot \left\{ \iint\limits_{S_w} 1 - \frac{0.12 + \left(u_{fsp}^{272.15} - 0.12\right) \cdot \exp\{0.0567[T(r,z,\tau) - 272.15]\}}{u_{fsp}^{293.15} - 0.001[T(r,z,\tau) - 293.15)]} dS_w \right\}$$
$$@T_{w-fre-avg} \le T(r,z,\tau) \le 272.15\text{K}. \tag{40}$$

Figure 5 presents the calculated change of the logs' icing degree $\Psi_{ice-bw-avg}$ and the derivative Ψ_{ice-bw}/dT according to Eqs. (40) and (25), respectively, during the 30 h freezing process of the studied pine and spruce logs. The graphs show that the change of these variables is happening according to complex dependences on the freezing time.

It can be seen that the values of $\Psi_{ice-bw-avg}$ increase gradually from 0 to 0.487 for P1 and to 0.498 for S1 at the end of the 30 h freezing (**Figure 5**—left). These values of $\Psi_{ice-bw-avg}$ mean that 1–0.487 = 0.513 relative parts (i.e., 51.3%) of the bound water in P1 and 1–0.498 = 0.502 relative parts (i.e., 50.2%) of the bound water in S1 remains in a liquid state in the cell walls of the wood at the end of the 30th h of the logs' freezing process when the calculated average logs' mass temperatures are equal to −27.59°C for P1 and −26.82°C for S1.

It can be pointed that during the first 2.42 h for P1 and 2.75 h for S1 of the freezing process the whole amount of the bound water in the logs is in a liquid state and because of that the icing degree $\Psi_{ice-bw-avg}$= 0 and also Ψ_{ice-bw}/dT= 0 K^{-1}.

From 2.42th h for P1 and from 2.75th h for S1 the crystallization of the bound water in the peripheral layers of the logs starts. This causes a jump in the change of

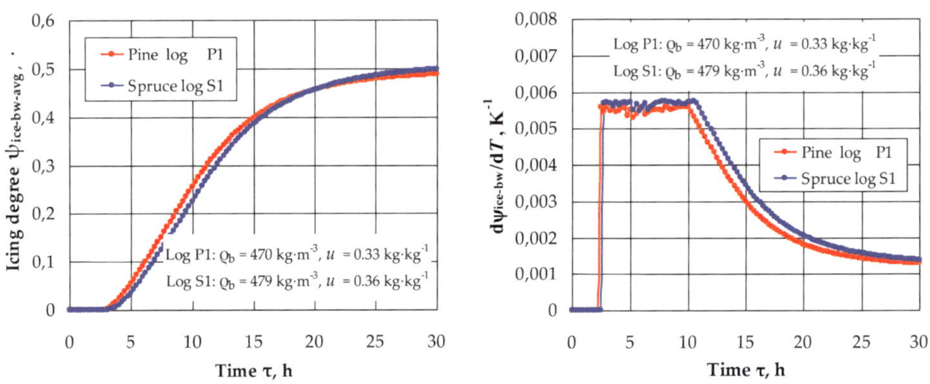

Figure 5.
Change in the relative icing degrees $\Psi_{ice-bw-avg}$ (left) and their derivatives Ψ_{ice-bw}/dT (right) during the refrigeration of the studied logs.

Ψ_{ice-bw}/dT from 0 to maximal values, equal to 0.0056 K^{-1} for P1 and 0.0057 K^{-1} for S1.

These maximal values remain practically unchanged until reaching 10.00th h for P1 and 10.75th h for S1, when the crystallization of the whole amount of the free water in the logs ends and the freezing of the bound water even in the logs' center starts.

The further decrease of the temperature in the freezer (refer to **Figures 3** and **4**) causes a gradual crystallization of the bound water in the logs. Because of that the derivatives Ψ_{ice-bw}/dT decrease slowly, and at the end of the 30 h freezing, they reach the value of 0.0013 K^{-1} for P1 and 0.0014 K^{-1} for S1.

4.4 Change of the thermal energies $Q_{fr\text{-}bw}$ and $Q_{Lat\text{-}bw}$

Figure 6 presents according to Eqs. (34) and (38) the calculated change of the thermal energies $Q_{fr\text{-}bw}$ and $Q_{Lat\text{-}bw}$ during the 30 h refrigeration of the studied logs.

It can be seen that the change of the energies $Q_{fr\text{-}bw}$ and $Q_{Lat\text{-}bw}$ is happening according to complex curvilinear dependences on the freezing time. The change of $Q_{fr\text{-}bw}$, depending on the freezing time, is similar to that of the icing degree $\Psi_{ice-bw-avg}$, and the change of $Q_{Lat\text{-}bw}$ is similar to that of the derivative Ψ_{ice-bw}/dT.

At the beginning of the logs' refrigeration process, when $\Psi_{ice-bw-avg} = 0$, both energies $Q_{fr\text{-}bw}$ and $Q_{Lat\text{-}bw}$ are also equal to 0. After that $Q_{fr\text{-}bw}$ increases gradually from 0 to 6.224 kWh·m^{-3} for P1 and 6.925 kWh·m^{-3} for S1 at the end of the 30 h freezing (see **Figure 6**, left). The larger value of $Q_{fr\text{-}bw}$ for S1 is caused from the larger amount of the frozen bound water in S1 in comparison with P1 at the end of the freezing process.

As it was mentioned above, the fiber saturation point of the pine wood is equal to 0.30 kg·kg^{-1}, and of the spruce wood, it is 0.32 kg·kg^{-1} [8, 10]. According to **Figure 5**, left, the relative icing degree, $\Psi_{ice-bw-avg} = 0.487$ for P1 and $\Psi_{ice-bw-avg} = 0.498$ for S1 at the end of the 30 h freezing. This means that the amount of the crystallized bound water is equal to $0.487 \times 0.30 = 0.146$ kg·kg^{-1} in P1 and it is equal to $0.498 \times 0.32 = 0.159$ kg·kg^{-1} in S1. Because of this not only the value of $Q_{fr\text{-}bw}$ for S1 is larger than that of P1, but also the maximal value of $Q_{Lat\text{-}bw} = 0.431$ kWh·m^{-3} for S1 is larger than the maximal value of $Q_{Lat\text{-}bw} = 0.405$ kWh·m^{-3} for P1. After reaching the maximal values, the energy $Q_{Lat\text{-}bw}$ decreases gradually and at the end of the 30 h freezing obtains a value of 0.232 kWh·m^{-3} for P1 and of 0.268 kWh·m^{-3} for S1.

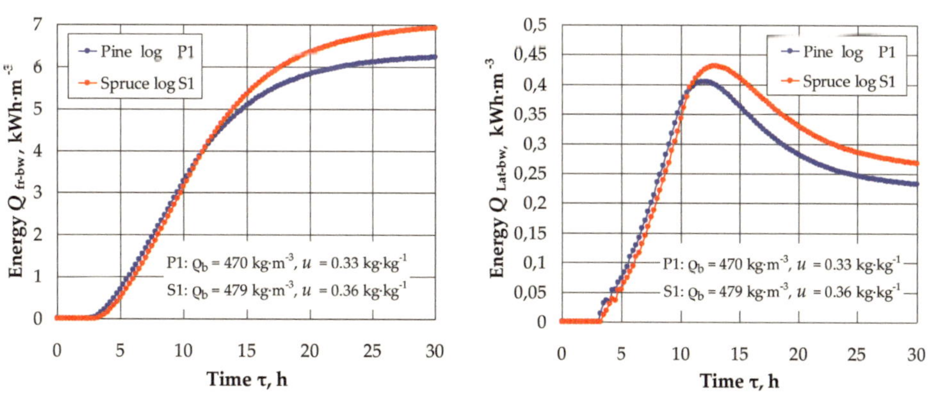

Figure 6.
Change in $Q_{fr\text{-}bw}$ (left) and $Q_{Lat\text{-}bw}$ (right) during refrigeration of the studied logs.

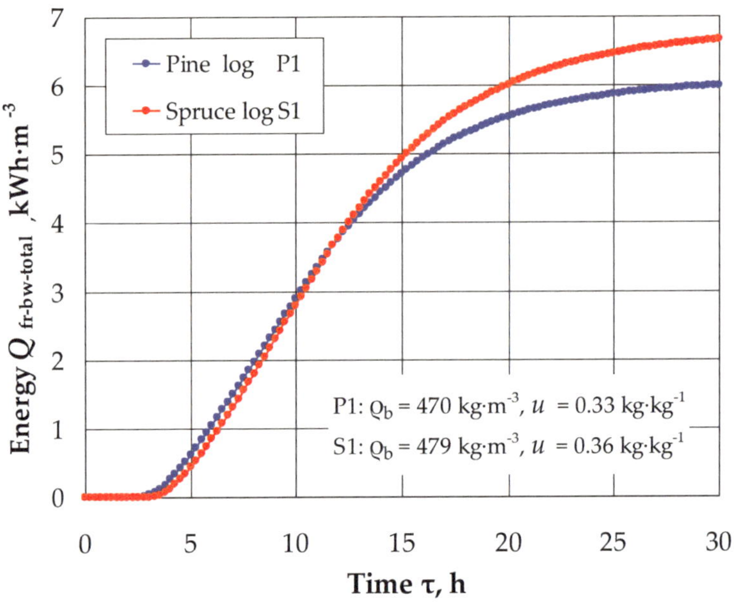

Figure 7.
Change in $Q_{fr\text{-}bw\text{-}total}$ during the refrigeration of the studied logs.

Figure 7 presents the change of the thermal energy $Q_{fr\text{-}bw\text{-}total}$ during the 30 h refrigeration of the studied logs, which is calculated according to Eq. (32). The change of this energy is similar to that of the energy $Q_{fr\text{-}bw}$ (see **Figure 6**, left). At the end of the 30 h freezing, the energy $Q_{fr\text{-}bw\text{-}total}$ reaches the following values: 5.992 kWh·m^{-3} for P1 and 6.657 kWh·m^{-3} for S1.

5. Discussions

This paper presents a methodology for mathematical modeling and research of two mutually connected problems: 2D nonstationary temperature distribution in logs subjected to refrigeration and change in two important energy characteristics of the bound water in logs during its freezing—thermal energy of the phase transition of the bound water in 1 m^3 wood from liquid into solid state, $Q_{fr\text{-}bw}$, and latent thermal energy of the bound water, $Q_{Lat\text{-}bw}$, which is released in 1 m^3 of the logs during water crystallization.

Mathematical descriptions and an approach for computing of the energies $Q_{fr\text{-}bw}$ and $Q_{Lat\text{-}bw}$ during the freezing of the bound water at temperatures below −1°C have been carried out. These descriptions are introduced in our own 2D nonlinear mathematical model of the 2D heat distribution in logs during their refrigeration at convective boundary conditions. The model was transformed in a form suitable for programming with the help of explicit schemes of the finite difference method, which excludes the necessity of any simplifications of the model.

A software program for numerical solution of the mathematical model and computation of 2D nonstationary change of the temperature in logs subjected to refrigeration and of the thermal energies $Q_{fr\text{-}bw}$ and $Q_{Lat\text{-}bw}$ has been prepared in Fortran, which has been input in the calculation environment of Visual Fortran Professional developed by Microsoft.

With the help of the program, computations for the determination of the energies $Q_{fr\text{-}bw}$ and $Q_{Lat\text{-}bw}$ and their difference, $Q_{fr\text{-}bw\text{-}total} = Q_{fr\text{-}bw} - Q_{Lat\text{-}bw}$, have been completed as an example for the case of a pine log P1 and a spruce log S1 with a diameter of 0.24 m and length of 0.48 m subjected to 30 h refrigeration in a freezer at approximately −30°C. Practically, $Q_{fr\text{-}bw\text{-}total}$ represents the energy needed for crystallization of that part of the bound water, which amount depends on u_{fsp}, t_{m-fr}, and τ_{fr}.

The logs subjected to refrigeration were with the following combinations of their initial temperature t_{w0}, basic density ρ_b, and moisture content u: t_{w0} = 25.2°C, ρ_b = 470 kg·m^{-3}, and u = 0.33 kg·kg^{-1} for log P1 and t_{w0} = 23.5°C, ρ_b = 479 kg·m^{-3}, and u = 0.36 kg·kg^{-1} for log S1.

It has been determined that the values of the energies $Q_{fr\text{-}bw}$,$Q_{Lat\text{-}bw}$, and $Q_{fr\text{-}bw\text{-}total}$ of the studied logs change according to complex relationships depending on the freezing time and after 30 h freezing of the logs reach the following values: $Q_{fr\text{-}bw}$ = 6.224 kWh·m^{-3}, $Q_{Lat\text{-}bw}$ = 0.232 kWh·m^{-3}, and $Q_{fr\text{-}bw\text{-}total}$ = 5.992 kWh·m^{-3} for P1 and $Q_{fr\text{-}bw}$ = 6.925 kWh·m^{-3}, $Q_{Lat\text{-}bw}$ = 0.268 kWh·m^{-3}, and $Q_{fr\text{-}bw\text{-}total}$ = 6.657 kWh·m^{-3} for S1.

6. Conclusions

Good adequacy and precision of the model toward the results from wide own experimental studies allow the carrying out of various calculations with the model, which are connected to the nonstationary temperature distribution and energy characteristics of logs from different wood species during their refrigeration. The mathematical model, after its connection with other our model of the logs' defrosting process [9, 10], could be input into the software of programmable controllers for optimized model-based automatic control [8, 20, 21, 35] of thermal treatment of frozen logs in the production of veneer.

The approach for the computation of the thermal energies of the bound water in logs during their refrigeration could be used for the creation of analogous models for the computation of the temperature distribution and the energy required for the refrigeration of different capillary porous materials (fruits, vegetables, meet, meet products, etc.).

Nomenclature

c	specific heat capacity, J·kg^{-1}·K^{-1}
D	diameter, m
E	exponent, —
L	latent heat, J·kg^{-1}, or length, m
q	internal heat source, W·m^{-3}
Q	thermal energy, kWh·m^{-3}
R	radius, m
r	radial coordinate: $0 \le r \le R$, m
S	shrinkage, %, or aria of ¼ of log's longitudinal section, m^2
T	temperature, K
t	temperature, °C
u	moisture content, kg·kg^{-1} = %/100
z	longitudinal coordinate: $0 \le z \le L/2$, m
α	heat transfer coefficients between log's surfaces and the surrounding air medium, W·m^{-2}·K^{-1}
λ	thermal conductivity, W·m^{-1}·K^{-1}

ρ	density, $kg{\cdot}m^{-3}$
σ	root square mean error (RSME), °C
τ	time, s
Δr	step along the coordinates r and z for solving of the model, m
$\Delta\tau$	step along the time coordinate for solving of the model, s
Ψ	relative icing degree of logs or relative degree of solidification of the metal, —
@	at

Subscripts

avg	average (for relative icing degree or for root square mean error)
b	basic (for wood density, based on dry mass divided to green volume)
bw	bound water
cr	crystallization
fr	freezing
fre	end of freezing
fsp	fiber saturation point
fw	free water
gen	generalized (for specific heat capacity)
ice	ice (for logs' icing degrees)
Lat	latent heat
m	medium (for cooling substance)
vM	volume of the metal
nfw	nonfrozen water
0	initial
p	parallel to the wood fibers
r	radial direction
total	total (for specific energy needed for freezing of the bound water)
v	volume
w	wood
we	wood effective (for specific heat capacity)
wL	wood with liquid water in it
wS	wood with solid state of water (ice) in it
wUfsp	wood at $u = u_{fsp}$
wUnfw	wood at $u = u_{nfw}$

Superscripts

272.15	at $T = 272.15$ K, i.e., at $t = -1$°C
293.15	at $T = 293.15$ K, i.e., at $t = 20$°C

Author details

Nencho Deliiski* and Natalia Tumbarkova
University of Forestry, Sofia, Bulgaria

*Address all correspondence to: deliiski@netbg.com

References

[1] Vorreiter L. Holztechnologisches Handbuch. Vienna: Fromm; 1949. 2080 p

[2] Chudinov BS. Theoretical research of thermo physical properties and thermal treatment of wood [thesis for DSc.]. Krasnojarsk, USSR: SibLTI; 1966 (in Russian)

[3] Kollmann FF. Technologie des Holzes and Holzwerkstoffe. I. Bd., 2 Auflage. Berlin/Göttingen, Heidelberg: Springer; 1951. 2233 p

[4] Kollmann FF, Côté WA Jr. Principles of Wood Science and Technology. I. Solid Wood. New York: Springer-Verlag, Berlin, Heidelberg; 1984. 592 p

[5] Shubin GS. Drying and Thermal Treatment of Wood. Moscow, USSR: Lesnaya Promyshlennost; 1990. 337 p. (in Russian)

[6] Požgaj A, Chovanec D, Kurjatko S, Babiak M. Structure and Properties of Wood. 2nd ed. Bratislava: Priroda a.s; 1997. 485p. (in Slovak)

[7] Trebula P, Klement I. Drying and Hydro-Thermal Treatment of Wood. Slovakia: Technical University in Zvolen; 2002. 449 p. (in Slovak)

[8] Deliiski N, Dzurenda L. Modelling of the Thermal Processes in the Technologies for Wood Thermal Treatment. Slovakia: Technical University in Zvolen; 2010. 224 p. (in Russian)

[9] Deliiski N. Transient heat conduction in capillary porous bodies. In: Ahsan A, editor. Convection and Conduction Heat Transfer. Rijeka, Croatia: InTech Publishing House; 2011. pp. 149-176. DOI: 10.5772/21424

[10] Deliiski N. Modelling of the Energy Needed for Heating of Capillary Porous Bodies in Frozen and Non-frozen States. Saarbrücken, Germany: Lambert Academic Publishing, Scholars' Press; 2013. 116 p. Available from: http://www.scholars-press.com//system/covergenerator/build/1060

[11] Steinhagen HP. Computerized finite-difference method to calculate transient heat conduction with thawing. Wood and Fiber Science. 1986;**18**(3): 460-467

[12] Steinhagen HP. Heat transfer computation for a long, frozen log heated in agitated water or steam—A practical recipe. Holz als Roh- und Werkstoff. 1991;**49**(7–8):287-290. DOI: 10.1007/BF02663790

[13] Steinhagen HP, Lee HW. Enthalpy method to compute radial heating and thawing of logs. Wood and Fiber Science. 1988;**20**(4):415-421

[14] Khattabi A, Steinhagen HP. Numerical solution to two-dimensional heating of logs. Holz als Roh- und Werkstoff. 1992;**50**(7-8):308-312. DOI: 10.1007/BF02615359

[15] Khattabi A, Steinhagen HP. Analysis of transient non-linear heat conduction in wood using finite-difference solutions. Holz als Roh- und Werkstoff. 1993;**51**(4):272-278. DOI: 10.1007/BF02629373

[16] Khattabi A, Steinhagen HP. Update of numerical solution to two-dimensional heating of logs. Holz als Roh- und Werkstoff. 1995;**53**(1):93-94. DOI: 10.1007/BF02716399

[17] Deliiski N. Modelling and automatic control of heat energy consumption required for thermal treatment of logs. Drvna Industrija. 2004;**55**(4):181-199

[18] Deliiski N. Computation of the 2-dimensional transient temperature

distribution and heat energy consumption of frozen and non-frozen logs. Wood Research. 2009;**54**(3):67-78

[19] Deliiski N, Dzurenda L, Tumbarkova N, Angelski D. Computation of the temperature conductivity of frozen wood during its defrosting. Drvna Industrija. 2015; **66**(2):87-96. DOI: 10.5552/ drind.2015.1351

[20] Hadjiski M, Deliiski N. Cost oriented suboptimal control of the thermal treatment of wood materials. IFAC-PapersOnLine. 2015; **48**(24):54-59. DOI: 10.1016/j.ifacol. 2015.12.056

[21] Hadjiski M, Deliiski N. Advanced control of the wood thermal treatment processing. Cybernetics and Information Technologies, Bulgarian Academy of Sciences. 2016; **16**(2):179-197. DOI: 10.1515/ cait-2016-0029

[22] Deliiski N, Tumbarkova N. A methodology for experimental research of the freezing process of logs. Acta Silvatica et Lignaria Hungarica. 2016; **12**(2):145-156. DOI: 10.1515/aslh-2016-0013

[23] Deliiski N, Tumbarkova N. An approach and an algorithm for computation of the unsteady icing degrees of logs subjected to freezing. Acta Facultatis Xylologiae Zvolen. 2017; **59**(2):91-104

[24] Deliiski N, Tumbarkova N. An approach for computing the heat sources in logs subjected to freezing. Acta Silvatica et Lignaria Hungarica. 2018;**14**(1):35-49. http://aslh.nyme.hu/ index.php?id=29435&L=0

[25] Salcudean M, Abdullah Z. On the numerical modelling of heat transfer during solidification processes. International Journal for Numerical Methods in Engineering. 1988;**25**: 445-473. DOI: 10.1002/nme.1620250212

[26] Dantzig J. Modelling liquid-solid phase changes with melt convection. International Journal for Numerical Methods in Engineering. 1989;**28**(8): 1769-1785. DOI: 10.1002/nme. 1620280805

[27] Hu H, Argyropoulos S. Mathematical modelling of solidification and melting: A review. Modelling and Simulation in Materials Science and Engineering. 1996;**4**:371-396. DOI: 10.1088/0965-0393/4/4/004

[28] Mihailov E, Petkov V. Cooling parameters and heat quantity of the metal during continuous casting of blooms. International Review of Mechanical Engineering. 2010;**4**(2):176-184

[29] Hrčka R. Model in free water in wood. Wood Research. 2017;**62**(6):831-837

[30] Stamm AJ. Wood and Cellulose Science. New York: Ronald Press; 1964

[31] Siau JF. Transport Processes in Wood. NewYork: Springer; 1984

[32] Kanter KR. Investigation of the thermal properties of wood [thesis]. Moscow, USSR: MLTI; 1955 (in Russian)

[33] Telegin AS, Shvidkiy BS, Yaroshenko UG. Heat- and Mass Transfer. Moscow: Akademkniga; 2002. 456 p. (in Russian)

[34] Table Curve 2D. Available from: h ttp://www.sigmaplot.co.uk/products/ tablecurve2d/tablecurve2d.php

[35] Hadjiski M, Deliiski N, Grancharova A. Spatiotemporal parameter estimation of thermal treatment process via initial condition reconstruction using neural networks. In: Hadjiski M, Atanasov KT, editors. Intuitionistic Fuzziness and Other Intelligent Theories and Their Applications. Cham, Switzerland: Springer International Publishing AG; 2019. pp. 51-80. DOI: 10.1007/978-3-319-78931-6

Chapter 8

Technique and Technology of Whole-Body Cryotherapy (WBC)

Alexander Baranov, Oleg Pakhomov, Alexander Fedorov, Vladimir Ivanov, Andrew Zaitsev and Ruslan Polyakov

Abstract

Whole-body cryotherapy (WBC) is a highly effective treatment method of a number of serious diseases. The therapeutic effect of WBC is achieved by stimulating cold receptors of the patient's skin, which provide supercooling of the skin surface to the level of −2°C. To achieve such a temperature of the skin surface, it is necessary to ensure heat removal with intensity not less than 3500 W/m^2. Such a heat flux can remove gas with the temperature not higher than −130°C. Procedures lasting less than 2 minutes do not form therapeutic effect. Procedures lasting more than 3 minutes are dangerous for the patient's health. WBC procedures are carried out in single- and multi-seat devices. Due to the compact placement of the patient in the WBC area, the share of useful heat load on the cryostatting system is up to 70%. In multi-seat installations, the useful heat load share is not more than 50%. During the WBC procedure, consumption of liquid nitrogen per patient is 3.77 kg. For the effective use of WBC technology, it is necessary to determine the general requirements for the power of cooling systems and the temperature of cryostatting of the WBC area.

Keywords: whole-body cryotherapy, WBC, supercooling of the skin surface, heat removal, useful heat load, WBC minimum procedure, cooling gas temperature

1. Introduction

Equipment for whole-body cryotherapy (WBC) has been used in clinics around the world for over 40 years [1, 2]. Despite this, until today there is no universally accepted concept describing the mechanism for achieving the healthcare effect of this physiotherapeutic procedure, and the physical conditions of safety and effectiveness of cryogenic cooling of the patient's skin surface have not been determined [3–6]. Temperature of the cooling gas and the duration of its contact with the patient's skin, being the most important technological parameters of WBC, vary over a wide range. The requirements for the power supply capacity of equipment for the implementation of WBC technology are not defined. In such conditions, manufacturers of devices for WBC procedures gradually increase the value of the minimum gas temperature in the WBC cab. Over 40 years of cryotherapeutic system production, the gas temperature declared by manufacturers of devices for WBC has doubled from 98 K in 1978 [1, 2] to 192 K [4–7]. By increasing the operating temperature of the equipment, manufacturers significantly reduce the cost of its production. For 40 years, the cost of devices for group WBC has decreased by 30 times. Low prices for equipment provide a high level of sales, so the trend of increasing operating temperature of WBC devices persists. An increase in the temperature level is accompanied by a decrease in the power of systems for cryostatting the WBC zone. The newest installations are equipped with refrigerators with a specific power of the electric driver of not more than 1 kW/m^3. At a temperature level of 170 K, a refrigerator with such a power has a heat-removing capacity of not more than 400 W/m^3, which is comparable with the physiological heat release of a patient under thermal comfort conditions (150 W) [7].

Unreasonable changes in WBC technology affect the effectiveness of the procedures. Recently, more and more articles appear, the authors of which express doubt that cryotherapy can provide the healthcare effects described in papers published before 1990 [7, 8]. The reason that many modern WBC systems are not able to provide the conditions for obtaining the healthcare effects described in the last century [1, 2] is the increase in gas temperature in the working zone of new installations. This can be seen even from the titles of the articles [1, 8]. The temperature increase from −170°C (102 K) to −110°C (163 K) changes the absolute value of the temperature by 1.6 times, which cannot but affect the intensity of heat removal, the degree of supercooling of the patient's body surface, etc. From a thermophysical point of view, it is obvious that from 1978 to 2018 the technology, which is commonly referred to as WBC, has qualitatively changed. And, judging by contemporary publications, this qualitative change had a negative impact on the healthcare effectiveness of the procedures, which until recently were successfully used to treat a number of severe diseases: rheumatoid arthritis, bronchial asthma, psoriasis, etc. [9, 10].

In such conditions, the determination of cause–effect relationships between the WBC technological parameters and the magnitude of the healthcare effect acquires high scientific and social significance. Formation of the thermophysical theory of WBC creates a scientific basis for restoring the production of effective cryotherapeutic installations at the modern technical level.

2. Historic reference

The WBC method is based on providing the total contact of the patient's skin surface with a cryogenic gas. With a contact duration of up to 3 minutes and a gas temperature of less than 140 K, the WBC procedure provides a number of positive effects that are used in treatment practice [11, 12]. The most demonstrative and controlled sign of the WBC effectiveness is the duration of analgesic action, which can last 6–8 hours [7]. The analgesic effect of WBC was first described and used in treatment practice by a Japanese doctor Yamauchi. For the WBC procedures, a special installation was made, called "Cryotium" by the author of the method [2]. The "Cryotium" design was analogously a refrigeration chamber for long-term storage of perishable products. The chamber was separated from the environment by the lock chamber (**Figure 1**), which was supposed to reduce the loss of cold air from the main chamber.

Given the relatively large size of the chamber, several patients were undergoing the WBC procedures simultaneously (from 5 to 12). To obtain cryogenic temperatures, liquid nitrogen was fed to the "Cryotium" heat exchangers instead of freon. The "Cryotium" design appeared by chance. Yamauchi believed that in order to obtain the maximum treatment outcome, the maximum decrease in temperature should be used. This condition became the basis of the design. To reduce the cost of manufacturing "Cryotium," Japanese engineers used the insulating structure and heat exchangers of the serial refrigeration chamber. The temperature regime in the chamber volume was

Figure 1.
Multi-seat cab for WBC.

determined by the requirement of the inadmissibility of air condensation on the surface of a heat exchanger. The temperature of the outer surface of the heat exchanger T_{HC} must be higher than the condensation temperature T''_A:

$$T_{HC} > T''_A = 81\ K. \tag{1}$$

The removal of heat from air to the surface of the heat exchanger was carried out by natural convection.

With natural convection, the calculated temperature gradient between the gas and the heat-removing surface is 20 K:

$$T_{A-HC} = T_A - T_{HC} \approx 20\ \text{K}. \tag{2}$$

Minimum possible air temperature in the cab:

$$T_A = T''_A + \Delta T_{A-HC} \approx 101\ \text{K}(-172^{\circ}\text{C}). \tag{3}$$

The value of the air temperature during the WBC sessions specified by the method's author [1] is the lowest possible temperature that could be achieved in this cab design. It is important to note that in "Cryotium" the temperature was maintained by choosing the pressure of liquid nitrogen (LN) vapor in the heat exchanger tubes (**Figure 2**). The boiling point of LN depends on pressure; by increasing the vapor pressure to a level of $P \geq 0.2$ MPa, it is possible to ensure the fulfillment of condition (1) without using the temperature control systems. The lack of a temperature control system has provided "Cryotium" with unique operational advantages over modern WBC devices. Heat exchangers filled with liquid nitrogen successfully dealt with an increase in heat load when patients entered, and the correct choice of operating pressure prevented air condensation.

Thus, the "Cryotium" design determined the WBC technology. Perhaps, that is why the author of the method did not give any reason for the WBC temperature regime in his works. The ratios of the boiling points of nitrogen and air, as well as the design features of the device in which the procedures were performed, have randomly created the conditions for a safe and highly efficient procedure.

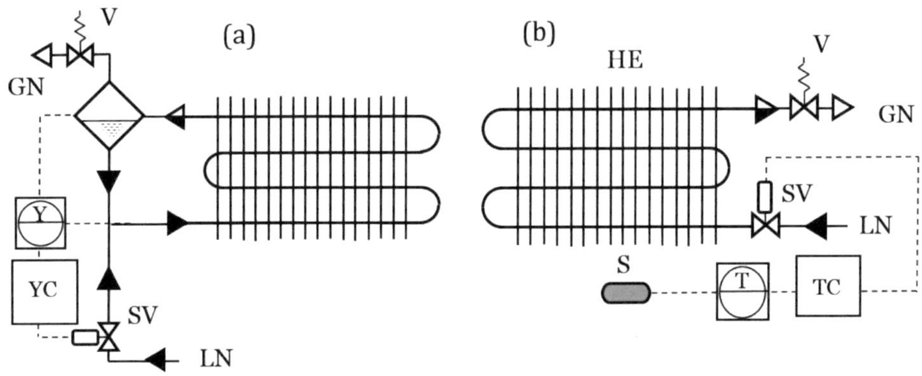

Figure 2.
The scheme of supplying liquid nitrogen to the heat exchangers of the installations for the WBC: (a) "Cryotium"; (b) "KR-2005."

Yamauchi used the WBC method for the treatment of rheumatoid arthritis [1]; the technique was so effective that it quickly spread to the countries of Western Europe. In Poland and Germany, devices similar to "Cryotium" were put in production. European manufacturers have tried to reproduce the Japanese installation on the base of available information but, for unknown reasons, have changed the basic operating principle of "Cryotium." This is translated into an increase in the minimum operating temperature from −170 to −160°C [13]. A slight increase in temperature led to a whole chain of changes in cooling technology, which caused a gradual decrease in the efficiency of European (Polish and German) installations for WBC. As already shown above, the temperature level of −170°C was maintained in "Cryotium" without using a temperature control system, only through relief of excess pressure in the LN vapor line (**Figure 2a**).

The liquid level controller (YC) in the heat exchangers of the "Cryotium" installation according to the sensor signals of the Y level controls the operation of the solenoid valve (SV), through which LN enters the system. The cryogenic liquid enters the heat exchanger (HE) tubes, where it partially evaporates due to the supply of heat from the procedural room air. Vaporization reduces the flow density in the tubes; the vapor–liquid mixture is pushed out from the top of the heat exchanger to the liquid separator (LS). In this apparatus, liquid and vapor are separated. The liquid flows into the lower section of the heat exchanger (HE) and again participates in the removal of heat. LN vapors accumulate at the top of the liquid separator (LS). The vapor pressure is controlled by a safety valve (V), which opens at a pressure of 0.22 MPa. The vapor pressure determines the LN boiling point and the temperature of the tubes of the heat exchanger (HE), which must meet condition (1). The air temperature in the main "Cryotium" cab at the presence of patients rises to −170°C. In the pauses between the procedures, when the heat load on the cooling system is reduced by 10 times [7], the air temperature in the main cab approaches the temperature of the heat exchanger tubes:

$$\Delta T_{A-HC} \rightarrow 5\ \text{K};\quad T_A \rightarrow T_A^{''} + \Delta T_{A-HC} \approx 86\ \text{K}\left(-187^{\circ}\text{C}\right). \qquad (4)$$

The air temperature in the cab remains at the minimum possible level. In European installations, the air temperature in the cab is controlled by the temperature controller (TC), which, by signals from the temperature sensor (S), opens the liquid nitrogen supply valve (SV). In order to maintain the temperature at −160°C, the TC limits the supply of LN to the heat exchanger (HE) in the period when there are no patients in the cab. The LN level decreases until the sensor (S) registers the set temperature. The upper sections of the heat exchanger (HE), through which the nitrogen passes in the vapor state, are heated to a temperature close to the air temperature in the chamber. The temperature of the inner tube surfaces exceeds the LN boiling point by more than 20 K:

$$T_{HC} \to T_A \approx 110 \text{ K}; \quad T_{HC} > T''_A + 20 \text{ K}. \tag{5}$$

When patients enter the main cab, relatively warm air enters from the lock chamber ($T_s = 210$ K). Because of this, the air temperature in the cab increases by 50 K or more. The temperature controller opens the SV and resumes the supply of LN to the heat exchanger (HE). However, it takes some time to fill the tubes of the heat exchanger; moreover, the temperature of the upper sections of the heat exchanger (HE) exceeds the boiling point of nitrogen by more than 20 K. Because of this, the LN film boiling occurs at which the intensity of heat removal to the liquid is much lower; therefore, the vapor–liquid mixture passes the heat exchanger (HE) and is discharged through valve (V) into the environment. Under such conditions, the heat exchanger does not cope with heat generation from the surface of the patient's bodies and cannot restore the specified temperature mode of the chamber until the patients enter the lock chamber. The WBC procedure takes place at a temperature significantly higher than the nominal −160°C [7]. After the patients leave, the thermal load on the cryostatting system is reduced 10 times, the air temperature drops to the nominal level, and the temperature controller stops the supply of LN to the heat exchanger (HE).

The increase in the air nominal temperature in the cab for WBC from −170°C to −160°C fundamentally changed the temperature algorithm of the procedure. The transition to the LN film boiling regime caused a significant overrun of the cryoagent. The operational drawbacks of the nitrogen cooling system and the uncertainty of the air temperature requirements in the main procedural cab created conditions for use in the WBC cryostatting system of three-stage chillers and steam cycles on gas mixtures. Refusing LN resulted in an increase in the nominal temperature in the main cab to −110°C.

Specialists in the field of WBC did not only pay attention to this but also actively promoted the "modernization" of cryotherapy equipment [13–15]. The ability to refuse to use LN and significantly reduce the costs of WBC procedures turned out to be so attractive that the specialists "did not notice" that the efficiency of the procedures in the "nitrogen-free" installations was 10 times lower than in "Cryotium" [7]. At the beginning of the twenty-first century, "Criohome" "cryotherapeutic" devices with a nominal temperature of −85°C was used for WBC procedures, i.e., the tendency to increase the temperature persists. Since 1985, the Russian direction of devices for WBC has been developing independently, based on the use of single-seat installations with a nitrogen cooling system (cryosaunas). The temperature in the cab of a single-seat cryosauna during the whole procedure is no higher than −130°C. The conditional constancy of temperature fundamentally changes the degree of supercooling of the skin surface; therefore, cryosaunas ensure the effectiveness of WBC at the level of the original technology implemented in "Cryotium" but with less energy loss. The current state of WBC in Europe is a consequence of the 40-year use of the method in the absence of a reliable concept of the method for obtaining the cryotherapeutic effect and the uncertainty of the technological requirements to specialized equipment [3–5]. In such conditions, manufacturers of WBC installations have flooded Europe with installations that, by their therapeutic efficacy, do not differ from traditional hypothermia. The popularization of the thermophysical theory of WBC will stop the regression of cryotherapy in Europe and the world.

3. Thermophysical theory of WBC

The WBC thermophysical theory was formulated at St. Petersburg National Research University of Information Technologies, Mechanics and Optics (ITMO University) in order to overcome the uncertainty of the technological requirements for specialized devices for WBC. In developing the theory, information on the conduct of WBC procedures and their effectiveness was used [1, 2, 7, 13]. As a criterion for optimizing the WBC technology, it is reasonable to use the duration of the analgesic effect of cryotherapy. The duration of the analgesic effect or effective time (WBC ET) is easy to determine in practice. To carry out computational experiments at ITMO University, a method was developed for calculating the WBC ET [16], which made it possible to perform studies on the optimization of the WBC technology in the mode of numerical experiment.

To calculate the WBC ET, a formula is proposed that relates the positive effect with the degree of approaching the skin surface temperature (T_s) to the temperature of the cryogenic damage onset (T_{cr} = 270.5 K), as well as with the area (f_s) and the duration (τ_{max}) of body surface contact with cryogenic gas:

$$\tau_E = f_s \int_{\tau=0}^{\tau_{max}} \frac{A}{(T_s - T_{cr})^2} \partial\tau, \qquad (6)$$

where A is an empirical constant, providing the calculation of the WBC ET in seconds, A = 1200.

The maximum duration of WBC (τ_{max}) is determined taking into account the requirements of the patient's hypothermic safety, which limit the permissible changes in body temperature on the surface (T_s) and on the inner boundary of the subcutaneous fat layer (T_f) (**Figure 3**).

Compliance with the established limitations of changing the value of T_s and T_f protects the patient from the danger of frostbite and frigorism, respectively:

$$T_s \geq 271 \text{ K } (-2°\text{C}); \; T_f > 309 \text{ K } (36°\text{C}).$$

Figure 3 shows a graphical representation of the patient's body shell (BS). BS is the outer layer of the body, the mass of which is 30% of the total body mass. BS consists of three types of tissues: epithelium 1, adipose tissue 2, and muscle tissue 3.

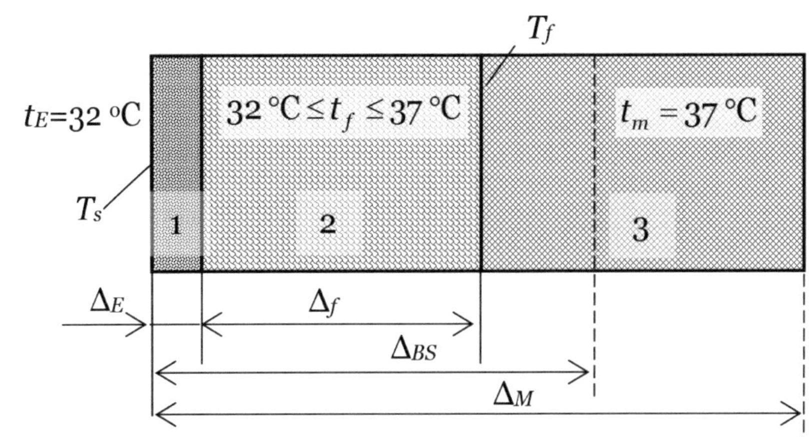

Figure 3.
Physical model of the human body shell (BS).

Layers 1 and 2 endure significant hypothermia without any harm; patient's safety is ensured when the violation of the normal temperature distribution does not extends beyond the inner boundary of the BS [13]. In the normal state t_E = 32°C, within the fat layer, the BS temperature rises 32°C ≤ t_f ≤ 37°C; the temperature of the muscle layer is equal to the human body core (BC) temperature t_m = t_{BC} = 37°C. It has been assumed that the thickness of layers 1 and 2 for an average patient has the following values: Δ_E = 2 mm; Δ_f = 2 mm.

The simplified physical model of a human BS has become the basis of a mathematical model.

4. Mathematical model of the human body shell

BS is a relatively thin surface layer. The calculated thickness of the BS of an average human is Δ_{BS} = 16 mm. In this case, the effective diameter of the body is 280 mm [7]. This allows us to describe the processes of heat transfer through the BS tissues by one-dimensional energy equation:

$$\rho \frac{\partial h}{\partial \tau} = \frac{\partial q_x}{\partial x} + q_v, \tag{7}$$

where h means the tissue enthalpy, kJ/kg; τ means the time, sec; q_x means the heat flux through BS, W/m^2; x means the coordinate along which heat is transferred, m; and q_v means the heat of metabolism, W/m^3.

The energy equation allows you to simulate thermal processes associated with significant changes in the temperature of the study object. The energy equation describes well the processes of changing the state of aggregation; this provides the mathematical model with certain advantages compared to models built based on the heat transfer equation [13]. When replacing derivatives with difference approximants, it is possible to obtain an algebraic expression suitable for automated calculations:

$$\rho \frac{\Delta h}{\Delta \tau} = \frac{\Delta q_x}{\Delta x} + q_v. \tag{8}$$

Solving Eq. (8) with respect to the value of enthalpy at the new time layer $\boldsymbol{h}'$, we get:

$$h' = h + \frac{(\Delta q_x + q_v \Delta x)\Delta \tau}{\Delta x \, \rho}, \tag{9}$$

where $\Delta h = h' - h$, h means the substance enthalpy value on the current time layer; Δx means the step of area elements along the x coordinate, m; and $\Delta\tau$ means the time step. Eq. (9) allows you to calculate the enthalpy of the nodal points of the simulated object under known boundary and initial conditions.

The structure of the cooling object and temperature distribution in the BS layers are described above (**Figure 3**). Thermophysical properties of the human body shell tissues are shown in **Table 1**.

Due to the high water content (φ), all human BS tissues have a high heat capacity, which ensures the accumulation of a significant amount of heat. The heat accumulated in the tissues protects the organs of the body core (BC), heart, lungs, kidneys, and liver from frigorism at a sharp decrease in ambient temperature. Thermal balance of the BS area element

Tissue	ρ, kg/m³	φ, %	c, J/(kg·K)	λ, W/(m·K)	q_v, W/m³
Skin	1093	72.0	3600	0.35	10,996
Muscles	1041	80.0	3458	0.475	7277
Adipose tissue	916	20	2250	0.21	—

Table 1.
Properties of the human BS tissues [17].

$$\Delta h = h' - h = \frac{(\Delta q_x + q_v \Delta x)\Delta\tau}{\Delta x \rho}, \quad (10)$$

determined by the ratio of the intensity of the heat fluxes transferred by the thermal conductivity of the tissue along the x coordinate Δq_x and the heat released by internal sources q_v. The intensity and direction of heat transfer by thermal conductivity depend on the temperature distribution along the x coordinate:

$$\Delta q_x = q_{i+1} + q_{i-1}; q_{i+1} = \lambda \frac{T_{i+1} - T_i}{\Delta x}; q_{i-1} = \lambda \frac{T_{i-1} - T_i}{\Delta x}. \quad (11)$$

At BS boundaries, heat transfer is described by boundary conditions. For the outer boundary, the intensity of the convective heat removal is calculated:

$$i = 1; \quad q_{i-1} = \alpha(T_g - T_i), \quad (12)$$

where α means the heat transfer coefficient at the natural convection of gas or liquid, $\alpha = f(T_g, T_i)$, and W/(m^2·K); T_g means the temperature of the heat-removing medium, K.

On the inner BS boundary, the temperature of the tissues during WBC does not change and is equal to the body core temperature (0):

$$i = n; \quad T_{i+1} = T_{BC} = const. \quad (13)$$

Eq. (10) describes the change in the heat content of tissues over time $h_i = f(\tau)$, and in order to form the trial of the WBC technology issues, a similar dependence for temperature should be obtained $\mathbf{T}_i = f(\tau)$. At each time step, the temperature value is calculated from the known value of the enthalpy of the array elements:

$$T_i = f(h_i). \quad (14)$$

The described algorithm of computations forms the mathematical model of the human BS, which is suitable for studies of processes of the therapeutic effect of low-temperature liquids and gases.

5. Thermophysical bases of achievement of the WBC healthcare effect

The mathematical model of the human BS allowed us to perform a numerical experiment to study the physical bases of the WBC healthcare efficacy. Quite often [4, 5] WBC is compared with the cold water immersion procedures. The basis for this comparison is that cryogenic gas and cold water remove a significant amount of heat from the body surface. Moreover, under conditions of natural convection, the coefficient of heat transfer from a source of heat to gas is usually 10 times lower than when heat is removed by water [7]. In reliance on this information, attempts to replace WBC with cheaper water procedures have been ongoing for 40 years [18]. Proponents of such a replacement do not take into account the fact that the WBC ET is more than 360 minutes, and water treatments provide pain relief for a maximum of 30 minutes. Such a difference in efficiency should be based on the fundamental differences between the results of heat removal by liquid and gaseous heat carrier.

Simulation of the BS surface cooling process with water with a temperature of 273 K and a cryogenic gas with a temperature of 140 K has allowed revealing such a difference (**Figure 4**), which reduces to the level of the minimum surface temperature of the cooling object. In cold water, the minimum surface temperature of the cooling object is at least 5.5°C. As the temperature difference between the water and the cooled surface decreases, the intensity of convective heat removal to water decreases to less than 950 W/m^2 [19]. The intensity of heat supply from the inner layers of the body to the surface, on the contrary, approaches the level of 880 W/m^2. Due to the small difference in heat fluxes at the boundary of the cooling object, the rate of temperature decrease $T_{i=1}$ reduces to 0.01 K/s; therefore, an increase in the duration of water hypothermia leads only to overcooling the patient's body. The estimated duration of the analgesic effect after water hypothermia is 31 minutes [16].

The use of a cryogenic gas with a temperature of 140 K for cooling the BS surface gives a completely different picture. The graph of surface temperature changes (**Figure 3**) is almost a straight line, which means that the temperature decreases almost without the rate change:

$$\partial T_{i=1}/\partial \tau \sim const. \tag{15}$$

The minimum temperature value $T_{i=1}$ = 271 K was obtained due to the termination of the numerical experiment on the patient's safety conditions during WBC: $T_s \geq 271$ K (−2°C); $T_f > 309$ K (36°C).

In the case of cooling with a cryogenic gas, the BS surface is supercooled to the minimum acceptable level. WBC technology is based on the use of this hypothermia to stimulate the cold receptors of the skin. Expression (6) for calculating the duration of positive effects contains a term that allows us to illustrate the intensity of stimulation of cold receptors by changing the temperature of the BS surface. This parameter of the WBC procedure is called the intensity of stimulating action (ISA):

$$I_{SA} = A/(T_{i=1} - T_{kr})^2. \tag{16}$$

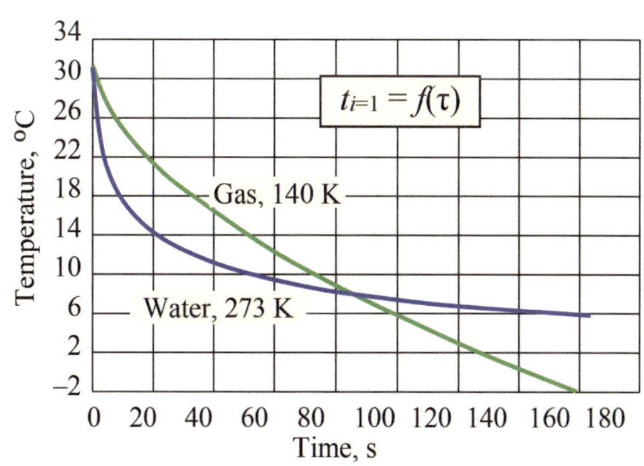

Figure 4.
Dependence of skin temperature on time.

The dependence graph $I_{SA} = f(T_{i=1})$ (**Figure 5**) shows how the receptor signal increases with the skin surface supercooling. At the lowest possible temperature of the skin surface under conditions of water hypothermia (5°C), I_{SA} = 21 s/s, while at WBC the maximum value is 225 times higher than I_{SA} = 4800 s/s. Differences in the intensity of stimulation of cold receptors determine the therapeutic benefits of WBC.

Data on the amount of heat Q_{HC} removed from a BS surface unit by a heat carrier (HC) and heat flux intensity has the fundamental importance for the development of WBC technology q_{HC}. The total heat removal is determined by integrating the instantaneous values $q_{HC}q_{HC}$, which were calculated by Eq. (12):

$$Q_{HC} = \int_{\tau=0}^{\tau_{\max}} q_{HC} \partial\tau. \tag{17}$$

The calculated values are given in **Table 2**. In cryogenic gas, heat removal was 440 kJ/m^2, which is 10% more than in cold water. The result obtained is significantly less than that supposed by some WBC popularizers who estimate heat removal from the patient's body at 1250–2500 kJ/m^2 [20].

At the same time, the result obtained is significantly more than can be removed from a WBC device by using 2 kg of liquid nitrogen for one patient's cooling [6, 10].

The assessment of the power of the specific heat flux, which the BS surface gives to the heat carrier, has essential practical importance. **Table 1** shows the maximum

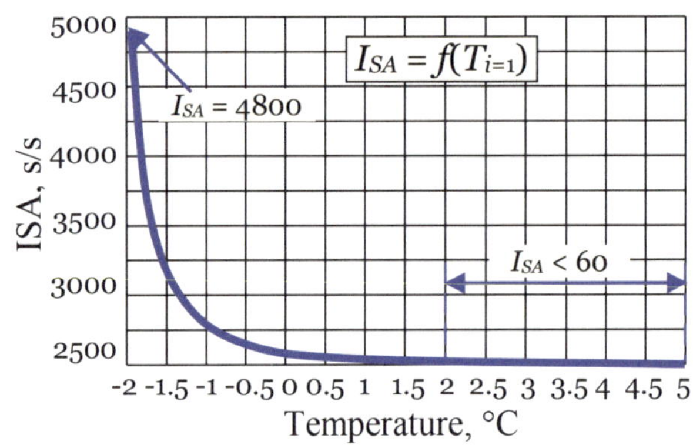

Figure 5.
Dependence of ISA value on body surface temperature.

Results	Heat carrier	
	gas	**water**
Cooling time, τ_{max}, sec	159	177
Minimum surface temperature of the object, T_s, °C	−2.0	5.5
The minimum temperature at the inner boundary of the fat layer, T_f, °C	309.2	309.0
Heat removed by heat carrier from the body surface, Q_{HC}, kJ/m^2	440	410
Heat removed through the inner boundary of the fat layer, Q_f, kJ/m^2	10.2	12.5
Heat flux from the body surface at the beginning of the cooling process, q_{HC}^{max}, kW/m^2	3.5	11.3
Heat flux from the body surface at the end of the cooling process, $q_{HC}^{\min}$, kW/m^2	2.3	0.95

Table 2.
The results of a numerical experiment on the simulation of heat removal by water with a temperature of 273 K and gas with a temperature of 140 K [16].

values, at the beginning of the procedure, and the minimum values, at the time of completion of the cooling powers of the heat flux to the cold water and the gaseous heat carrier. For designing WBC devices, it is useful to know the mean value of the heat flux, which the heat carrier must remove in a single procedure, 2.9 kW/m2. This value is 29 times greater than the nominal calorific capacity of the human body; therefore, it is often challenged by manufacturers of WBC devices [13]. Estimation of the heat reserve in BS tissues before and after the procedure shows that the heat flux to the heat carrier is provided by the heat capacity of the body shell tissues (**Figure 6**).

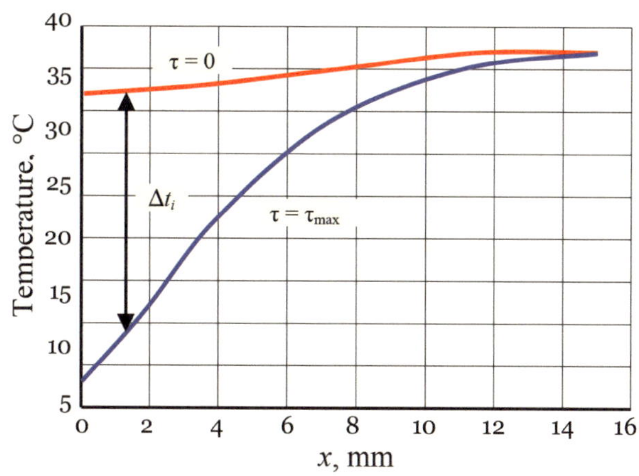

Figure 6.
The temperature of the body shell before and after the WBC procedure.

Lowering the surface temperature of the BS creates conditions for increasing heat transfer with thermal conductivity from the deep to the periphery of the body. As a result, there is a change in the distribution of the tissue temperature throughout the entire thickness of the BS. The amount of heat removed from different BS tissues is determined by the enthalpy difference before and after the WBC procedure. Taking into account the constancy of the heat capacity of the tissues in the temperature range from −2 to 40°C [17], the amount of heat removed can be calculated from the temperature difference:

$$\Delta T_i = T_i^{\tau=0} - T_i^{\tau=\tau_{\max}}. \tag{18}$$

The amount of accumulated heat removed from one area element:

$$Q_{Ai} = \Delta T_i \Delta x \, \rho_i c_i. \tag{19}$$

The total amount of heat released due to supercooling of each of the three types of BS tissues is the sum of portions of heat released in the area elements of this tissue layer:

$$\begin{aligned} &\textit{epithelial layer}, \quad 1 \le i \le n_E, \quad Q_{AE} = \textstyle\sum_{i=1}^{n_E} \Delta T_i \Delta x \, \rho_i c_i; \\ &\textit{fat layer}, \quad n_E + 1 \le i \le n_F, \quad Q_{AF} = \textstyle\sum_{i=n_E+1}^{n_F} \Delta T_i \Delta x \, \rho_i c_i; \\ &\textit{muscular layer}, \quad n_F + 1 \le i \le n_{BS}, \quad Q_{AM} = \textstyle\sum_{i=n_F+1}^{n_{BS}} \Delta T_i \Delta x \, \rho_i c_i, \end{aligned} \tag{20}$$

where n_E, n_F, and n_{BS} means the number of area elements located between the outer surface and the inner boundary of the epithelial layer, the fat layer, and the patient's body shell, respectively.

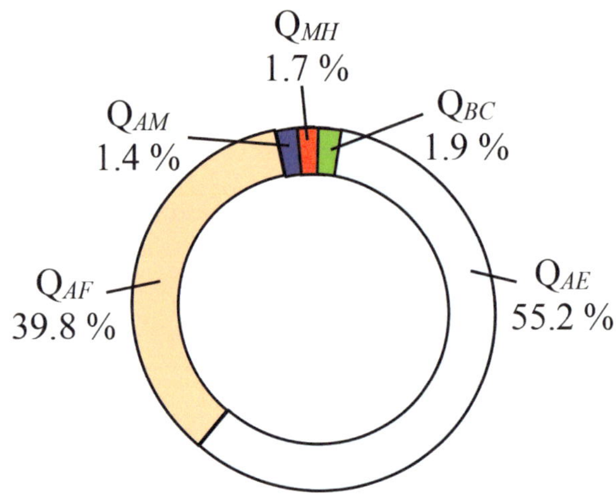

Figure 7.
Sources of the heat gained the WBC procedure.

Part of the heat removed was obtained from internal sources in the epithelial and muscle layers, heat of metabolism Q_{MH}. This heat was calculated by a known value q_v (**Table 1**):

$$Q_{MHE} = \tau_{\max}\Delta x(n_E - 1)q_{VE};\ \ Q_{MHM} == \tau_{\max}\Delta x(n_{BS} - n_F - 1)q_{VM}, \tag{21}$$

where q_{VE} and $q_{VM}q_{VM}$ means the specific heat of metabolism of epithelium and muscles, respectively, W/m^3.

Some of the heat removed came from the patient's body core; the amount of heat gained can be determined by numerical integration and instantaneous values of the heat flux transferred by thermal conductivity through the inner boundary of the body shell:

$$Q_{BC} = \int_{\tau=0}^{\tau_{\max}} q_{n_{i+1}} \partial\tau. \tag{22}$$

The histogram on **Figure 7** gives an idea of what is the source of heat removed from the surface of the patient's body shell. The main share of the heat of 55.2% was gained due to supercooling the epithelial layer. The heat gained by supercooling the fat layer Q_{AF} is 39.8%. The supply of heat from the body core Q_{BC} and internal sources Q_{MH} in the body shell tissues is less than 2%. This supply of heat is gained by supercooling the muscle layer Q_{AM}.

The calorific capacity of the body does not play any role in the formation of the heat load on the cooling system of the WBC device, which is determined by the heat storage capacity of the body shell tissues. The safety of the WBC procedures is ensured by the correct choice of the contact duration of the body surface with a cryogenic gas. The thermal control system of the body does not affect the safety of procedures.

6. Selecting optimal gas temperature in the WBC zone

In practice, there are two options for carrying out WBC procedures in multi-seat and single-seat installations [7, 21–23]. The cooling conditions in these installations differ significantly; therefore, the technology of group and individual WBC should be developed separately.

Contrary to the popular belief [7, 13], GWBC and IWBC provide effects on only a fraction of the skin area. In a group installation, the contact area of the cooling gas with the patient's body in a multi-seat cab is up to 70.5% of the total surface area of the body. In an individual cab, the contact area reaches 66% [7]. Temperature regimes of GWBC and IWBC are fundamentally different.

The GWBC technology was influenced by the design of the device for performing the procedures (**Figure 1**). Using a low-temperature food storage chamber for WBC procedures, Japanese engineers and doctors were forced to carry out WBC procedures in groups. The dimensions of the chamber were too large for individual procedures. This forced solution is contrary to the general practice of physiotherapy; treatment is always carried out individually.

Systems for implementing technology I, individual cryosaunas, were developed 20 years after multi-seat installations [7] with consideration of the experience of their operation. Modern installations for IWBC use a nitrogen cooling system (NCS), so they quickly reach a given temperature level and allow you to adjust the temperature of the gas in the WBC zone.

It is impossible to develop universal recommendations on selecting the optimal temperature of the gaseous heat carrier for GWBC and IWBC, since in multi-seat and single-seat installations the algorithm for changing the temperature of the cooling gas during the procedure is different.

To conduct a preliminary analysis of the effect of gas temperature in the WBC zone on the magnitude of the positive effect achieved, it can be assumed that the procedure takes place in isothermal conditions:

$$0 < \tau < \tau_{max}; \quad T_1 = \text{const}. \tag{23}$$

It is impossible to implement WBC in the isothermal mode, since it takes some time for the patient to enter the low-temperature zone and exit from it. However, the study of WBC processes in ideal temperature conditions allows us to formulate the general technological conditions of efficiency.

To determine the optimal gas temperature in the WBC zone, the calculated values of the WBC ET obtained by Eq. (6) were used. Simulation of the BS cooling process under conditions of natural gas convection with a temperature from 90 to 190 K allowed us to plot the dependence of the ET value on the gas temperature in the WBC zone (**Figure 8**).

When isothermally cooling the surface of the patient's body, the maximum value of ET (325 min) is achieved at a temperature of 140 K. At temperatures below 140 K, the WBC efficiency gradually decreases. At a temperature of 100 K, the value of WBC ET is almost three times lower than the maximum [7]; therefore, when conducting WBC procedures, it is advisable to use a gas with a temperature from 120 to 140 K [16]. At temperatures above 140 K, the WBC efficiency rapidly decreases. At a temperature of 160 K, the WBC ET of the procedures is 10 times lower than the maximum value and is close to the results achieved during water procedures. The results of the computational experiment on simulation of cooling the body surface with gas with a temperature of 160 K (−110°C) ideally coincide with the results of tests performed by doctors in sports medicine [8], which in comparing the therapeutic effect of WBC procedures at a temperature of −110°C and water baths with a temperature of 8°C, did not reveal any advantages of the WBC. The results obtained have clear thermophysical reasons. As the temperature of the gaseous heat carrier increases, the intensity of heat removal from the BS surface decreases, and the safe cooling time increases.

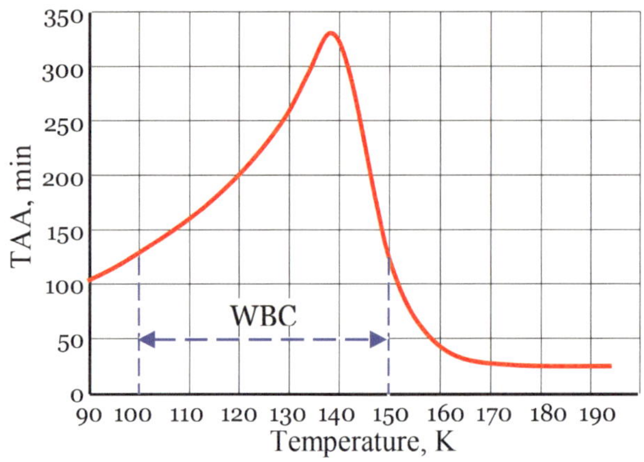

Figure 8.
The estimated duration of the effect of WBC at different gas temperature.

When the gas temperature is above 150 K, the danger of supercooling of the body core ($T_f \rightarrow 309$ K) occurs before the surface of the body shell is supercooled. The reason for the termination of the WBC procedures becomes a violation of the condition $T_f \geq 309$ K. At the same time, the temperature of the body shell surface remains at a sufficiently high level of $T_s \geq 275$ K (2°C), due to which the cold receptors of the skin do not experience significant irritation and the accumulation of a positive WBC effect is extremely slow. The picture described is identical to what is observed at the time of completion of the water hypothermia procedure. Under conditions of isothermal cooling of the body with gas with a temperature of 160 K (−110°C), the estimated duration of the WBC procedure is 207 sec. During this time, the BS surface temperature drops only to 275 K. At this BS surface temperature, the I_{SA} value is 80 times less than the maximum value (**Figure 5**). Under actual conditions, the WBC procedures in installations with a minimum temperature of 160 K (−110°C) do not ensure the constancy of the gas temperature, so the BS surface temperature after the procedure is much higher than the calculated one and is 15–20°C [24]. Such a temperature on the surface of the skin can be obtained using water baths with a temperature of 8°C; therefore, the doubts of some authors [11, 18, 25] on the advisability of using cryogenic technologies are fully justified.

According to the results of simulating the process of cooling the BS surface with a cryogenic gas, it can be argued that for effective procedures the gas temperature in the WBC zone should be not lower than 140 K.

7. Selecting the optimal duration of the WBC procedure

The author of the WBC method, Yamauchi, limited the exposure of the body contact with a cryogenic gas to a period of 180 sec [1, 2]. The minimum air temperature in the Japanese installation was −175°C. According to the contemporary idea that WBC technology is based on metered supercooling of the body shell, the choice of the cooling exposure should be related to the temperature of the gas in the WBC zone. Using the assumption of the constancy of the gas temperature in the WBC zone, it is possible to determine the maximum duration of cooling at different gas temperatures. Computational experiments on the mathematical model of the human BS showed that with an increase in the heat carrier temperature from 90 to 190 K, the safe duration of a patient's stay in the WBC zone increases from 54 to

237 sec [3]. At a temperature of 140 K, the safe exposure time for cooling is 161 sec. The practice of using WBC has shown that, along with the maximum duration of cooling, it is necessary to limit the minimum duration of stay of patients in a cryotherapeutic installation [7].

The reasons for this limitation are explained by the graph of dependence $ET = f(\tau)$ (**Figure 8**). The graph shows that there is a fairly long period in the WBC procedure when a positive effect is not formed. At a gas temperature of 140 K, this phase of the procedure accounts for almost 80%, but 93% of the positive effect is formed after its completion. The reason for the low efficiency of WBC at the beginning of the procedure is the relatively high skin temperature (T_S) (**Figure 8**), which drops to 275 K (2°C) only by the end of the first phase of the procedure. As it can be seen from the graph of dependence $I_{SA} = f(T_S)$ (**Figure 5**), at a skin temperature of $T_S > 275$ K, the intensity of the WBC stimulating effect is negligible $I_{SA} < 60$. The first phase of the WBC procedure reduces the surface temperature of the skin to a temperature of $T_S = 275$ K, so it is called the cooling time (τ_{cool}) (**Figure 9**). It is obvious that the duration of the WBC procedure must be longer than the duration of the cooling time but less than the time of the violation of safety conditions $\tau_{\text{cool}} < \tau < \tau_{\max}$. The second, effective, phase of the procedure ensures the formation of the main positive result, the longer the duration of the effective phase, the greater the effect of the procedure.

$$\tau_{EP} = \tau_{\max} - \tau_{cool} \tag{24}$$

The calculated dependences of the WBC safe exposure ($\tau_{\max}$) and the duration of the cooling time (τ_{cool}) on the gas temperature (T_g) (**Figure 10**) show that increasing the gas temperature from 90 to 150 K increases the effective phase of the procedure. A further increase in temperature increases the duration of the cooling time. At temperatures above 160 K, the estimated duration of the cooling time exceeds the safe WBC exposure. Even with isothermal cooling, it is impossible to provide an effective WBC when using gas with a temperature $T_g > 150$ K; in real conditions the gas temperature should be no higher than 140 K.

Numerical experiments on a mathematical model of the human body shell allowed to formulate general ideas about the technological foundations of effective WBC. When developing technological recommendations on the design of installations for the implementation of GWBC or IWBC methods, it is necessary to take into account the algorithm for changing the temperature of the gas in contact with the patient's body surface.

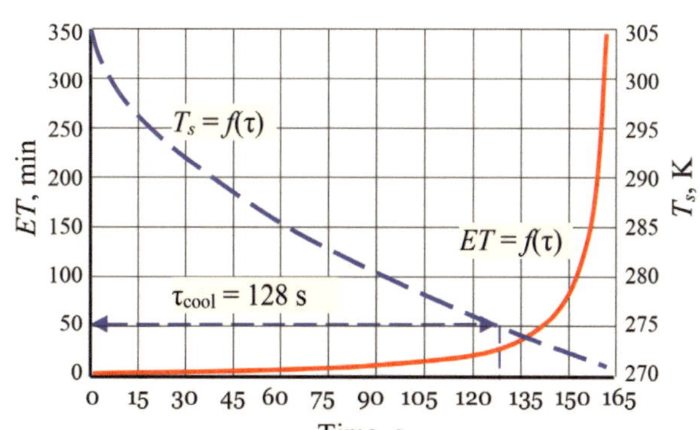

Figure 9.
The change of the body temperature surface T_s *and the value of* ET *during the WBC procedure with gas temperature of 140 K.*

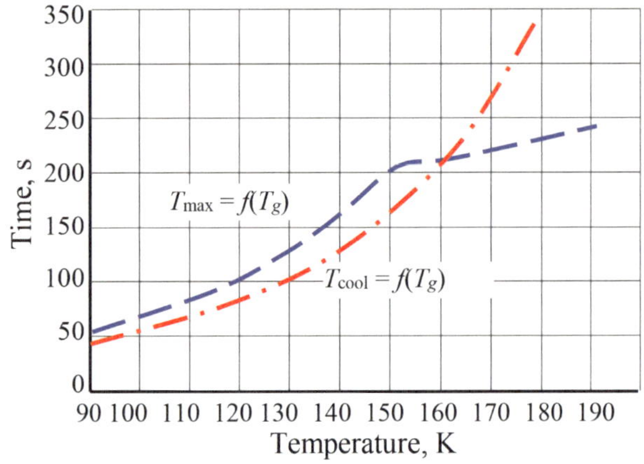

Figure 10.
The dependence of the cooling phase duration and safe exposure WBC on the gas temperature.

8. Algorithm for changing the cooling gas temperature during the individual and group WBC procedures

Installations for GWBC consist of two or three heat-insulated rooms with different air temperatures [7]. Patients pass from the treatment room to the chamber with the minimum temperature (main chamber, MC) and back through the lock chambers (LC). In most modern installations, the temperature in the main chamber is maintained at 160 K and in the lock chamber means at 210 K. At the time of entry (exit) of patients into the LC or MC, warmer air enters from adjacent volumes. Because of this, the air temperature in the MC volume increases by at least 25 K. From the body surface of each patient, 3.5 to 4.5 kW of heat is released into the MC volume. Taking into account these factors, the actual GWBC temperature regime depends not only on the choice of the nominal temperatures in MC and LC but also on the power of the cooling system. Another uncertainty factor is the duration of stay of patients in the main cab. There are different opinions about the advisability of pre-cooling the body surface at an intermediate temperature of 210 K. Some researchers believe that a gradual decrease in temperature increases subjective comfort and safety [7]. In other works it is proposed to reduce the time of stay of patients in LC to a minimum [13]. Given all the reasons presented, it is obvious that it is extremely difficult to simulate the GWBC process. The temperature of the cooling gas varies according to a complex schedule, which consists of at least eight stages (**Figure 11**).

The algorithm of changing the gas temperature in IWBC is much simpler (**Figure 11**). The patient enters the cab filled with atmospheric air, which is quickly replaced by vapors of liquid nitrogen with a temperature not higher than 140 K. The time to reduce the gas temperature in the IWBC cab to the optimum level depends on the power of the cooling system and is at least 20 sec.

Taking into account the results of simulating the WBC process under conditions of a constant gas temperature, it can be argued that the GWBC procedures using the algorithm shown in **Figure 11** do not provide significant therapeutic outcomes. To restore the effectiveness of GWBC, it is necessary to significantly reduce the minimum air temperature in the main treatment cab. Experiments on a mathematical model of the body shell showed that the effectiveness of GWBC reaches the optimal level when the air temperature in the main cab drops to 130 K. However, modern installations for GWBC cannot maintain the temperature at this level, since they use

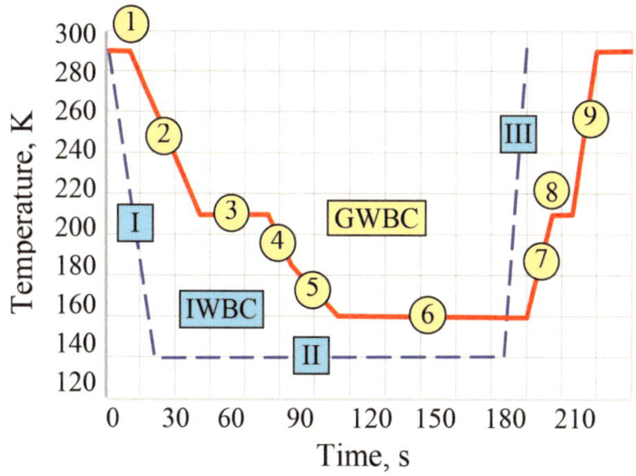

Figure 11.
Algorithm for changing the temperature of the of the cooling gas with the technology GWBS and IWBS.

compression cooling systems on gas mixtures [7]. To lower the temperature, it is necessary to use other heat transformation cycles in the cooling system, the power of which will allow compensating for the heat load associated with WBC procedures.

9. The heat load on the cooling system of the WBC zone

When designing cooling systems of the WBC zone, it is necessary to adequately estimate the power of the heat fluxes that need to be compensated. It was shown above that during the WBC procedure, 440 kJ/m^2 of the heat is released from the patient's body surface, and the mean heat flux from the body to the cryogenic heat carrier varies from 3.5 to 2.3 kW/m^2 (**Table 2**). Taking into account the surface area of the body (1.6 m^2 [7]), the heat input from one patient will be 700 kJ; the mean power of the heat input is 4.6 kW. It is necessary to spend at least 2.7 kg of liquid nitrogen only to remove the heat released from a patient's body surface with a gas with the temperature of T_g = 140 K:

$$G_{LN} = \frac{Q_{HC}}{\left[r_{LN} + c_{LN}\left(T_g - T''_{LN}\right)\right]'} \tag{25}$$

where r_{LN} means the heat of vaporization of nitrogen, r_{LN} = 199 kJ/kg;c_{LN} means the heat capacity of nitrogen vapor, c_{LN} = 1.002 kJ/kg; and T''_{LN} means the boiling point of liquid nitrogen at atmospheric pressure,T''_{LN} = 78 K.

Estimated nitrogen flowrate for removal of the heat from the body surface is 2.7 times higher than in modern nitrogen-cooled WBC installations [13]. To restore the WBC effectiveness, it is necessary to provide cryotherapy installations with sufficiently powerful cooling systems.

The heat input from the patients Q_{HC} is only the useful part of the heat load on the cooling system. In addition, it is necessary to compensate for the heat input from the walls of the thermal fencing of the WBC zone and Q_{TI} the heat that the warm gas fluxes bring from the adjacent volumes (lock chamber or environment) to the volume of the treatment cab Q_{GC}. The total heat load is defined as the sum of heat received from different sources:

$$Q_{\Sigma} = Q_{HC} + Q_{TI} + Q_{GC}. \quad (26)$$

The energy efficiency of the installation design for WBC can be estimated by the share of the useful load on the cryostatting system, the coefficient of thermal efficiency:

$$\eta_H = Q_{HC}/Q_{\Sigma}. \quad (27)$$

To estimate the expenditure of energy and select the optimal technology for WBC procedures, it is necessary to conduct a numerical experiment on a mathematical model of a cryotherapeutic device.

10. Mathematical model of cryotherapeutic device

Given the variety of design solutions used in the manufacture of devices for WBC, the mathematical model should have the most generalized form. It is necessary to stop considering particular design features and focus on the fundamental issues. It becomes possible with a one-dimensional model of the WBC zone (**Figure 12**). The model considers the processes occurring in a volume unit of the WBC zone.

The surface of the patient's body 1 is cooled with a gaseous heat carrier 2 which fills the volume of the thermal fencing 3. The heat flux is removed from the patient's body surface q_{HC}. The heat flux is supplied to the gas 2 from the surface of the thermal fencing 3 q_{TI}. The movement of gas fluxes creates an additional source of the heat load q_{GC}. To keep the gas temperature in the WBC zone at a constant level, it is necessary to ensure the removal of total heat q_{Σ} to the cryostatting system. To form a one-dimensional model, it is necessary to relate all system indicators to the volume unit of the WBC zone. The specific heat transfer surface of the thermal insulation f_3 and the patient's body f_1 are determined taking into account the volume of the treatment cab V_3:

$$f_3 = F_3/V_3; \; f_1 = nF_1/V_3, \quad (28)$$

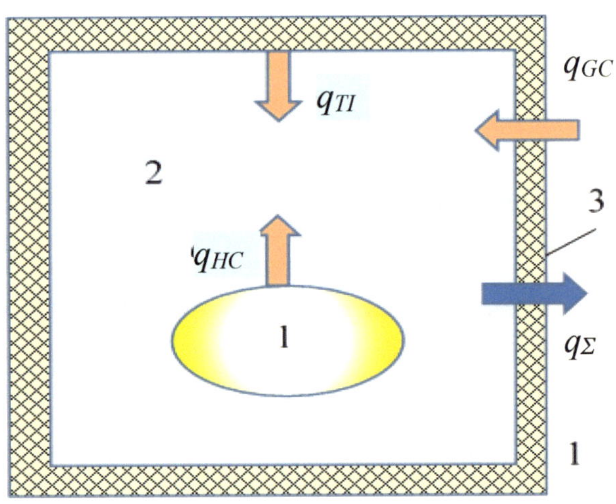

Figure 12.
Heat fluxes in the WBC zone.

where F_3 means the total area of the internal surface of the cab's thermal fencing, n means the number of patients in the cab, and F_1 means the surface area of a patient's body.

The specific heat input from the patient's body and the thermal fencing is calculated considering the temperatures of their surfaces:

$$q_{TI} = \alpha_{3-2}\, f_3(T_3 - T_2); q_{HC} = \alpha_{1-2}\, f_1(T_1 - T_2), \tag{29}$$

where α_{3-2} and α_{1-2} means the heat transfer coefficients from the thermal fencing and the patient's body, respectively; T_1 and T_3 means the temperatures of surfaces of the body and the fencing; and T_2 means the temperature of the heat carrier gas.

Specific characteristics of devices designed for the implementation of GWBC and IWBC technologies have large differences. In multi-seat installations, the patient accommodation density is 0.4–0.7 person/m^3, and the specific volume of free space V_0 is at least 98%. The specific surface area of the patient's body is 0.6–1.0 m^2/m^3; the thermal fencing area of the WBC zone is 2.4–3.0 m^2/m^3. Low compactness of accommodation of patients in multi-seat installations is necessary so that they can move from one low-temperature chamber to another.

In single-seat cryosaunas, the patient accommodation density reaches 2.0 persons/m^3, the specific surface area of the patient's body is 3.2 m^2/m^3, the thermal fencing area of the WBC zone is 6.4 m^2/m^3, and the specific free space volume is 84% [7]. High compactness of the patient accommodation is ensured by the fact that the patient does not move during the procedure; therefore, the cab size is comparable to the size of the patient's body.

The heat input with gas fluxes is determined by the intensity of convective mass transfer of warm gas to the volume of the WBC zone. The heat input by gas convection across the boundary of the WBC zone is determined from the expression:

$$q_{GS} = c_p g_G (T_g - T_2), \tag{30}$$

where g_G means the specific transfer of the gas mass into the volume of the WBC zone, kg/(m^3 sec); c_p means the heat capacity of the gas, kJ/(kg K); and T_g means the temperature of the gas entering the WBC zone.

Large heat flows with gas fluxes are supplied into the WBC zone as patients enter and exit. For example, a multi-seat lock chamber and a cab of a single-seat cryosauna are filled with atmospheric air at the moment patients enter. 93 kJ/m^3 of heat enters the lock cab with atmospheric air. When the temperature recovers to the nominal level, the air density in the lock cab increases by 40%; this is accompanied by supplying additional air from the atmosphere, which contributes another 27 kJ/m^3 of heat. In one procedure, 120 kJ/m^3 of heat transferred by gas convection enters the lock chamber.

The basis of the mathematical model of the WBC zone is a one-dimensional energy equation:

$$\rho \frac{\partial h}{\partial \tau} = \frac{\partial q_x}{\partial x} + q_v, \tag{31}$$

where q_v means the heat from internal sources:

$$q_v = q_\Sigma + q_{HC} + q_{TI} + q_{GC}. \tag{32}$$

In the ideal case, $q_v = 0$, since $q_\Sigma = -(q_{HC} + q_{TI} + q_{GC})$, i.e., the cooling system compensates for the heat input per volume unit of the WBC zone.

To account for material balance in the mathematical model of the WBC zone, the continuity equation is used:

$$\frac{\partial \rho}{\partial \tau} + \frac{\partial g_x}{\partial x} = 0. \tag{33}$$

So, the transfer of heat by the thermal conductivity of gas is small; expression (30) is simplified and can be transformed by replacing the derivatives with differential approximants:

$$\rho \frac{\Delta h}{\Delta \tau} = q_v; h' = h + \frac{q_v \Delta \tau}{\rho}. \tag{34}$$

The numerical solution of the continuity (Eq. (33)) allows to take into account the input of gas mass to compensate for the change in density:

$$g_x' = g_x + \frac{(\rho' - \rho)\Delta x}{\Delta \tau}. \tag{35}$$

Eqs. (29), (34) and (35) allow to analyze the processes occurring in the WBC zone during the implementation of individual or group technology. To perform a computational experiment, it is necessary to adopt an algorithm for changing the temperature of the cooling gas for IWBC and GWBC.

Formulation of a temperature algorithm for the IWBC process is relatively simple. let us take the time of filling the zone with a cryogenic gas (**Figure 11**) τ_I = 20 sec, τ_{II} = 150 sec, and τ_{III} = 10 sec and the gas temperature in the isothermal phase II T_{II} = 140 K. The specific heat transfer surfaces of heat sources are f_1 = 3.2 m^2/m^3, and f_3 = 6.4 m^2/m^3; the specific free space volume is 84% [7]. When simulating the IWBC process, heat fluxes from various sources (28) and (29) are calculated, and the integral heat input is determined:

$$Q_{HC} = f_2 \int_{\tau=0}^{\tau=\tau_{max}} q_{HC} \partial\tau; Q_{TI} = f_3 \int_{\tau=0}^{\tau=\tau_{max}} q_{TI} \partial\tau; Q_{GC} = \int_{\tau=0}^{\tau=\tau_{max}} q_{GC} \partial\tau. \tag{36}$$

By Eqs. (26) and (27), the total heat load on the cooling system and the coefficient of thermal efficiency are calculated. It is assumed that the cooling system covers all types of heat load, so the specific power of the refrigerator electric drive can be determined by the heat load and the value of the coefficient of performance at the current temperature level:

$$N_5 = q_\Sigma / \varepsilon_5; \varepsilon_5 = f(T_g), \tag{37}$$

where ε_5 means the coefficient of performance and the ratio of the heat removed to expenditure of energy in the refrigerator at the temperature level of 140 K, ε_5 = 0.25 W/W.

For the instantaneous values of the calculated power of the system refrigerator, the specific expenditure of energy for cooling the IWBC zone per procedure is calculated:

$$Q_5 = \int_{\tau=0}^{\tau=\tau max} N_5 \partial\tau \tag{38}$$

Let us determine the specific values of the liquid nitrogen flowrate per procedure:

$$g_{LN} = \frac{q_\Sigma}{\left[r_{LN} + c_{LN}\left(T_g - T''_{LN}\right)\right]'}; G_a = \int_{\tau=0}^{\tau=\tau_{\max}} g_a \partial\tau. \quad (39)$$

The results of the numerical experiment are summarized in **Table 3**. In the experiment on simulating the GWBC process, the time algorithm presented on the graph (**Figure 11**) was used; the nominal gas temperature in the main cab was 130 K. Energy indicators for the main and lock chambers were calculated.

The energy efficiency of the technology was estimated by the total energy expenditures in the main and lock chambers. The results of the numerical experiment on simulating the GWBC process are summarized in **Table 3**.

The data in **Table 3** show that WBC procedures require the removal of large amounts of heat from the low-temperature zone. Specific heat input to the IWBC zone is Q_Σ = 2012 kJ/m3. Given the short duration of the procedure ($\tau_{\max}$ = 180 sec), the mean heat load on the cooling system of the IWBC zone was 11.8 kW/m^3. Considering that this heat load must be removed at a temperature level of 140 K, the estimated power of the cooling system of the low-temperature zone was 45.3 kW/m^3. Refrigerators of such power are rather expensive; therefore, it is economically feasible to use the nitrogen cooling system of the WBC zone. Specific liquid nitrogen flowrates are 7.503 kg/m^3.

Indicators	IWBC	GWBC		
		Cab	Lock chamber	Total
Features of WBC zone				
f_1, m^2/m^3	3.2	0.62	0.62	
f_2, m^2/m^3	6.4	2.4	2.4	
V_0, %	0.84	0.97	0.97	
Heat input to the WBC zone				
q_Σ^{max}, kW/m3	33	7.43	3.97	
Q_Σ, kJ/m^3	2012	422	144	566
Q_{GC}, kJ/m3	92	142	96	238
Q_{HC}, kJ/m3	1427	246	33	279
Q_{TI}, kJ/m3	493	33	14	47
Q_Σ/τ_{max}, kW/m3	11.8	2.76	1.18	3.94
η	0.71	0.58	0.23	0.49
The expenditure of electrical energy and liquid nitrogen flowrate for cooling				
N_5^{max}, kW/m3	136	11.04	4.72	15.76
Q_5, kW/hour/m^3	2.15	0.50	0.07	0.57
Q_5/τ_{max}, kW/m3	45.3	11.04	4.72	15.76
G_{LN}, kg/m^3	7.503	1.56	0.42	2.02
G_{LN}/τ_{max}, g/(sec·m3)	0.04	0.0029	0.009	0.0038

Table 3.
Energy features of devices for IWBC and GWBC.

The energy indicators of the GWBC zone are much lower (**Table 3**). The specific heat input is Q_Σ = 566 kJ/m^3, the mean heat load on the cooling system is 15.6 kW/m^3, and the power of the cooling system is determined by the sum of the inflows of heat into the main cab and lock chamber.

Due to the low compactness of the accommodation of patients in the treatment area, the GWBC thermal efficiency coefficient was 0.49. Under conditions of a single-seat installation, the thermal efficiency coefficient was 0.71, which indicates a more rational expenditure of energy. This is clearly illustrated by the histogram of the structure of the heat load on the cooling system of the IWBC and GWBC zones (**Figure 13**).

In single-seat installations, the heat storage capacity of the thermal fencing makes a significant contribution to the heat load, due to which the share of heat removed from thermal insulation reaches 24%. At the beginning of each procedure, a single-seat cab is filled with atmospheric air, which heats the inner surface of the thermal insulation. When implementing the GWBC technology, the heat load from the insulation is insignificant means of 9%, but the convective heat supply is 24%. The negative impact of convective heat transfer is determined by a large share of the free space in the low-temperature zone.

The data in **Table 3** do not allow giving an unambiguous preference for a particular technology. This is due to the fact that all indicators are related to the volume unit of the WBC zone, while the technological task of the process is to cool the surface of the patient's body shell. If we calculate the specific heat load values and the expenditure of energy for cooling a unit of the shell surface (**Table 4**), the advantages of the IWBC technology become indisputable. According to all energy indicators, the IWBC technology is 1.5 times more efficient than the GWBC process.

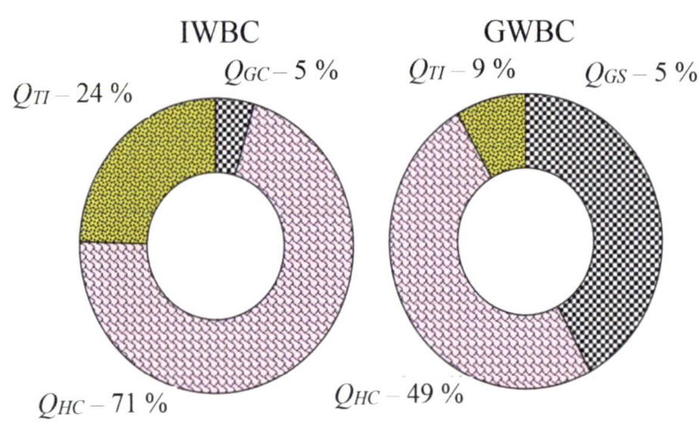

Figure 13.
The structure of the heat load on the cooling system zones IWBC and GWBC.

Indicators	IWBC	GWBC
Q_Σ/f_1, kJ/m2	629	913
Q_5/f_1,Q_5/f_1, kW/hour/m2	0.67	0.92
G_{LN}/f_1, kg/m2	2.34	3.24
$G_{LN}/f_1 \cdot F1$, kg/m2	3.77	5.18

Table 4.
Energy features of devices for IWBC and GWBC.

11. Conclusion

The performed analysis of the healthcare and energy efficiency of the two options for the implementation of the WBC technology allows us to reasonably give preference to individual procedures that not only combine high healthcare efficiency with relatively low expenditure of energy but also to a greater extent correspond to the traditional principle of individuality of therapeutic techniques.

The effectiveness of WBC technology depends on the choice of the duration of contact with cryogenic gas. The minimum duration of WBC procedure at the optimum gas temperature (−130°C) is 120 s. Meanwhile, one should remove 440 kJ/m^2 with an average intensity of at least 2.4 kW/m^2 and spend not less than 1.7 kg/m^2 of liquid nitrogen on heat removal. The electric drive of the cooling system of WBC zone should have an average power of at least 9.3 kW/m^2, and in the case of using nitrogen cooling system, the cryoagent consumption should be not less than 2.4 kg/m^2.

Other declarations

The research was supported by the Ministry of Education and Science of the Russian Federation (Project 11.4942.2017/6.7).

Author details

Alexander Baranov*, Oleg Pakhomov, Alexander Fedorov, Vladimir Ivanov, Andrew Zaitsev and Ruslan Polyakov
Saint Petersburg National Research University of Information Technologies, Mechanics and Optics, Saint Petersburg, Russian Federation

*Address all correspondence to: abaranov@corp.ifmo.ru

References

[1] Yamauchi Y, Yamauchi T, Miura K. The analgesic effects of −170°C whole body cryo-therapy on rheumatoid arthritis (R.A.); curable. Pain. 1987;**30** (Supp. 1):S261

[2] Yamauchi T, Yamauchi Y, Miura K, Cooper A. Clinical effects of −170°C whole body cryotherapy (W.B.C.T.) on steroid dependant chronic diseases. Journal of Steroid Bio-Chemistry. 1986; **25**(Supp. 1):25

[3] Bouzigon R, Grappe F, Ravier G, Dugue B. Whole- and partial-body cryostimulation/cryotherapy: Current technologies and practical applications. Journal of Thermal Biology. 2016;**61**: 67-81

[4] Carrard J, Lambert A, Genné D. Transient global amnesia following a whole-body cryotherapy session. BML Case Reports. 2017;**2017**

[5] Greenwald E, Christman M, Penn L, Brinster N, Liebman T. Cold panniculitis: Adverse cutaneous effect of whole-body cryotherapy. JAAD Case Reports. 2018;**4**:344-345

[6] Rose C, Edwards KM, Siegler J, Graham K, Caillaud C. Whole-body cryotherapy as a recovery technique after exercise: A review of the literature. International Journal of Sports Medicine. 2017;**38**:1049-1060. DOI: 10.1055/s-0043-114861

[7] Baranov AY. Low-temperature installations for medical purposes. In: Part 1. Apparatus for General Cryotherapy Treatment: (Textbook). St. Petersburg: University ITMO; IHBT; 2016. p. 178

[8] Costello JT, Donnelly AE, Karki A, Selfe J. Effects of whole body cryotherapy and cold water immersion on knee skin temperature. International Journal of Sports Medicine. 2014;**35**:35-40

[9] Cámara-Lemarroy CR, Azpiri-López JR, Vázquez-Díaz LA, Galarza-Delgado DA. Abdominal aortic dissection and cold-intolerance after whole-body cryotherapy: A case report. Clinical Journal of Sport Medicine. 2017;**27**: e67-e68

[10] Polidori G, Cuttell S, Hammond L, et al. Should whole body cryotherapy sessions be differentiated between women and men? A preliminary study on the role of the body thermal resistance. Elsevier. Medical Hypotheses. 2018;**120**:60-64

[11] Krueger M, Costello JT, Achtzehn S, Dittmar KH, Mester J. Whole-body cryotherapy (−110°C) following high-intensity intermittent exercise does not alter hormonal, inflammatory or muscle damage biomarkers in trained males. Cytokine. 2018;**113**:277-284. DOI: 10.1016/j.cyto.2018.07.018

[12] Piras A, Campa F, Toselli S, Rocco DM, Raffi M. Physiological responses to partial-body cryotherapy performed during a concurrent strength and endurance session. Applied Physiology, Nutrition, and Metabolism. 2018. DOI: 10.1139/apnm-2018-0202. [Epub ahead of print]

[13] Zagrobelny Z. Krioterapia Miejscowa I Ogolnoustrojowa. Wroclaw: Urban & Partner. 2003. pp. 189 s. 218

[14] Fricke R, Grapow G, Nobbe T, Knauer G. Ganzkörperkältetherapie bei −110°C von zwei minuten bewirkt steigerung von muskelkraft und leistung. Physikalische Medizin Rehabilitationsmedizin Kurortmedizin. 1999;**9**:144

[15] Fricke R. Ganzkörperkältetherapie in einer kältekammer mit temperaturen um −110°C. Physikalische Medizin, Balneologie, Med. Klimatologie. 1989, 18. Jahrgang: pp. 1-10

[16] Baranov AY, Malysheva TA. Modeling unsteady heat transfer in cryomedicine. Journal of International Academy of Refrigeration. 2000;**2**:38-41

[17] Tsyganov DI. Cryomedicine: Processes and Devices. Monograph. Moscow: SAINS-PRESS; 2011. p. 304

[18] Poppendieck W, Faude O, Wegmann M, Meyer T. Cooling and performance recovery o trained athletes: A meta-analytical review. International Journal of Sports Physiology and Performance. 2013;**8**:227-242

[19] Vieira A, Bottaro M, Ferreira-Junior JB, Vieira C, Cleto VA, Cadore EL, et al. Does whole-body cryotherapy improve vertical jump recovery following a high-intensity exercise bout? Open access Journal of Sports Medicine. 2015;**6**: 49-54

[20] Chernyshev IS. Cryogenic Aerotherapy in Modern Medicine/ Toolkit. Moscow: Med-Cryonica LLC; 2010. p. 26

[21] Stanek A, Cholewka A, Gadula J, Drzazga Z, Sieron A, Sieron-Stoltny K. Can whole-body cryotherapy with subsequent kinesiotherapy procedures in closed type cryogenic chamber improve BASDAI, BASFI, and some spine mobility parameters and decrease pain intensity in patients with ankylosing spondylitis? BioMed Research International. 2015;**2015**: 404259

[22] Rymaszewska J, Ramsey D. Whole body cryotherapy as a novel adjuvant therapy for depression and anxiety. Archives of Psychiatry and Psychotherapy. 2008;**10**(2):49-57

[23] Bouzigon R, Ravier G, Dugué B, Grappe F. The use of whole-body cryostimulation to improve the quality of sleep in athletes during high international competitions. British Journal of Sports Medicine. 2014;**48**:572

[24] Makhaniok AA, Levin ML, Dragun VL. Thermophysigal aspects gas cryotherapy. Founded by the national academy of sciences of Belarus. Physico-Technical Series. 2011;**3**:S 59-S 71

[25] Westerlund T, Oksa J, Smolander J, Mikkelsson M. Thermal responses during and after whole-body cryotherapy. Journal of Thermal Biology. 2003:601-608

Chapter 9

Comparative Evaluation of Cryogenic Air Separation Units from the Exergetic and Economic Points of View

Stefanie Tesch, Tatiana Morosuk and George Tsatsaronis

Abstract

The industrial use of cryogenic air separation units started more than 120 years ago. Cryogenic air separation processes produce pure nitrogen, oxygen, and argon, as well as other noble gases. In cryogenic air separation units, the produced amounts of nitrogen and oxygen vary between 200 and 40,000 Nm^3/h and 1000 and 150,000 Nm^3/h, respectively. Different configurations of this process lead to various amounts of gaseous and liquid products. In addition, the purity of the products is affected by the schematic. Oxygen in gaseous or liquid form is typically used in the metallurgical (e.g., steel) industry, in chemical applications (as oxidizer), in power plants (for oxy-fuel combustion processes), as well as in the medical and aerospace sectors. Nitrogen in gaseous or liquid form is used as inert or flushing gas in the chemical industry and as a coolant for different applications. In this chapter, different schematics of air separation units are analyzed. An exergetic analysis is applied in order to identify the thermodynamic inefficiencies and the processes that cause them. Finally, the systems are evaluated from the economic point of view.

Keywords: air separation unit, exergetic analysis, economic analysis

1. Introduction

The industrial use of air separation units for producing pure oxygen, nitrogen, argon, helium, and other noble gases is established well. The main products of air separation units are oxygen, nitrogen, and argon in liquid and/or gaseous state. Air separation processes are classified as cryogenic and noncryogenic air separation units, which mainly differ regarding the production capacity and the purity of the products. Cryogenic air separation units are used to gain large amounts of products with high purity. Modern air separation units produce up to 6000 t/d of oxygen and 10,000 t/d of nitrogen [1]. In multi-train air separation plants, production rates of up to 30,000 t/d oxygen are possible. In comparison to that, the first ASU plant had a production rate of 0.1 t/d [2]. In the noncryogenic air separation processes, adsorption, chemical, polymeric membranes, and ion transport membranes are used. However, these processes produce only small amounts of products.

The applications of pure oxygen, nitrogen, and argon are wide. Approximately 55% of the produced oxygen is used in the metallurgical industry [3] and 25% in the chemical industry for production of ethylene glycol. Oxygen is required for water and waste water treatment for welding and cutting, as well as an oxidizer [4].

In the chemical and metallurgical industries, nitrogen is mainly used as inert or flushing gas. It is also used for temperature control purposes in chemical reactions. Finally, nitrogen is used in the medicine, cryotherapy, and food industry.

Argon is used as an extinguishing working fluid, packaging gas in the food industry, filling gas for light bulbs, carrier gas for gas chromatography, inert and cutting gas in the laser technology, etc.

In this chapter, two different schematics for an air separation unit are analyzed with external and internal compression [5–7]. The systems are evaluated using exergetic and economic analyses.

2. State of the art

Cryogenic air separation is a fully developed process for the separation of air into its components for large production rates and high purity. **Table 1** gives an overview of the different types of air separation processes and their typical production rates and purities.

Cryogenic air separation systems consist of at least four blocks (**Figure 1**): air compression and purification, main heat exchanger, cryogenic distillation column, and product compression (internal or external).

Dustless air (the typical composition of dry air is given in **Table 2**) enters the air compression block and is compressed in a multistage compressor with interstage cooling. In the subsequent purification block, all chemical components within the air, which will freeze during the liquefaction of air, have to be removed. Particular attention should be given to the contents of water and carbon dioxide: must be <0.1 ppm for H_2O and <1 ppm for CO_2 [10]. Thermal swing adsorption (TSA) or pressure swing adsorption (PSA) is used for the purification, which consists of two vessels filled with granular adsorbents. The compressed air enters one bed, while the second bed is regenerated. For the regeneration, a so-called waste nitrogen stream is used. This stream is a side product stream of the column block. Depending on the used adsorption process, the waste nitrogen stream is either heated or pressurized in order to desorb the adsorbed impurities from the bed.

Component	Capacity Nm^3/h	Purity mol%	Separation method	Load range %
	1–1000	<99.5	Membrane	30–100
Nitrogen	5–5000	<99.99	Pressure swing adsorption	30–100
	200–400,000	Any with residual concentrations down to ppb[1] range	Cryogenic air separation	60–100
	100–5000	<95	Vacuum pressure swing adsorption	30–100
Oxygen	1000–150,000	Any with residual concentrations down to ppb range, oxygen content mostly >95	Cryogenic air separation	60–100
Argon			Cryogenic air separation	

[1]ppb: parts per billion.

Table 1.
Production range of cryogenic and non-cryogenic air separation processes (data adopted from [8]).

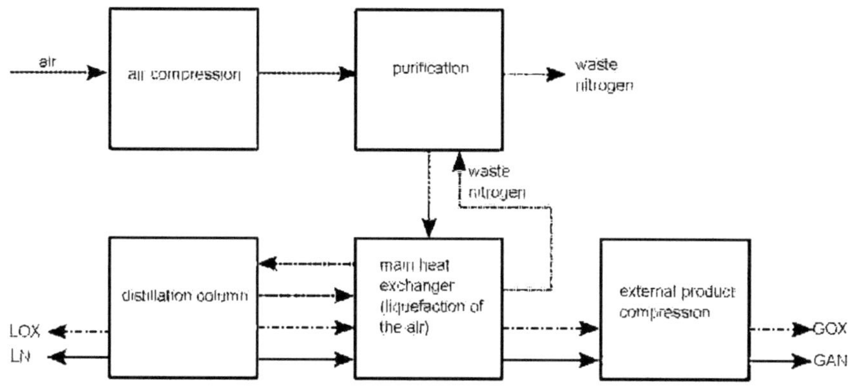

Figure 1.
General structure of an air separation unit.

Component	Volume fraction
Nitrogen	78.08 vol.%
Oxygen	20.95 vol.%
Argon	0.93 vol.%
Carbon dioxide	400 vppm*
Neon	180 vppm
Helium	5 vppm
Methane	1.8 vppm
Krypton	1.1 vppm
Hydrogen	0.5 vppm
Nitrous oxide	0.3 vppm
Carbon monoxide	0.2 vppm
Xenon	0.09 vppm

**vppm: volume parts per million.*

Table 2.
Composition of dry air (adopted from [9]).

For the liquefaction of air, a temperature of −172°C is required. In the main heat exchanger (multi-stream heat exchanger; typically, plate and fin heat exchanger design), the air is cooled and partially liquefied. The heat transfer processes within the main heat exchanger are quite complex, due to the large number of streams in different passages and the high number of channels and interactions. Detailed analyses are reported in [11–13].

The partially liquefied air leaves the main heat exchanger and enters the distillation column block, which is a double-column system [14]. It consists of a high-pressure column (operation pressure is around 5–6 bar) and a low-pressure column (operation pressure is around 1.3 bar). The condenser of the high-pressure and the reboiler of the low-pressure column are thermally coupled. The different boiling points of nitrogen and oxygen lead to the production of gaseous nitrogen at the top of the high-pressure column and an oxygen-enriched mixture at the bottom of the high-pressure column. The gaseous nitrogen is partially or totally liquefied in the condenser. A part is fed back to the high-pressure column as reflux; a second part leaves the system as liquid nitrogen and a third part enters the low-pressure column as reflux. At the top of the low-pressure column also gaseous nitrogen is gained, which is fed to the main heat exchanger. The liquid and gaseous oxygen leave the column system at the bottom of the low-pressure column either before or after the reboiler. The two gaseous product streams are fed to the main heat exchanger in order to cool and partially liquefy the air. In addition, a waste nitrogen stream leaves the low-pressure column and is also heated within the main heat exchanger.

The air separation process can be extended in order to obtain specified product requirements. The integration of an additional cooling cycle is also possible [15].

Air separation units can be distinguished regarding the kind of product compression [16]:

- *Internal compression* or "pumped LOX cycle," where the product oxygen is produced at an elevated pressure by using a pump and heating high-pressure liquid oxygen in the main heat exchanger against high-pressure air.

- *External compression* or "low-pressure GOX cycle," where the product oxygen is taken as a gas from the bottom of the low-pressure column and is subsequently compressed to the required pressure using a compressor.

Nowadays, most of the air separation plants use the internal compression of oxygen [17, 18]. The internal compression has several advantages from the thermodynamic and safety points of view. From the thermodynamic point of view, the increase of the oxygen pressure requires less power if it is pressurized in liquid state instead of gaseous state.

In addition, there are safety-related problems in conjunction with the oxygen compressors, which lead to higher costs, lower efficiency, and reliability in comparison to air and/or nitrogen compressors [17, 19]. The internal compression of oxygen has a second advantage from the safety viewpoint. Due to the fact that hydrocarbons accumulate in the bottom of the column, an explosion could occur. Therefore, in air separation plants where only gaseous products are produced, a small amount of liquid oxygen has to be withdrawn from the bottom to decrease the potential of hazards [16]. In contrast, in air separation units with internal compression, the liquid oxygen is continuously withdrawn from the sump and thus decreases the potential of hazards.

In literature, different systems have been studied from the energetic and exergetic points of view.

In [20], schematics of air separations units are evaluated, which differ regarding (a) the kind of product compression (internal or external) and (b) the amount of produced gaseous oxygen. The specific power consumption varies between 0.464 and 0.639 kW/Nm^3. A specific power consumption of 0.38 kWh/Nm^3 for a large-scale air separation unit located in China is reported in [18].

An air separation unit with a nitrogen liquefaction block is analyzed from exergetic point of view in [15]. The nitrogen liquefaction block is the subsystem with the highest exergy destruction. The total exergy destruction ratio is 51%. In [21], two cryogenic air separation units are analyzed from the exergetic point of view. The paper evaluates a two- and a three-column system (as part of an integrated gasification combined cycle), which produces one gaseous oxygen and three gaseous nitrogen streams at different pressure levels (88, 25, and 1.3 bar). The highest exergy destruction is reported for the preprocessing feed subsystem (air compressors, interstage cooler, and purification system), which amounts to 47 and 54% for the two- and three-column system, respectively. An air separation unit with an internal compression unit is analyzed from the energetic, exergetic, and economic points of view in [22]. The exergetic analysis shows that the air compression and distillation blocks have the highest exergy destruction. An exergetic analysis is applied to an air separation unit that produces gaseous oxygen and nitrogen in [23]. The results demonstrate that the air compression system causes 38.4% of the total exergy destruction, while the distillation system is responsible for 28.2% of the total exergy destruction. A double-column and a single-column air separation unit are analyzed from the exergetic point of view in [24]. The paper discusses the effect of the air pressure on the exergy destruction within the main heat exchanger.

Information about noncryogenic processes can be found in [1, 2, 25–28].

3. Evaluation methods

The exergy-based methods are powerful tools for identifying thermodynamic and cost inefficiencies, as well as environmental impacts and risks associated with the inefficiencies within energy conversion systems [29, 30]. This evaluation method includes the following analyses:

- conventional exergetic analysis
- advanced exergetic analysis
- exergoeconomic analysis
- exergoenvironmental analysis
- exergy-risk-hazard analysis

3.1 Exergy analysis

In order to apply an exergetic analysis, the reference environment needs to be defined. In this chapter, the average European conditions are chosen: $T_0 = 15\,^{\circ}\mathrm{C}$ and $p_0 = 1.0134$ bar.

For each stream, the mechanical, thermal, and chemical exergies are calculated according to [29, 31]. For all components, the exergies of fuel and product are defined, and the exergy destruction is calculated using exergy balances for the k-th component and the overall system:

$$\dot{E}_{F,k} = \dot{E}_{P,k} + \dot{E}_{D,k} \tag{1}$$

$$\dot{E}_{F,tot} = \dot{E}_{P,tot} + \dot{E}_{D,tot} + \dot{E}_{L,tot} \tag{2}$$

Here, $\dot{E}_{F,k}$ is the exergy of fuel of the k-th component; $\dot{E}_{P,k}$ represents the exergy of product of the k-th component; $\dot{E}_{D,k}$ is the exergy destruction within the k-th component; $\dot{E}_{F,tot}$ is the exergy of fuel of the overall system, and $\dot{E}_{P,tot}$ is the exergy of product of the overall system; $\dot{E}_{D,tot}$ represents the exergy destruction within the overall system; $\dot{E}_{L,tot}$ is the exergy loss from the overall system.

The exergetic efficiency is

$$\varepsilon = \frac{\dot{E}_P}{\dot{E}_F} \tag{3}$$

The following exergy destruction ratios are used in the analysis according to [29].

$$y_k = \frac{\dot{E}_{D,k}}{\dot{E}_{F,tot}} \tag{4}$$

$$y_k^* = \frac{\dot{E}_{D,k}}{\dot{E}_{D,tot}} \tag{5}$$

3.2 Economic analysis

The economic analysis is performed according to the total revenue requirement (TRR) method [29]. First, the purchased equipment costs (PEC) and the bare module costs (C_{BM}) have to be estimated for all components. Afterward, the fixed capital investment (FCI) and total capital investment (TCI) are determined. The FCI consists of the direct and indirect costs, whereas the direct costs are further divided into onsite and offsite costs. The total revenue requirement (TRR) consists of the sum of the levelized carrying charges (CC_L), the levelized operating and maintenance costs (OMC_L), and the levelized fuel costs(FC_L) .

4. Process description

System Case A (CA) is a conventional air separation unit with two distillation columns, a nitrogen liquefaction block, and an external compression unit of the product. Case B (CB) is an air separation unit with two distillation columns and an internal compression unit [5–7]. The flowsheets of Cases A and B are given in **Figures 2** and **3**. The key values for the simulations are based on [15, 32].

4.1 Air compression and purification block (ACPB)

In both systems, the dustless air is compressed (in a two-stage compression process with interstage cooling) to approximately 6 bar and purified in the adsorption block (AD). In Case A, the "pure" air enters the main heat exchanger (MHE). In Case B, the "pure" air is divided into two parts: Stream 11 enters the main heat exchanger, while stream 17 is further compressed in the booster air compressor (BAC or AC3). The three air streams (streams 12, 14, and 18) are fed into the main heat exchanger.

In Case A, the heat exchanger 3 (HE3) is also assigned to the air compression and purification block because this component is required in order to heat the waste nitrogen to 170°C, which is required for the desorption of the water vapor and carbon dioxide from the adsorption beds. A temperature between 150 and 200°C is required for the desorption [32, 33].

In Case A, the air compression and purification block consists of AC1, AC2, IC1, IC2, AD, and HE3. In Case B, the following components belong to the air compression and purification block: AC1, AC2, IC1, IC2, AD, and AC3.

4.2 Main heat exchanger (MHE)

The main heat exchanger is the core component of an air separation unit where the cleaned air is cooled to −173.4°C and partially liquefied using the streams leaving the column block. In Case A, these are two gaseous nitrogen streams, one gaseous oxygen stream, and a waste nitrogen stream. In Case B, the three air streams are cooled using one gaseous nitrogen, one gaseous oxygen, and one waste nitrogen stream. Stream 15 (28.1% of the air mass flow rate, i.e., stream 11) leaves the MHE at a temperature of −120°C and is expanded in the expander (EXP1).

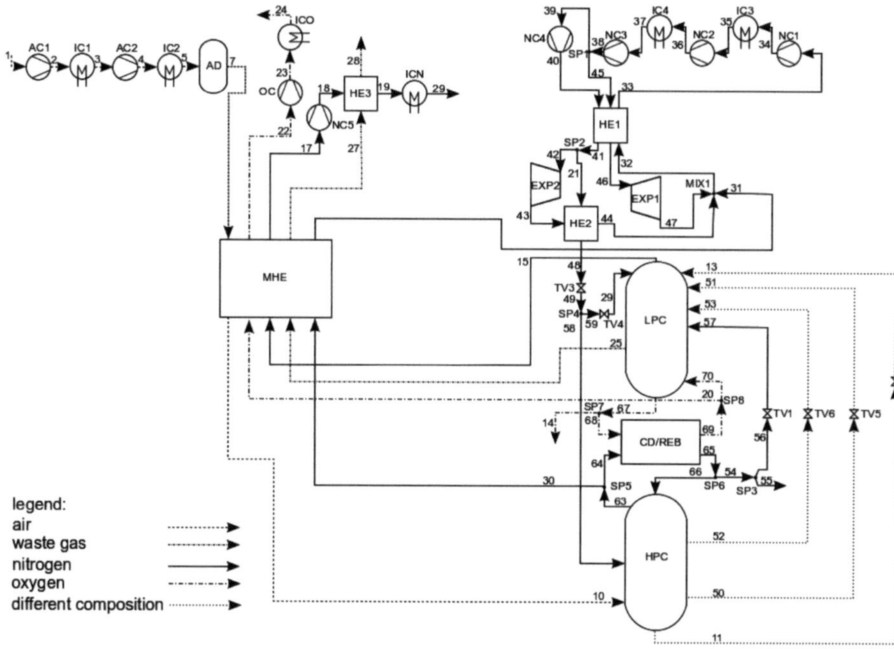

Figure 2.
Flowsheet of case A.

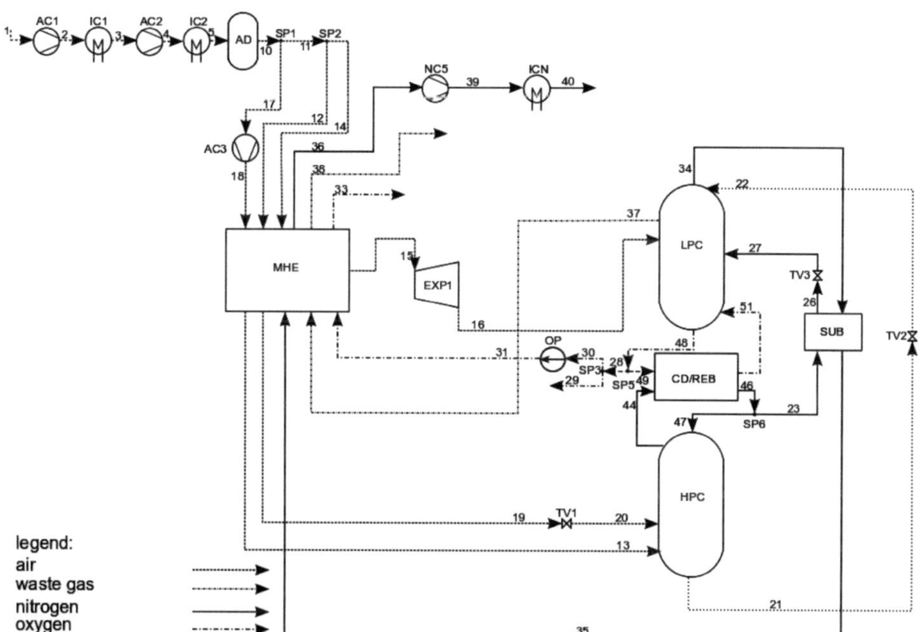

Figure 3.
Flowsheet of case B.

The share of this mass flow rate is slightly higher in comparison to the data available in literature. As reported in [32], a mass portion of 10–20% at a temperature of −100 to −130°C is common for this stream. In [24], it is mentioned that the air is divided at a temperature level of −140°C and fed to the expander. After the expander, stream 16 enters the low-pressure column (LPC) at an intermediate sieve tray.

The air stream at high pressure leaves the MHE (stream 19) and is expanded within a throttling valve (stream 20). In Case A, the oxygen and nitrogen streams are heated to 15°C, and the waste nitrogen leaves the main heat exchanger at the temperature of 33°C, which results in a minimal temperature difference of 2 K. In Case B, the liquid oxygen stream is vaporized and heated within the main heat exchanger and leaves it also at 15°C. The waste nitrogen is heated to 170°C, which results in a minimal temperature difference of 2.7 K for the MHE.

4.3 Column block (CB)

The column block consists of a low-pressure column and a high-pressure column (LPC and HPC) that are thermally coupled by the condenser and reboiler. Gaseous nitrogen is gained at the top of the high-pressure and low-pressure column. In Case A, the nitrogen stream is condensed only partially, while in Case B, it is totally condensed. In both systems, a part of the liquid nitrogen is fed back to the column as reflux (CA: stream 66, CB: stream 47). In Case A, the remaining stream is again divided into two parts, while stream 55 leaves the systems as a product stream and stream 54 is throttled and fed to the low-pressure column. From the bottom of the high-pressure column, an oxygen-enriched stream (CA: stream 11, CB: stream 21) is withdrawn, throttled, and fed to the low-pressure column. In Case A, the liquid and gaseous oxygen leaves the column block at the bottom of the low-pressure column either before or after the reboiler. In Case B, the oxygen stream is withdrawn only in liquid state.

In Case A, additional side streams are fed from the high-pressure column to the low-pressure column. In Case B, a subcooler is used which is introduced in order to decrease the liquid fraction of stream 27 after the throttling, which in turn increases the liquid nitrogen reflux to the low-pressure column and has a positive effect on the purity of the gaseous nitrogen stream. In the subcooler, the top product stream of the low-pressure column is heated before it is fed to the MHE.

The column block consists of the components HPC, LPC, CD/REB, TV1, TV2, TV5, and TV6 in Case A. In Case B, it contains the HPC, LPC, CD/REB, SUB, TV2, and TV3.

4.4 Product postprocessing block (PPPB)

In Case A, the gaseous oxygen and nitrogen, after leaving the main heat exchanger, are compressed to 20 bar within the oxygen (OC) and nitrogen compressors (NC). Afterward, both streams are cooled to the ambient temperature and leave the systems. In Case B, the liquid oxygen is pressurized to 20 bar using the oxygen pump (OP). The gaseous nitrogen is compressed after leaving the main heat exchanger and is cooled to ambient temperature.

The product postprocessing block consists of the components NC5, OC, ICN, and ICO in Case A or OP, NC5, and ICN in Case B.

4.5 Nitrogen liquefaction block (NLB)

Stream 31 leaves the MHE and is fed to the nitrogen liquefaction block where it is mixed with streams 44 and 47, which are recycled streams within the nitrogen liquefaction block to stream 32. This stream is heated in the heat exchanger 1 (HE1) and compressed in a three-stage compression process with interstage cooling to 38 bar. Afterward, stream 38 is divided into two streams (stream 39 and 45). Stream 45 is cooled within HE1, while stream 39 is first compressed to 45 bar in NC4 and is then cooled within HE1. At the outlet of the HE1, stream 41 is split into streams 42 and 21. Streams 46 and 42 are expanded in expanders 1 and 2 (EXP1 and EXP2), respectively, which are connected with the nitrogen compressors 3 and 4

(NC3 and NC4). Stream 21 is cooled in the HE2 by stream 43, which is afterward mixed with streams 47 and 31. The cooled stream leaves the HE2 as stream 48 and is throttled and split into streams 58 and 59. Stream 58 is directly fed to the HPC, while stream 59 is throttled again and then enters the LPC.

In Case A, the nitrogen liquefaction block is formed by the following components: HE1, HE2, NC1, NC2, NC3, NC4, IC3, IC4, EXP1, EXP2, and MIX1.

Some components shown in **Figures 2** and **3** cannot be assigned to any of the afore-mentioned blocks. These components from the remaining block are in Case A TV3 and TV4, and in Case B EXP1 and TV1.

5. Simulation

The two systems were simulated using Aspen Plus [34]. For the equation of state, the Peng-Robinson-equation was selected. The general assumptions made for the simulation are given in **Table 3**.

Variable	Unit	Case A	Case B
Air			
Inlet temperature	°C	15	
Inlet pressure	bar	1.013	
Mass flow rate	kg/s	16.39	33.50
Composition	mol/mol	x_{N_2} = 0.7720 x_{O_2} = 0.2080 x_{Ar} = 0.0905 x_{H_2O} = 0.0102 x_{CO_2} = 0.0003	
Turbomachines			
Isentropic efficiency (compressor)	%	84	
Isentropic efficiency (expander)	%	90	
Isentropic efficiency (pump)	%	70	
Mechanical efficiency	%	99	

Table 3.
General assumptions for the simulation of the systems.

6. Results and discussions

6.1 Energy analysis

The total power consumption is calculated as 17.5 MW for Case A and 15.9 MW for Case B, while the specific power consumption is 2.31 kWh/Nm^3_{GOX} and 2.11 kWh/Nm^3_{GOX} for Cases A and B, respectively. Due to the fact that the amount of produced gaseous oxygen is the same in both systems, the specific power consumption per gaseous oxygen decreases from Case A to Case B. The production rates of the product streams, as well as their purities, are given in **Figures 4** and **5**. The mass flow rates of the gaseous and liquid oxygen are kept constant for both systems.

Figure 6 gives an overview of the specific power consumption per produced oxygen obtained from the literature. The large deviations in the results obtained for Cases A and B are related to several reasons:

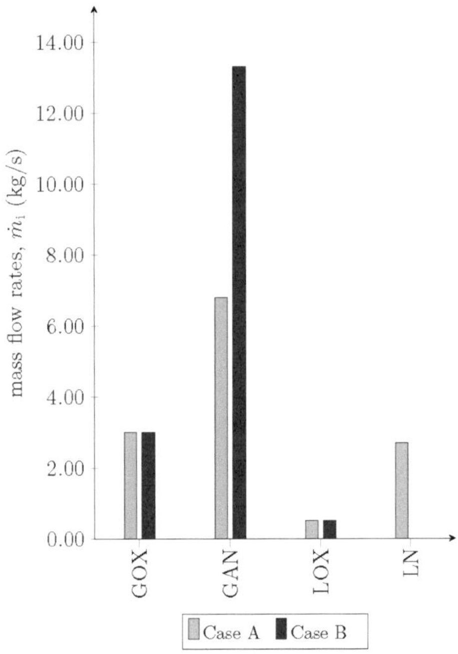

Figure 4.
Mass flow rates of the product streams.

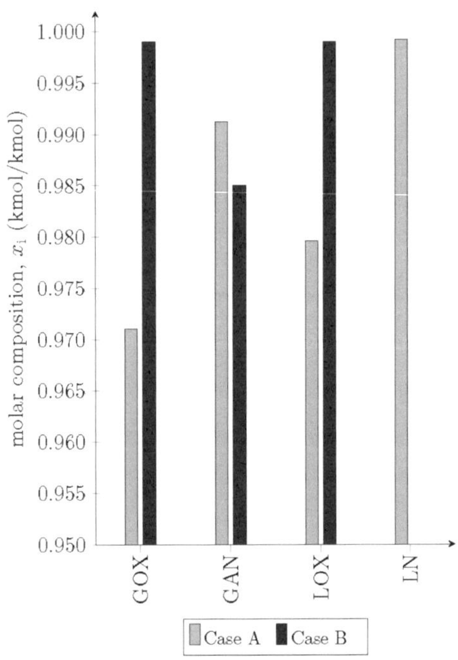

Figure 5.
Purity of the product streams.

- In Cases A and B, the gaseous oxygen and nitrogen streams leave the systems at 20 bar. For the data obtained from the literature, it is not clearly indicated whether the product streams leave the system at atmospheric pressure or at a higher pressure level. Solely in [38], it is mentioned that the oxygen leaves the system at atmospheric pressure.

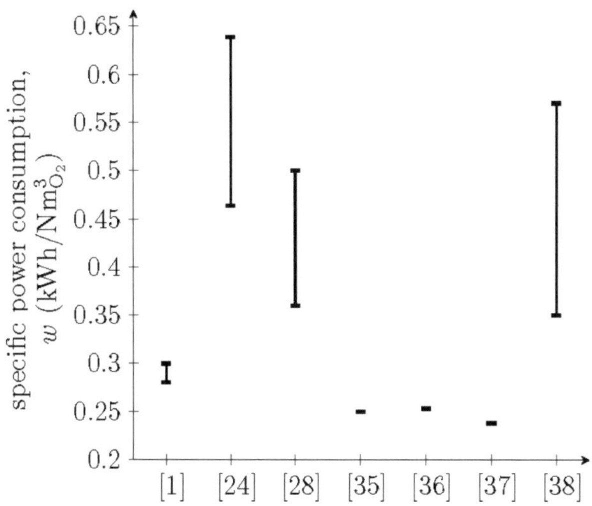

Figure 6.
Specific power consumption obtained from literature [35–37].

- In some publications, the specific power consumption is calculated per produced oxygen, which also includes the liquid oxygen stream but is not identified as this.
- The production of oxygen and nitrogen with high purity and the additional production of liquids have a very large influence on the power consumption. In Case A, the additional cooling cycle significantly affects the overall power consumption.
- Another reason is the size of the air separation units. Both analyzed systems are small-scale plants, which tend to have higher specific power consumption in comparison with large-scale units.

The power consumption/generation of the turbomachines in Cases A and B is given in **Figure 7**. In Case A, NC1 and NC2 have the highest power consumption. In Case B, the compressor with the highest power consumption is NC5. This compressor requires more power in comparison to NC5 in Case A because the mass flow rate of the gaseous nitrogen is twice as high as in Case B. The differences in the power consumption of the components AC1 and AC2 are also related to the higher air mass flow rate in Case B in comparison to Case A.

6.2 Exergy analysis

The results of the overall system are given in **Table 4**. The exergetic efficiency of Cases A and B amounts to 28.4 and 31.1%, respectively. In [39], an exergetic efficiency of 26.6% is reported for a single air separation unit, which is in the same range as the results obtained for Cases A and B.

The difference in the exergy of fuel between Cases A and B (**Table 4**) is related to the slightly different total power consumption. The exergy of product for both overall systems is the same. Due to the fact that the amount of gaseous and liquid oxygen is identical in both systems, the significantly higher amount of gaseous nitrogen in Case B compensates the product stream of the liquid nitrogen, which is not available in Case B. The exergy loss is significantly higher in Case B. This is due to the fact that the mass flow rate of the waste nitrogen stream is significantly higher in Case B than in Case A in order to produce the same amount of liquid and gaseous oxygen.

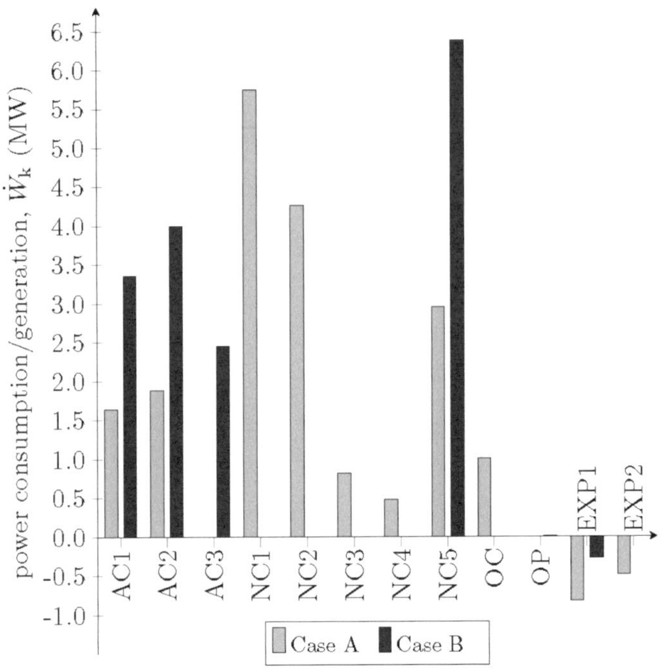

Figure 7.
Power consumption/generation.

	$\dot{E}_{F,tot}$	$\dot{E}_{P,tot}$	$\dot{E}_{D,tot}$	$\dot{E}_{L,tot}$	ε_{tot}
	MW	MW	MW	MW	%
Case A	17.6	5.0	12.4	0.2	28.4
Case B	16.1	5.0	10.1	0.9	31.1

Table 4.
Results obtained from the exergetic analysis of the overall system, Cases A and B.

A graphical representation of the exergy streams is given in **Figures 8** and **9**. For each component, the inlet and outlet exergy streams associated with a material stream, the power for the turbomachines, and the exergy destruction are shown. The high-pressure and low-pressure columns, the condenser/reboiler, sub cooler (in Case B), and some throttling vales are summarized as column block (CB). **Figure 8** shows that the components within the nitrogen liquefaction block have the highest exergy destruction in Case A. The exergy destruction ratio of this block accounts for 60.8% of the total exergy destruction. In Case B (**Figure 9**), the ICN has the highest exergy destruction among all components.

6.3 Economic analysis

The estimated bare module costs (**Table 5**) for the component blocks are based on data from the following sources:

- column and expanders [40]
- heat exchangers and pumps [41]
- compressors [42]

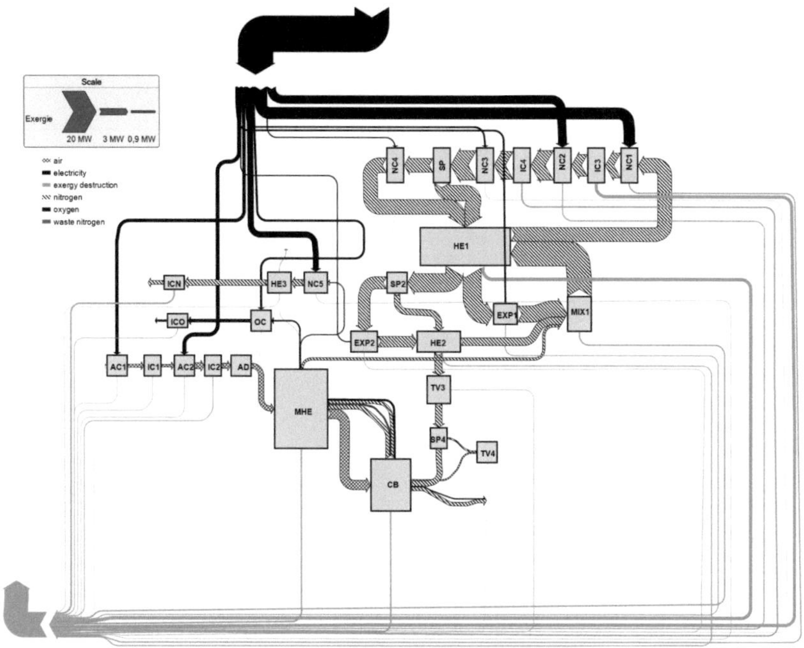

Figure 8.
Sankey diagram Case A.

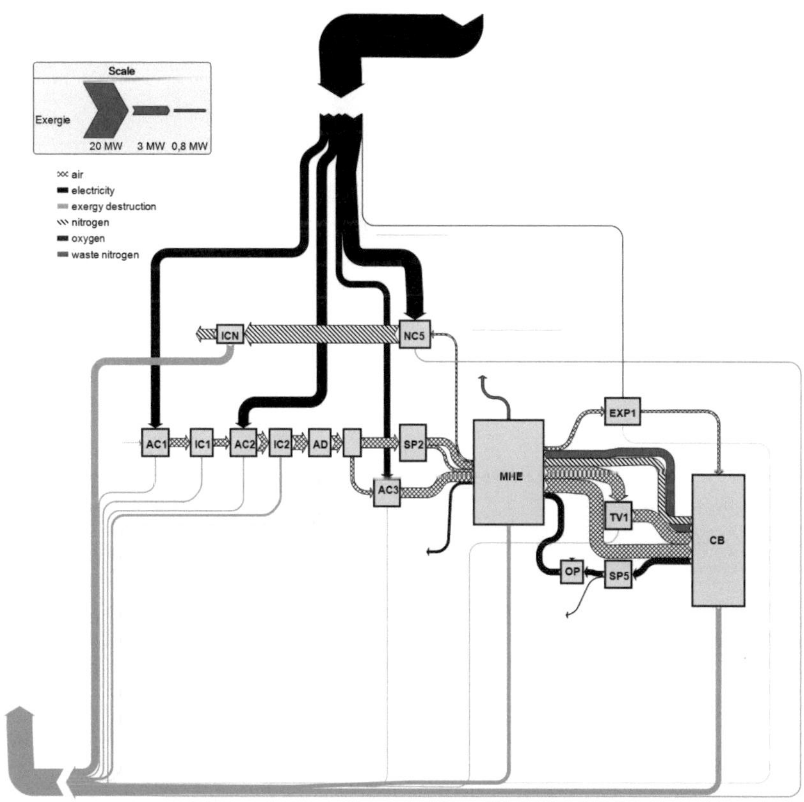

Figure 9.
Sankey diagram Case B.

	$C_{BM,ACPB}$	$C_{BM,MHE}$	$C_{BM,CB}$	$C_{BM,NLB}$	$C_{BM,PPPB}$	$C_{BM,rest}$	$C_{BM,tot}$
	10^3US$	10^3US$	10^3US$	10^3US$	10^3US$	10^3US$	10^3US$
Case A	3013	1246	12,798	8882	2487	0	28,426
Case B	5050	2350	18,491	0	2622	435	28,948

Table 5.
Bare module costs for the component blocks (reference year 2015).

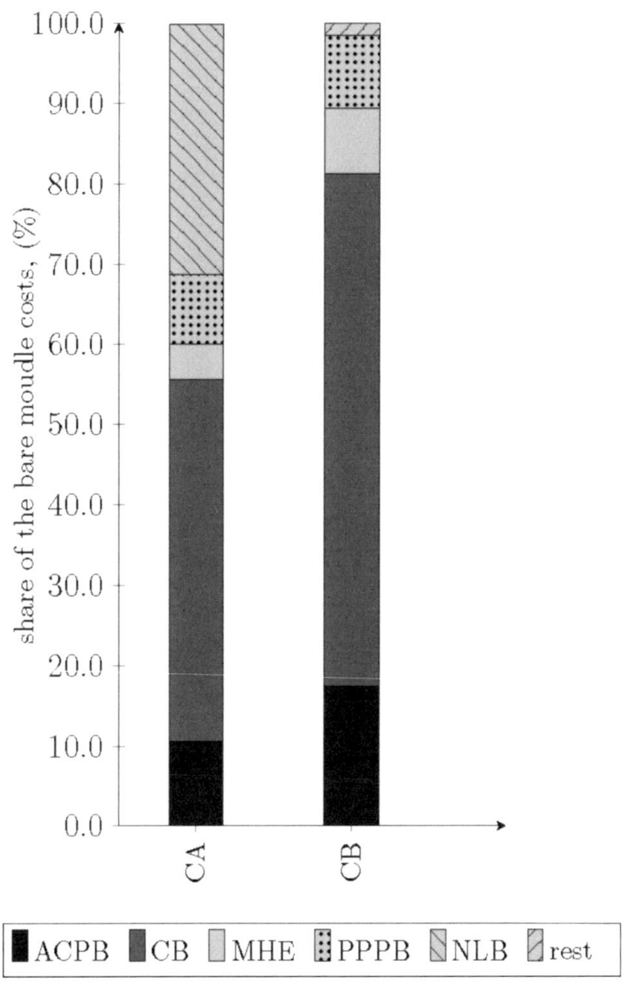

Figure 10.
Distribution of the bare module costs among the component blocks.

The bare module costs are slightly higher (1.8%) for Case B than in Case A. The distribution of the bare module costs among the component blocks is given in **Figure 10**.

In both systems, the column block has the highest costs: 45% for Case A and 64% for Case B. In Case A, the nitrogen liquefaction block exhibits the second highest share, 31%. In Case B, the air compression and purification block has the second highest bare module cost, which amounts to 17.4% of the total sum.

The FCI and TCI, as well as the specific investment costs, are shown in **Table 6**. Due to the fact that the fixed and capital investment costs are calculated based on the bare module costs, the FCI and TCI are slightly higher for Case B in comparison to Case A.

	FCI	TCI	Specific investment costs
	10^6 US$	10^6 US$	10^3 US$/kW$_{\dot{E}_P}$
Case A	40.2	46.9	9.39
Case B	40.9	47.7	9.57

Table 6.
Fixed, total, and specific investment costs.

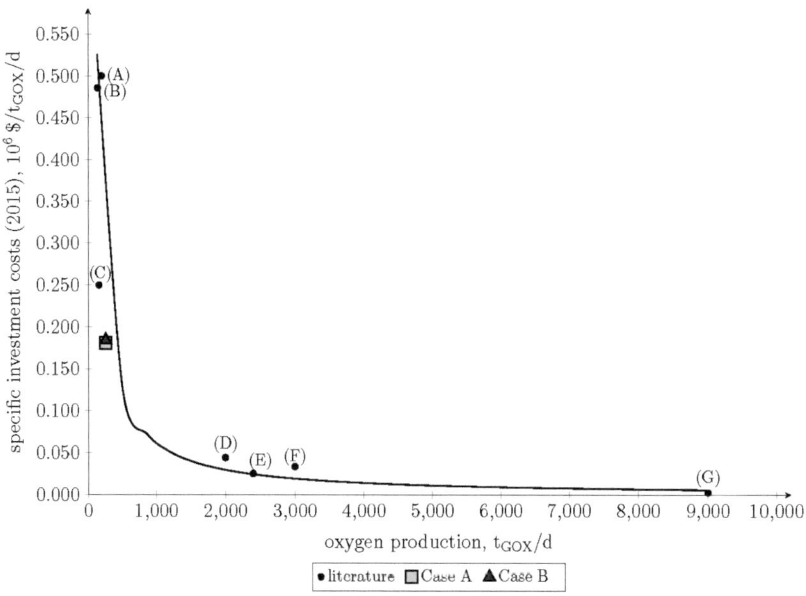

Figure 11.
Specific investment costs depending on the oxygen production (values obtained from (A) [43], (B) [44], (C) [45], (D) [46], (E) [47], (F) [48], (G) [49]).

	CC_L	OMC_L	FC_L	TRR
	10^3 $/a	10^3 $/a	10^3 $/a	10^3 $/a
Case A	5506	1953	17,771	25,232
Case B	5608	1989	16,182	23,779

Table 7.
Levelized carrying charges, operating and maintenance costs, fuel costs, and total revenue requirement.

In order to compare the investment costs with costs for real plants, the specific investment per gaseous oxygen is calculated which amounts to $0.184 \times 10^6\,\$/\frac{t_{GOX}}{d}$ and $0.181 \times 10^6\,\$/\frac{t_{GOX}}{d}$ for Cases A and B, respectively. **Figure 11** shows the specific investment costs as a function of the produced oxygen for different plants for the reference year 2015. As shown in this figure, the specific investment costs decrease for air separation plants with large production capacity, which corresponds to the economies of scale. The specific investment costs for Cases A and B are close to the curve; this shows that the cost estimation conducted here is reasonable. **Table 7** shows the results for the levelized carrying charges, operating and maintenance costs, fuel costs, and total revenue requirement. The fuel costs contribute 70 and 59% to the total revenue requirement for Cases A and B, respectively. Due to the lower TCI and the lower power consumption, the TRR is 6% lower for Case B in comparison to Case A.

7. Conclusion

This chapter discusses two different air separation units. Both systems are designed for a small production capacity—260 t/d of gaseous oxygen and 43 t/d of liquid oxygen. The results of the exergetic analysis show an exergetic efficiency of 28.4 and 31.1% for the system with external and internal compression unit, respectively. From the economic point of view, the bare module costs are slightly higher for the system with external compression unit (Case A) in comparison to the system with internal compression unit (Case B). The results for the total revenue requirement show a relatively large contribution of the fuel cost, which leads to a lower *TRR* in Case B in comparison to Case A. In addition, Case B has advantages from the safety point of view due to the use of an oxygen pump instead of an oxygen compressor.

Author details

Stefanie Tesch, Tatiana Morosuk* and George Tsatsaronis
Institute for Energy Engineering, Technische Universität Berlin, Berlin, Germany

*Address all correspondence to: tetyana.morozyuk@tu-berlin.de

References

[1] Castle WF. Air separation and liquefaction: Recent developments and prospects for the beginning of the new millennium. International Journal of Refrigeration. 2002;**25**(1):158-172

[2] Beysel G. Enhanced cryogenic air separation: A proven Process applied to Oxyfuel. In: 1st Oxy-fuel Combustion Conference; Cottbus, Germany; 8-11 September 2009. Available from: http://www.ieaghg.org/docs/oxyfuel/OCC1/Plenary%201/Beysel_ASU_1stOxyfuel%20Cottbus.pdf [Cited: 01 December 2017]

[3] Emsley J. Nature's Building Blocks: An A-Z Guide to the Elements. Oxford, United Kingdom: Oxford University Press; 2003

[4] Baron RF. Cryogenic Systems. 2nd ed. New York, USA: Oxford University Press; 1985

[5] Morosuk T, Tesch S, Tsatsaronis G. Concepts for regasification of LNG in industrial parks. In: Al-Megren HA, Altamimi RH, editors. Advances in Natural Gas Emerging Technologies. Rijeka, Croatia: InTech; 2017. pp. 27-53

[6] Tesch S, Morosuk T, Tsatsaronis G. Exergy-based methods for air separation. In: Proceedings of the 14th Cryogenics 2017 IIR International Conference; 15-19 May 2017; Dresden, Germany

[7] Tesch S, Morosuk T, Tsatsaronis G. Exergy evaluation of an air separation unit with pumped liquid oxygen. In: Proceedings of the 2nd International Conference on Cryogenics and Refrigeration Technologies (ICCRT 2018); 7-10 May 2018; Bucharest, Romania

[8] Haring HW. Industrial Gases Processing. Weinheim: Wiley-VCH; 2008

[9] Landesanstalt für Umwelt Baden-Württemberg (LUBW). Zusammensetzung der Luft; 2013. Available from: http://www4.lubw.baden-wuerttemberg.de/servlet/is/18340/ [Cited: 10 July 2017]. [Regional office for Environment Baden-Wuerttemberg. Composition of air, 2013. Available from: http://www4.lubw.baden-wuerttemberg.de/servlet/is/18340/ [Cited: 10 July 2017]]

[10] Jain R, Piscataway NJ. Pre-purification of air for separation. US Patent 005232474A. August 3, 1993

[11] Picon-Nunez M, Polley GT, Medina-Flores M. Thermal design of multi-stream heat exchangers. Applied Thermal Engineering. 2002;**22**(14):1643-1660

[12] Boehme R, Parise JAR, Pitanga MR. Simulation of multistream plate–fin heat exchangers of an air separation unit. Cryogenics. 2003;**43**(6):325-334

[13] Ghosh I, Sarangi SK, Das PK. An alternate algorithm for the analysis of multistream plate fin heat exchangers. International Journal of Heat and Mass Transfer. 2006;**49**(17-18):2889-2902

[14] The Linde Group. Cryogenic air separation—history and technological progress; n.d. Available from: http://www.linde-engineering.com/internet.global.lindeengineering.global/en/images/AS.B1EN%201113%20-%20%26AA_History_.layout19_4353.pdf [Cited: 30 September 2015]

[15] Cornelissen RL, Hirs GG. Exergy analysis of cryogenic air separation. Energy Conversion and Management. 1998;**39**(16-18):1821-1826

[16] Schmidt W. ASU reboiler/condenser safety. Presented at the European Industrial Gas Assocation Production Safety Seminar; February 2, 2006. Available from: http://www.

airproducts.com/~/media/downloads/white-papers/a/enair-separation-unit-reboiler-condenser-safety-whitepaper.pdf [Cited: 28 July 2015]

[17] Dawson BK, Siegmund SC, Yonggui Z. Flowsheet optimization for multi-product air separation units. In: Proceedings of the 1st Baosteel Annual Academic Conference; 27-28 May 2004; Shanghai, China

[18] XB Z, JY C, Yao L, Huang YH, Zhang XJ, Qiu LM. Research and development of large-scale cryogenic air separation in China. Journal of Zhejiang University Science A (Applied Physics & Engineering). 2014;**15**(5):309-322

[19] Jones D, Bhattacharyya D, Turton R, Zitney SE. Optimal design and integration of an air separation unit (ASU) for an integrated gasification combined cycle (IGCC) power plant with CO2 capture. Fuel Processing Technology. 2011;**92**(9):1685-1695

[20] Alsultannty YA, Al-Shammari NN. Oxygen specific power consumption comparison for air separation units. Engineering Journal. 2014;**18**(2):67-80

[21] Van Der Ham LV, Kjelstrup S. Exergy analysis of two cryogenic air separation processes. Energy. 2010;**35**(12):4731-4739

[22] Ebrahimi A, Meratizaman M, Akbarpour Reyhani H, Pourali O, Amidpour M. Energetic, exergetic and economic assessment of oxygen production from two columns cryogenic air separation unit. Energy. 2015;**90**:1298-1316

[23] Fu C, Gundersen T. Using exergy analysis to reduce power consumption in air separation units for oxy-combustion processes. Energy. 2012;**44**(1):60-68

[24] Taniguchi M, Asaaoka M, Ayuhara T. Energy saving air-separation plant based on exergy analysis. Kobelco Technology Review. 2015;**33**:34-38

[25] Smith AR, Klosek J. A review of air separation technologies and their integration with energy conversion processes. Fuel Processing Technology. 2001;**70**(2):115-134

[26] Lemcoff NO. Nitrogen separation from air by pressure swing adsorption. In: Dabrowski A, editor. Studies in Surface Science and Catalysis: Adsorption and its Applications in Industry and Environmental Protection. Vol. 120. Elsevier; 1999. pp. 347-370

[27] Sadeghzadeh Ahari J, Pakseresht S, Mahdyarfar M, Shokri S, Zamani Y, Nakhaei Pour A, et al. Predictive dynamic model of air separation by pressure swing adsorption. Chemical Engineering & Technology. 2006;**29**(1):50-58

[28] Dunbobbin BR, Brown WR. Air separation by a high temperature molten salt process. Gas Separation & Purification. 1987;**1**(1):23-29

[29] Bejan A, Tsatsaronis G, Moran MJ. Thermal Design and Optimization. New York, USA: Wiley; 1996

[30] Tsatsaronis G, Morosuk T. Understanding and improving energy conversion systems with the aid of exergy-based methods. International Journal of Exergy. 2012;**11**(4):518-542

[31] Morosuk T, Tsatsaronis G. A new approach to the exergy analysis of absorption refrigeration machines. Energy. 2008;**33**(6):890-907

[32] Agrawal R, Herron DM. Air liquefaction: Distillation. Encyclopedia of Separation Science. 2000:1895-1910

[33] Cryotec Anlagenbau GmbH. Personal Communication. 2017

[34] Aspen Plus V8.6. The Software is a Proprietary Product of AspenTech. 2014. Available from: http://www.aspentech.com

[35] Dowling AW, Balwani C, Gao Q, Biegler LT. Equation-oriented optimization of cryogenic systems for coal oxycombustion power generation. Energy Procedia. 2014;**63**:421-430

[36] Gils HC. Abschätzung des möglichen Lastmanagementeinsatzes in Europa. Tagungsband der 8. Internationale Energiewirtschaftstagung an der TU Wien. Wien, Austria; February 13-15, 2013. Available from: http://elib.dlr.de/83717/1/Gils_Lastmanagementpotenziale_IEWT2013.pdf [Cited: 09 April 2018]. [Gils HC. Estimation of the use of the possible load management in Europe. Proceedings of the 8th International Energy Economy Conference at the TU Wien. Wien, Austria; February 13-15, 2013. Available from: http://elib.dlr.de/83717/1/Gils_Lastmanagementpotenziale_IEWT2013.pdf [Cited: 09 April 2018]]

[37] Kollmann A. LoadShift: Lastverschiebung in Haushalt, Industrie, Gewerbe und kommunaler Infrastruktur—Potenzialanalyse für Smart Grids. 2015. Available from: https://nachhaltigwirtschaften.at/resources/e2050_pdf/reports/endbericht_201507e_loadshift_industrie.pdf [Cited: 09 April 2018]. [Kollmann A. LoadShift: Load transfer in household, industry, and municipal infrastructure—potential assessment for smart grids. 2015. Available from: https://nachhaltigwirtschaften.at/resources/e2050_pdf/reports/endbericht_201507e_loadshift_industrie.pdf [Cited: 09 April 2018]]

[38] Pfaff I, Kather A. Comparative thermodynamic analysis and integration issues of CCS steam power plants based on oxy-combustion with cryogenic or membrane based air separation. Energy Procedia. 2009;**1**(1):495-502

[39] Zhou H, Cai Y, Xiao Y, Mkhalel ZA, You B, Shi J, et al. Process configurations and simulations for a novel single-column cryogenic air separation process. Industrial & Engineering Chemistry Research. 2012;**51**(47):15431-15439

[40] Peters MS, Timmerhaus KD, West RE. Plant Design and Economics for Chemical Engineers. 5th Edition. McGraw-Hill Chemical Engineering Series. New York: McGraw-Hill; 2003

[41] Ulrich GD, Vasudevan PT. Chemical Engineering Process Design and Economics: A Practical Guide. 2nd ed. Durham, N.H: Process Pub; 2004

[42] Smith R. Chemical Process Design and Integration. West Sussex, United Kingdom: Wiley; 2005

[43] CHEManager. Air Liquide eröffnet Deutschlands größte Luftzerlegungsanlage. 2013. Available from: https://www.chemanager-online.com/newsopinions/nachrichten/air-liquide-eroeffnet-deutschlandsgroesste-luftzerlegungsanlage [Cited: 03 2017]. [CHEManager. Air Liquide starts the operation the largest air separation unit in Germany. 2013. Available from: https://www.chemanager-online.com/newsopinions/nachrichten/air-liquide-eroeffnet-deutschlandsgroesste-luftzerlegungsanlage [Cited: 03 April 2017]]

[44] Air Liquid SA. 2014. Available from: http://www.airliquide.pt/pt/novoinvestimento-de-cerca-de-40-milhoes-de-euros-no-brasil.html#.VsNA5ubDE9K [Cited: 16 Feburary 2016]

[45] RT. Giant air-separation unit starts up in Russia. 2007. Available from: https://www.rt.com/business/giant-air-separation-unitstarts-up-in-russia/ [Cited: 16 February 2016]

[46] Cockerill R. Cryogenmash achieves Russian first with onsite plant. 2009. Available from: http://www.gasworld.com/

cryogenmash-achieves-russianfirst-with-onsite-plant/4185.article [Cited: 03 April 2017]

[47] Chemietechnik. Linde baut Luftzerlegungsanlage in Indonesien. 2011. Available from: http://www.chemietechnik.de/texte/anzeigen/115070 [Cited: 03 April 2017]. [Chemietechnik. Linde builds air separation unit in Indonesia; 2011. Available from: http://www.chemietechnik.de/texte/anzeigen/115070 [Cited: 03 April 2017]]

[48] Business Standard Reporter. Matheson K-Air to set up air separation plant in Pune. 2013. Available from: https://www.business-standard.com/article/companies/matheson-k-air-to-set-up-air-separation-plant-in-pune-111030800081_1.html [Cited: 03 April 2017]

[49] Group TL. Press Releases. 2011. Available from: http://www.the-linde-group.com/de/news_and_media/press_releases/news_2011_05_23.html [Cited: 03 April 2017]

Chapter 10

Impact of Air-Conditioning Filters on Microbial Growth and Indoor Air Pollution

Amira Hassan Al-abdalall, Sarah Abdullah Al-dakheel and Hmidah Abdulhadi Al-Abkari

Abstract

Contemporary lifestyles dictate that people spend between 60 and 90% of their daily lives indoors. For those living in warm climates, air conditioning is thus considered a necessity. Air conditioners function by removing hot and humid air from building interior and replacing it with cooler air. Microorganisms are considered among the most important sources of poor quality of indoor air, and contamination of this air by microbial pollutants is being increasingly recognized as a public health problem and a probable cause of the so-called sick building syndrome. In this regard, microfiber glass panel filters are considered to provide an effective solution for air filtration and have been demonstrated to improve air quality in many applications. However, recent research has demonstrated that certain microorganisms are able to colonize panel filter surfaces. Studies on selected microbes isolated from the most commonly used filters have revealed that the bacterial and fungal moist masses carried on sponge-type filters are greater than those carried on polyester and high-efficiency particulate air (HEPA) filters. Moreover, microbial moist mass has been found to increase with increasing incubation time. In addition, recent research has shown that certain microorganisms, particularly fungi, can colonize the materials used in heating, ventilation, and air-conditioning systems (HVAC).

Keywords: colonization, air-conditioning system, filter, HEPA, indoor, sick building syndrome

1. Introduction

Microorganisms are among the most important sources of poor indoor air quality, and contamination of indoor air by microbial pollutants is being increasingly recognized as a public health problem and as one of the factors contributing to sick building syndrome. Bioaerosols, such as those comprising fungi, bacteria, and viruses, in indoor air can cause allergic and infectious diseases, respiratory problems, and hypersensitivity reactions. People who are sensitive to indoor environmental problems complain of a wide variety of symptoms, ranging from headache, tiredness, nausea, and sinus congestion to eye, nose, and throat irritations [1].

Although invisible to the naked eye, the atmosphere is populated by a diversity of microorganisms, including bacteria, fungi, algae, and protozoa. Researchers have estimated that the total bacterial count within the troposphere layer ranges from 1×10^3 to

1×10^5 cells/m^3. Dust is formed during the passage of organic and inorganic particles from external and internal resources, which subsequently aggregate and precipitate. House dust, for example, consists of cotton fibers, hair, bacteria, molds, and remaining paint particles [2, 3]. The findings of a previous study have indicated that the average number of fungi contaminating 820 indoor air-conditioning units was 1252 CFU/m^3, with range of 17–9100 CFU/m^3. In addition, Baxter [4] found that the average number of spores isolated from 85 buildings was 913 cells/m^3, ranging from 68 to 2307 cells/m^3. Daily and seasonal numbers of contaminant microorganisms in the air vary and depend primarily on environmental factors, such as vegetation, human activities, and seasonal fluctuations [5]. Most of these microorganisms are bacteria and fungi [6].

These microbial contaminants affect the residents of enclosed and humid buildings, particularly in the case of toxic hygrophytic fungi, such as *Phoma* sp., *Exophiala* sp., *Aureobasidium pullulans*, *Acremonium* sp., and *Sporobolomyces*, that are frequently isolated from the cooling pipes of air-conditioning systems. Gram-negative bacteria and their toxins are also isolated from leaks in air-conditioning pipes. Yang [7], for example, identified *Legionella pneumophila*, which is the causal agent of legionnaires disease, as a dominant bacterium in the water leaking from cooling systems. In addition, *Pseudomonas aeruginosa*, which has also been isolated from water leaking from air-conditioning systems, is an opportunistic bacterium responsible for several diseases. Many studies have proven that the heating, ventilation, and air-conditioning (HVAC) systems can become contaminated with organic pollutants, bacteria, and fungi, as well as by particulate matter derived from mice, insects, and nematodes. The bacteria and fungi colonizing these systems tend to saprophytic and thrive in areas that meet their environmental requirements [7]. Fungi have been proven to be a source of airborne contamination in air-conditioning systems [8], including *Alternaria*, *Aspergillus flavus*, *Aspergillus fumigatus*, *Aspergillus niger*, *Aspergillus ochraceus*, *Aspergillus versicolor*, *Botrytis cinerea*, *Cladosporium herbarum*, *Epiccocum purpurascens-sterilia*, *and Penicillium* spp., among which *A. fumigatus*, which has been isolated from air-conditioning filters, is responsible for many dangerous infections. With regard to bacteria, *Propionibacterineae*, *Staphylococcus*, *Streptococcus*, and *Corynebacterineae* (17, 17.5, 20, and 3%, respectively) have been detected in aeration pipes and air filters installed in indoor areas [9, 10]. In addition small percentages of species from the genera *Fusobacterium* and *Veillonella* (0.02 and 0.1%, respectively), which are associated with the mouth cavity and saliva, have also been identified as air-conditioning system contaminants [11–13].

With a view toward providing clean indoor air, several studies have been conducted to investigate measures that can be used to control the levels of microorganisms that colonize filtering, heating, ventilation, and air-conditioning systems. In this regard several types of air filters have been studied with the aim of preventing the penetration of particles. However, although high-efficiency particulate air (HEPA) filters are widely used in hospitals, *Aspergillus*-associated infections continue to occur [14]. Currently, most indoor air-conditioning systems contain internal filters that extract microorganisms from the air (**Figure 1**). However, these microbes often remain viable and can be returned to the surrounding atmosphere under certain circumstances, such as inefficient operation, during periods of maintenance, or due to temporary malfunction [15].

It is widely acknowledged that air-conditioning filters do not remove all the particles from the air. Even the use of HEPA filters will not completely eliminate the problem of microbial contamination, as this material will only retain particles of a minimum of 3 microns in size. Thus, dust particles with sizes smaller than 3 microns will pass through unhindered. Furthermore, when the filters become excessively wet, they can provide a fertile environment for the proliferation of molds and bacteria [16, 17].

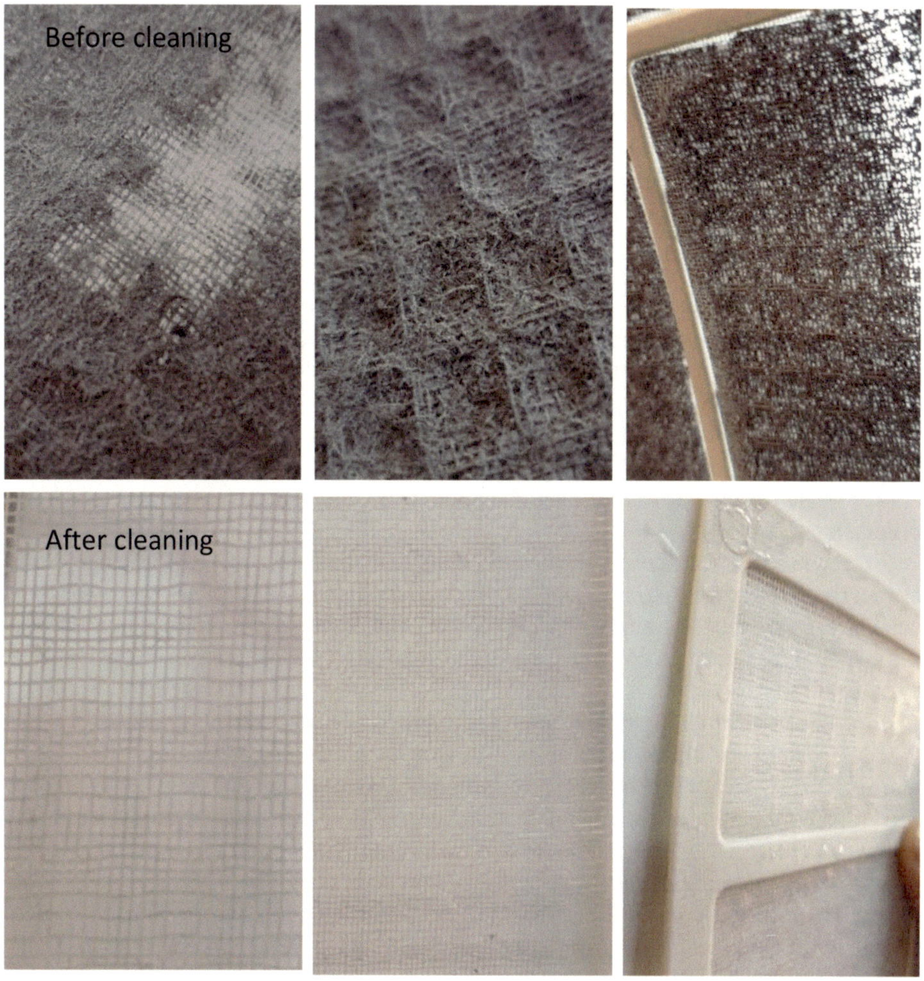

Figure 1.
Accumulated dust on discarded polyester filters.

In this chapter, several new topics related to environmentally sustainable buildings were presented, as it clearly describes the potential impact of HVAC systems on the indoor air quality with the aim to enhance the healthy buildings. The chapter is structured as follows: besides the Introduction (Section 1), Section 2 introduces the principle of air filtration, Section 3 is concerned with presenting the traditional air filters, Section 4.5 demonstrates a comparison between the most common and modern HVAC filters, Section 6 provides the impact of HVAC filters on indoor air quality, and Section 7 is concerned with presenting several results for research progress about the relationship between microbes and traditional filters and microbial colonization of the types of filters commonly used in air-conditioning systems.

2. Principle of air filtration

2.1 How filters work

There are five different collection mechanisms that determine air filtering performance: straining, interception, diffusion, inertial separation, and electrostatic attraction.

The first of these mechanisms applies mainly to mechanical filters and is influenced by particle size. **Figures 6** to **10** illustrate the five mechanical principles of particle capture and their contribution to the retention of particles of different sizes.

Straining (sieving) occurs when the opening between the media components (e.g., fibers, screen mesh, and corrugated metal) is smaller than the diameter of the particle the filter is designed to capture. This principle spans across most filter designs and is entirely related to the size of the particle, media spacing, and media density (**Figure 2**).

Interception occurs when a large particle, because of its size, collides with a fiber in the filter that an air stream is passing through (**Figure 3**).

Diffusion occurs when the random (Brownian) motion of a particle causes that particle to come into contact with a fiber. When a particle vacates an area within the media, by attraction and capture, it creates an area of lower concentration within the medium into which another particle diffuses, only in turn to be captured itself. To enhance the likelihood of this attraction, filters employing this principle operate at low media velocities and/or high concentrations of microfine fibers, glass, or otherwise (**Figure 4**).

Inertial separation is based on a rapid change in air direction and the principles of inertia to separate particulate matter from the air stream. Particles moving at a certain velocity tend to remain at that velocity and travel in a continuous direction. This principle is normally applied when there is a high concentration of coarse particulate matter and in many cases represents a pre-filtration stage prior to the passage of air through higher-efficiency final filters (**Figure 5**).

Electrostatic attraction is obtained by charging the media as a part of the manufacturing process (**Figure 6**). However, it plays a minor role in mechanical filtration. After fiber contact is made, smaller particles are retained on the fibers by a weak electrostatic force. The force may be created through a manufacturing process or be dependent upon airflow across media fibers. The force is eradicated as media fibers collect contaminants that act as an insulator to a charge. Electrostatic filters, which are composed of polarized fibers, may lose their collection efficiency over time or when exposed to certain chemicals, aerosols, or high relative humidity. A decrease in pressure in an electrostatic filter generally increases at a slower rate than it does in a mechanical filter of similar efficiency.

Inertial separation and interception are the dominant collection mechanisms for particles greater than 0.2 μm in size, whereas diffusion is dominant for particles less than 0.2 μm in size.

As mechanical filters become loaded with particles over time, their collection efficiency and reduction in pressure typically increase. Eventually, the decrease

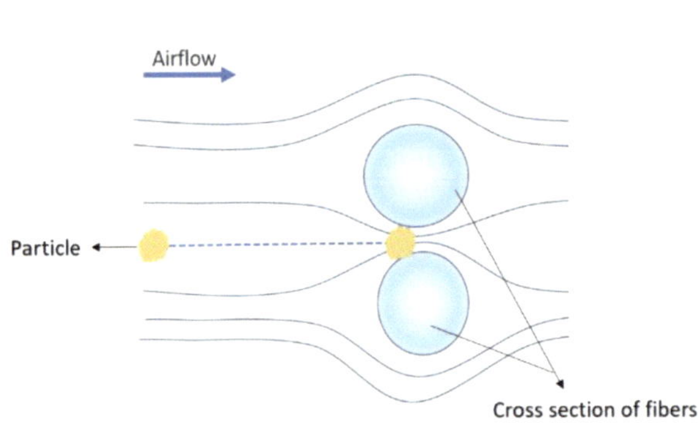

Figure 2.
Model of straining (sieving) mechanism, depends on the space between the fibers.

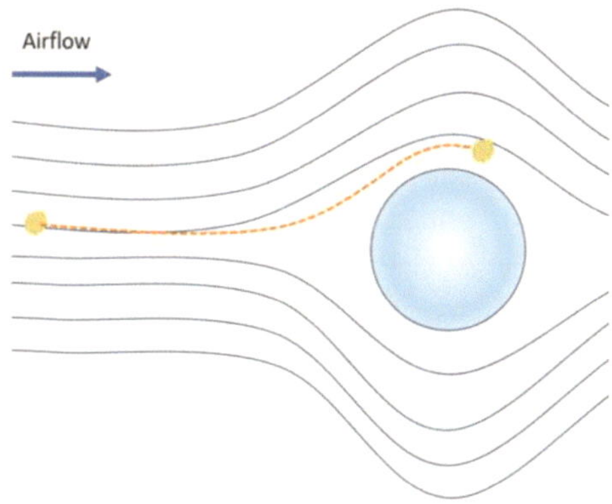

Figure 3.
Model of interception effect mechanism, depends on the collision between the fiber and the particle passing through the filter.

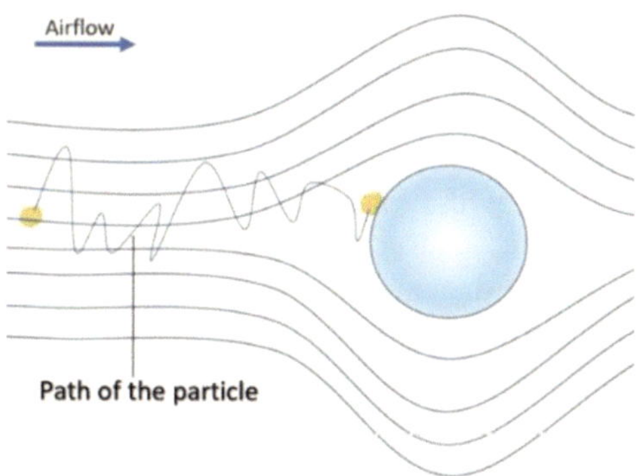

Figure 4.
Model of diffusion mechanism, depends on the motion of the particle causing contact with a fiber.

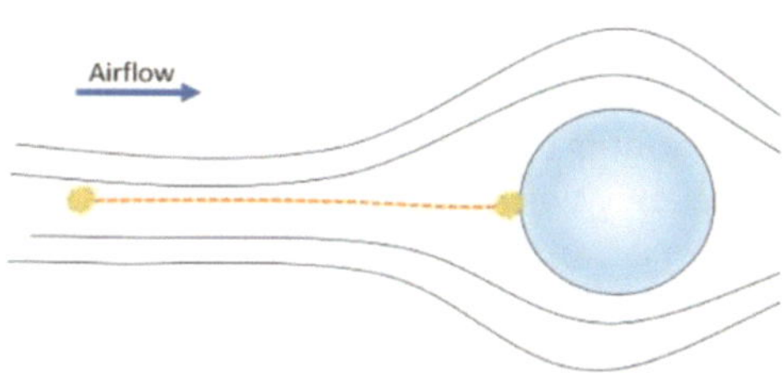

Figure 5.
Model of inertial separation mechanism, depends on the collision between the fiber and the small particles for reducing its velocity.

in pressure significantly inhibits airflow, and when this occurs, the filters must be replaced. For this reason, a decrease in pressure across mechanical filters is often monitored, as this can provide an indication of when the filters need to be replaced. Thus, unlike mechanical filters, a decrease in the pressure of electrostatic filters is a poor indicator of the need to change filters. When selecting an HVAC filter, these differences between mechanical and electrostatic filters should be borne in mind because they will have an impact on filter performance (collection efficiency over time), as well as on maintenance requirements (changeout schedules).

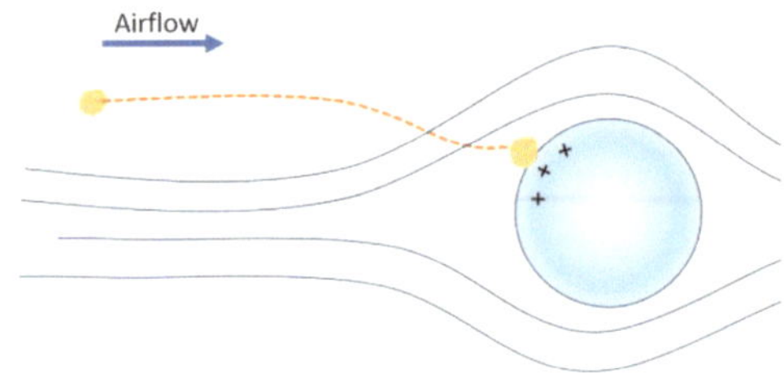

Figure 6.
Model of electrostatic attraction mechanism, depends on charging the fiber to retain the small particles by a weak electrostatic force.

3. Traditional types of household air filters

Humans consume approximately 30 liters of oxygen per hour. Hence, our requirement for air is relatively small: 0.15 m^3/h. However, because we also produce carbon dioxide, our bodies require approximately 5 m^3/h of fresh air in order to maintain carbon dioxide concentrations below life-threatening levels. When installing an air-conditioning system, it is advisable to determine the amount of air needed, and this will generally be set at between 15 and 20 m^3 per individual per hour. However, larger volumes of air might be necessary for managing temperature or drawing off polluted air.

Ensuring that air is free of dust and aerosols is not only important for maintaining buildings and their interior but also essential for maintaining the health and well-being of the human inhabitants.

This may be due to the higher foot traffic during business hours. The air output of these places is relatively high, and the cleaning of air-conditioning units may prove difficult, which could favor microbial growth and increased accumulation of dust on filters and in ducts. With respect to building contamination, it has been found that hospitals tend to have higher levels of contamination than other types of building examined. Given that hospitals are permanently inhabited by patients, this accordingly increases the potential for contamination and possibly infection by opportunistic pathogens [18, 19].

Al-Abdalall and Al-Abkari [20] examined the most commonly used filters incorporated in air-conditioning systems, namely, sponge, polyester, and HEPA, in order to assess the efficiency with which these filters can prevent the passage of fungi and bacteria. They accordingly found that complex filters were the most efficient in terms of purifying air, with efficiency rates up to 91.8% for bacteria and 100% for fungi. Sponge filters were deemed to be the least efficient filters, with estimated filtration rates of 2 and 50% for bacteria and fungi, respectively. This difference can probably be explained in terms of the passage of air through filters, with filters containing smaller pores being able to trap the larger cells of bacteria or fungi more efficiently. In other words, sponge filters are less efficient for air purification due to the large filter pores, whereas the filters of HVAC systems are able to capture particles smaller than 0.5 microns and prevent all particles with sizes greater than 3 microns from passing through [21].

In this regard, there are a number misconceptions concerning the relationship between filter efficiency and particle size, and in order to resolve this issue, a number of companies have developed certain filter-related standards based on particle counts at the most penetrating particle size (MPPS). The European Standard applies to HEPA ULPA filters used in the field of ventilation and for technical processes (e.g., for clean room technology or applications in the nuclear and pharmaceutical industries).

Many indoor air quality problems can be solved or avoided by cleaning or replacing air filters on a regular basis. Since using air filters is one of the most common methods of purifying air, it is recommended that filters be checked at 3-month intervals or that arrangements are made for a certified technician to change the filter at the beginning of each season.

Filters tend to become clogged, and once their holding capacity has been reached, particulate matter tends to be released downstream in the system and into the heat exchanger and can thereby cover the interior of the ductwork and the blower motor. This matter can subsequently cause problems and malfunctions in the mechanical and electrical parts of the system, resulting in high repair costs and even the need for replacement. The dispersed matter will circulate back into the house, potentially resulting in the proliferation of molds and other fungi. Dirty filters can also have a detrimental effect on energy consumptions due to impeded airflow, resulting in an increased in fan runtime.

4. Types of most common HVAC filters

4.1 Minimum efficiency reporting value (MERV)

According to the American Society of Heating, Refrigerating, and Air-Conditioning Engineers (ASHRAE) Standard 52.2-2007 [22], the performance of an air filter is determined by measuring the particle counts on both the upstream and the downstream sides of the air filter device being tested. Through provided capture efficiency values for a range of particle sizes, it facilitates the selection of a filter that has the best efficiency with regard to removal of the target contaminant.

To simplify filter selection, the Standard defines a minimum efficiency reporting value. The MERV is a single number that simplifies the filter selection process by providing the specifier, or the user, with a single value of specification for filter selection. For most filters with mechanical-based filter operation, this number will most probably be a minimum value at installation and throughout the life of the filter. The particle size ranges specified by Standard 52.2 and an illustration of how to read an ASHRAE 52.2-2007 [22] test report are shown in **Tables 1** and **2** and **Figure 5**, respectively (**Figure 7**).

Range	Size (microns)
1	0.3–0.4
2	0.4–0.55
3	0.55–0.7
4	0.7–1.00
5	1.00–1.30
6	1.30–1.60
7	1.60–2.20
8	2.20–3.00
9	3.00–4.00
10	4.00–5.00
11	5.00–7.00
12	7.00–10.00

Table 1.
Particle size ranges of Standard 52.2.

Example particle	Size (μm)
Hair	20–200
Pollen	10–100
Spores	10–25
Toner	5–20
Oil fog	0.3–5
Bacteria	0.2–25
Tobacco smoke	0.01–1
Virus	0.002–0.05

Table 2.
Particle size ranges of common pollutants specified by Standard 52.2.

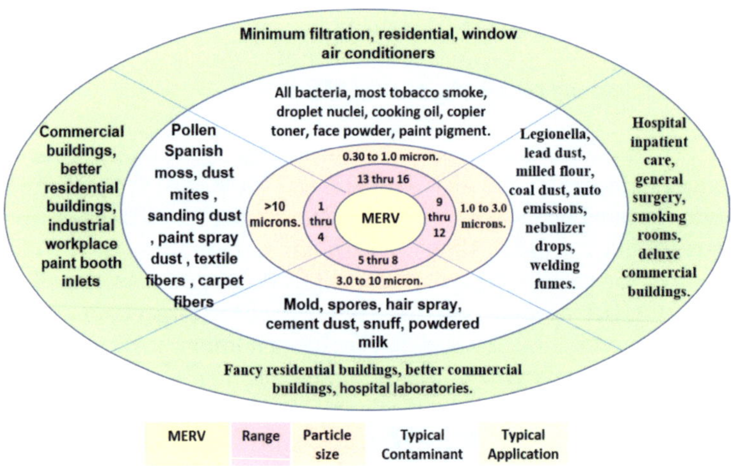

Figure 7.
How to read an ASHRAE 52.2-2007 test report.

Unfortunately, filters that use the principle of electrostatic attraction can "fool" the test by providing a high MERV during tests. However, due to the loss of electrostatic attraction during operation, a much lower value is obtained during application. Hence, the user may not be getting the particle removal efficiency that they originally specified.

Multiple studies have shown that coarse-fiber media (charged synthetic media), unlike fine-fiber media (fiberglass media), perform differently in real-life applications. Coarse-fiber media depends on an electrostatic charge to achieve the published filter efficiency. When atmospheric air, in which 99% of the particulate matter less than 1.0 micron in size, passes through a filter, the very fine particulate matter will dissipate the charge, and the filter rapidly loses efficiency.

Qian [23] isolated Streptophyta from dust samples collected from the filters of air-conditioning systems at a rate of 45%, whereas the rate in the indoor air was found to be only 2.4%, which provides an indication of the efficiency of HVAC filter systems (preventing particles sizes that are larger than 3 μm).

4.2 Types of most common HVAC filters

There are six types of filters, which are briefly described below.

4.2.1 Flat fiberglass filters

The main advantage of fiberglass filters is they are very cheap, easy to install, and readily available in stores. Accordingly, although they have a lifespan of only 1 month, replacing them on a monthly basis would not pose an inordinate financial burden. Unfortunately, they are not particularly effective in terms of trapping particles.

4.2.2 Pleated filters

These filters are more effective with regard to trapping dust than fiberglass filters. They can trap approximately 45% of airborne debris. Their MERV rating typically lies somewhere between 10 and 13. They also have a 1-month lifespan but tend to be more expensive than fiberglass filters.

4.2.3 Electrostatic air filters

These filters use electricity to attract charged particles, which are trapped internally. They are very efficient at trapping dust particles and debris and have a 6-month lifespan.

4.2.4 Washable filters

They are the most economical type of filter, which can be removed and cleaned as directed, dried, and then reinstalled. Furthermore, they do not need replacing at monthly intervals. These filters can prevent the passage of debris and tend to function better when dirty. However, they have a MERV rating of only 1–4.

4.2.5 Disposable filters

These filters comprise a cardboard frame and filter material. As their name implies, it is necessary to replace them when they become dirty.

4.2.6 HEPA filters

HEPA filters are considered the best type of filter because they trap even the smallest particles and keep premises smelling fresh. They can capture up to 97% of all particulate matter and remove all allergens from indoor air.

5. Modern HVAC filters

The type of air filter is the first factor people take into consideration before deciding on which air purifier to purchase.

Air filters and electrostatic filter cleaners are typically rated according to the minimum efficiency reporting value, commonly known as the MERV rating. The MERV scale is a measurement scale developed in 1987 by ASHRAE to rate the effectiveness of air filters, which determines efficiency in terms of the size of particle that the filter will capture. Values vary from 1 to 16, with a higher number indicating the greater efficiency of the filter in trapping airborne particles.

5.1 UV lights and filtration systems for cleaner indoor air

Recently, UV lights have been widely employed in the ducts of HVAC filtration systems. These lights facilitate effective and inexpensive control and solve the problem of microbial outgrowth in HVAC systems, eliminating up to 99.9% of the microorganisms and destroying airborne viruses, bacteria, and fungi. The types and quantities of microorganisms killed depend on the length of exposure and the output of the lamps. Nowadays, more advanced UV lights, such as air scrubbers, are employed, which can kill both airborne viruses and bacteria and those growing on surfaces.

There are two main types of UV lights used for HVAC systems, the most common of which are coil sterilization UV lights, which are installed near the return ducts, so as to kill mold that may grow on the air handler coil. These UV lights operate 24 h a day and eliminate the need for removing mold from the air handler coils. The second main type of UV lights is air sterilizer UV lights that function by sterilizing the air passing through the return ducts [24].

5.2 Activated carbon air filters

Also referred to as charcoal-impregnated air filters, these types of filters are used to effectively remove odors and fumes from the air during the air recirculating process.

Commercial activated carbon filters provide high-efficiency odor, fume, and gas removal and are fabricated using the finest quality coatings, including bulk air filter media and pads cut to size, pleats, panels, and high-density granular carbon packs.

Synthetic media substrates, such as non-woven polyester, are impregnated with finely ground coatings, including activated carbon, zeolite, or alumina, and a heat set to retain these coatings even when the activated carbon filter media is rinsed or vacuumed. Just as a sponge soaks up water, the media of activated carbon air filters absorb odors and fumes. Moreover, the odor-causing molecules are permanently removed from the air, rather than simply being masked with a different odor.

The rate of adsorption depends on the relationship between the pore structure, or surface area, and the shape of the contaminating molecules. Activated carbon filters are disposable air filters, and once they have become saturated with odors, fumes, or gases, after approximately 3 to 6 months of use, they must be replaced. The amount of activated carbon required will depend on the amounts of odors, fumes, or gases to be removed [24].

5.3 Deodorizing filters

Deodorizing air filters use acidified titanium, activated carbon, ceramic fiber, pulp, and other advanced materials that are prepared using a variety of rigorous refinement processes. They function by purifying the air and maintaining air fresh and show superior efficacy when used in conjunction with UV irradiation [24].

5.4 Antibacterial filters

Antibacterial filters are prepared by incorporating a bactericidal substance in the filter media. However, doubts remain regarding the effectiveness of these filters. One type is prepared by simply spraying the additive onto the surface of the filter medium, and therefore effective coverage is often not achieved, and not all of the filter layers will kill bacteria. A second type is prepared by application of a bacteriostatic agent, which does not kill the bacteria and may indeed promote the development of drug resistance among the bacteria. A third type may generate certain gaseous substances or odors that are potentially harmful to humans. Furthermore, it should be emphasized that it is difficult to capture bacteria on the windward side of the HEPA filters fabricated from inorganic materials, into which the bacteria may even penetrate. However, these microorganisms are only likely to survive under suitable conditions of temperature and humidity. Accordingly, the efficacy of these antibacterial filters remains inconclusive, and ASHRAE has recommended that HVAC systems incorporating antibacterial filters should be used with caution, so as not to produce any additional chemical pollution within indoor environments [24].

Although air filters often show excellent removal efficiency with regard to pathogens, the captured microorganism can remain viable within the filter and subsequently grow and become re-dispersed in the air flow, thereby generating a secondary source of pollutants and unpleasant odors. In an effort to resolve this issue, a large number of studies have been conducted with the objective of improving the effectiveness of air filters with antibacterial properties, and some of these studies have demonstrated that such air filters can be successfully prepared by incorporating inorganic nanoparticles and natural plant extracts [25].

Kim [26] also evaluated the efficacies of various functional filters coated with antimicrobial chemicals in deactivating representative microorganisms on filters or as bioaerosols. Specifically, they examined the effectiveness of functional filters coated with different chemicals, including ginkgo and sumac; Ag-apatite and guanidine phosphate; SiO_2, ZnO, and Al_2O_3; and zeolite, using a model ventilation system to evaluate the efficiency in which bacteria (*Escherichia coli* and *Legionella pneumophila*), bacterial spores (*Bacillus subtilis* spore), and viruses (MS2 bacteriophage) were removed. Their result showed that although the functional filters could facilitate the biological removal of various bioaerosols, physical removal was minimal. Appropriate use of chemical-coated filter materials could reduce exposure to these agents.

5.5 Electrostatic and HEPA filters

Electrostatic filters: Static electricity attracts dirt and dust to vertical and overhead surfaces. The static is often generated when two surfaces rub together and are then separated. Electrostatic filters generate a static electrical charge on all particles in the air that passes through eight filter layers. The discharged particles are then attracted to collector plates with an opposite electrical charge. These filters have the advantage of being washable.

High-efficiency particulate air filters: HEPA filters have a strong particle-trapping capacity that facilitates the removal of a high percentage (99.97%) of airborne particles that pass through an air purifier and accordingly meet US government standards. This contrasts with the 60–90% efficiency of medium filters [27]. Furthermore, HEPA filters perform significantly better than electrostatic air cleaners, in which filtering is based on ionic processes. HEPA filters are therefore often used in medical facilities and in households in which the residents suffer from severe allergies.

5.6 Microbial filtration efficiency of HEPA filters

Photocatalysts are nanoscale metal oxide materials (commonly titanium dioxide) that are applied to substrate surfaces, forming a film after drying under the action of light. They have a strong catalytic degradation function and can be used to degrade hazardous atmospheric gases. They can also be used to effectively kill a variety of bacteria, with an antibacterial rate of 99.99%. Furthermore, toxins released during the degradation of bacteria and fungi can be rendered harmless. These catalysts also have a range of other properties, including deodorant and dirt removal functions.

Exposure to bioaerosols can cause various adverse health effects, including infectious and respiratory diseases and hypersensitivity. Consequently, controlling the exposure to bioaerosols constitutes an important aspect of disease control and prevention.

Photocatalytic oxidation systems use a UV light source and a titanium dioxide photocatalyst to produce oxidants that destroy gaseous contaminants. When the photocatalyst is irradiated with UV light at wavelength of 254–365 nm, a photon from the light excites a catalyst electron in the valence band to jump to the conduction band, leaving a hole. This photocatalytic oxidation process converts organic pollutants into carbon dioxide and water. Using this technique, pollutants, particularly volatile organic carbons, are preferentially adsorbed on a catalyst surface and oxidized. The hole generated by photocatalysis can further react with surrounding water to produce a hydroxyl radical (ÆOH), whereas the electron in the conduction band reacts with oxygen to yield a superoxide radical anion (ÆO2). These radicals can attack the cell membranes of microorganism, thereby causing the release of K^+ ions, RNA, proteins, and other important components and eventually resulting in cell death [28]. Given these properties, researchers have applied photocatalytic oxidation to many substrates and achieved impressive results, indicated by the affective control of test microbes.

To date, however, photocatalytic oxidation has yet to be applied to HEPA filters in HVAC systems [27]. HEPA filters have been mandated for use in the removal of airborne microorganisms in many codes adopted in the field of healthcare, including the American Institute of Architects Guidelines for Design and Construction of Hospital and Health Care Facilities (AIA Guidelines), the American Society of Heating, Refrigerating and Air-Conditioning Engineers standards, the Joint Commission on Accreditation of Healthcare Organizations Environment of Care standards, the Centers for Disease Control and Prevention (CDC) guidelines, and recommended practices [29]. Although HEPA filters can efficiently capture aerosolized microorganisms, the area downstream of the filter can become a breeding ground for microbes. Under conditions of suitable temperature and humidity, microbes retained within a filter can multiply using particulates adhered to the filter as a food source, and the microbial progeny can ultimately disperse into the filtered air [30, 31]. Thus, instead of being an apparatus to control air quality, these systems can potentially become a source of pathogens. Efforts have accordingly been made to eliminate the breeding ground problem. For example, Goswami [32] examined four microbial species (*Aspergillus niger*, *Penicillium citrinum*, *Staphylococcus epidermidis*, and *Bacillus subtilis*) that are representative of the genera most commonly detected in hospitals in Thailand, a country characterized by a hot and humid climate, with an average temperature of 27°C and average relative humidity ranging from 62 to 84% [33, 34].

Very high relative humidity not only reduces the probability of microorganisms coming into contact with hydroxyl radicals but also provides sites conducive to microbial survival. Excessive amounts of water can also occlude the reactive sites of filter surfaces and subsequently reduce the photocatalytic oxidation efficiency [32, 34, 35]. Hence, the effect of relative humidity has been investigated.

Chuaybamroong [27] examined the application of photocatalytic oxidation to HEPA filters for disinfection of airborne microorganisms. Experiments were conducted at two TiO_2 loadings on HEPA filters irradiated with UV-A under two relative humidities. They assessed the inactivation of two fungal (*Aspergillus niger* and *Penicillium citrinum*) and two bacterial (*Staphylococcus epidermidis* and *Bacillus subtilis*) isolates and found that, on average, 60–80% of microorganisms retained on a photocatalytic filter were inactivated, although in the case of *S. epidermidis,* 100% inactivation was observed.

These authors suggested that high humidity may induce the reactivation of organisms, whereas water may occupy most of the TiO_2 sites, leaving fewer available sites for microbial colonization. These conjectures are consistent with the opinions of [36], who noted that although the presence of water vapor enhances the likelihood of hydroxyl radical formation, at certain humidity levels, radical formation would not increase with increasing water vapor and even decrease due to the occlusion of adsorption site on the TiO_2 surface. Consequently, high humidity would decrease filter efficiency. Furthermore, Peccia [37] indicated that high levels of relative humidity may promote changes in the biopolymers within microbial cells, including cell wall components, or alter protein structure, thereby affecting DNA repair enzymes, and hence could protect the microorganism from desiccation and/or attenuate the incident UV irradiation.

6. Impact of HVAC filters on indoor air quality

Filtration technology is currently an integral aspect of air purification techniques that focus on particulate matter, the most common examples of which are fibrous filters [38]. The use of glass fiber filters is another mature filtration technique with a proven high efficiency (99.0%), similar to that of the HEPA filters [39]. Furthermore, wire mesh filters can also provide good filtration efficiency down to sizes of 2–10 μm [40].

On the basis of particle filtration efficiency, air filters can be divided into four types: prefilters, medium filters, HEPA filters, and ultralow particulate air (ULPA) filters [41]. The filtration efficiency of ULPA filters is greater than 99.999%, with particles of diameters down to 0.12–0.17 μm being effectively trapped [42]. Similarly, HEPA material has a strong ability for trapping particles and can remove 99.97% of particulate matter, smog, and bacteria with sizes down to 0.3 μm (**Figure 8**). In contrast, the efficiency of medium filters is only between 60 and 90% [27, 43]. Most air purifiers currently on the market incorporate HEPA filters, as these are internationally recognized as the most efficient filters, capturing particles of different diameters. HEPA filters are designed to be over 99.99% efficient and are used in a diverse range of situations, including in theaters, hospitals, respirators, and vehicles [44]. Furthermore, filtration based on the use of non-woven nano-fiber material is an emerging filtration technique with an extremely high efficiency that is comparable to, or even superior to, HEPA-based filtration in the smaller particle size range [45].

According to the Environmental Protection Agency (EPA), pollutant levels may be two to five times higher indoors than outdoors, which indicates that the poor quality of indoor air is mainly attributable to the inefficient circulation of air. In regions characterized by hot summer, there is a heavy reliance on air-conditioning systems for maintaining comfortable conditions during the hot summer months. Accordingly, windows tend to remain shut, and little fresh air enters into our homes and places of work. The EPA warns that high temperatures and humidity can increase the concentrations of certain pollutants, with young children and the elderly being at particular risk from the detrimental effects of indoor air pollution.

Mechanisms for trapping dust in air using a standard filter, killing almost of all airborne microbes using UV lamps, and removing fine particles (dust) and died microbes using a high-efficiency particulate air filter (**Figure 8**).

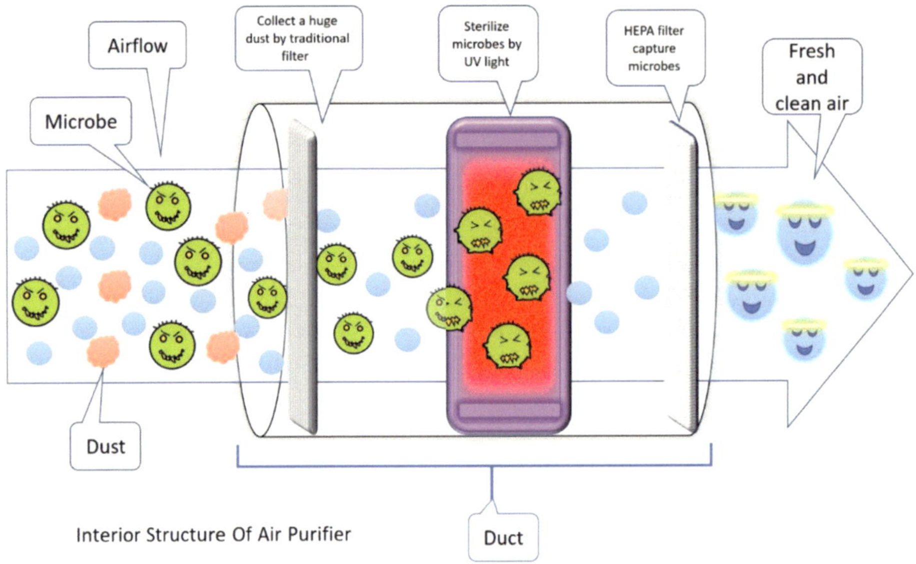

Figure 8.
A standard filter traps large dust particles in internal structure of an air purifier. HEPA filter captures microbes.

6.1 Air duct cleaning services

Given that dirty HVAC units have been proven to be less efficient, it is essential that air-conditioning ducts are periodically cleaned through employing an air duct cleaning service. Such cleaning should include complete care of the internal elements of the HVAC unit.

Ductwork would only be essential if there has been renovation, asbestos abatement, lead paint removal, or a significant accumulation of dust debris. Cleaning would be considered essential in the presence of the following: animal feces, mold, foul odors, noticeable debris, or pet hairs. Furthermore, if occupants suffer from an unexplained allergy, then it would be advisable to consider cleaning. Abe [46] noted that it is possible to remove fungi and bacteria from filters by washing with water and detergent; however, if the fan and heat exchanger are also contaminated, specialist cleaning would be required (**Figure 9**).

Kujundzic [47] mentioned that cleaning room air could contribute to reducing the levels of particulate matter within the home and that this can be achieved by using filters that retain the filtered particles.

6.2 Air-conditioning systems and mold

Mold is a pervasive problem, of which many property owners are fully unaware. In the Eastern Province of Saudi Arabia, the average annual humidity level is approximately 74%, but can be notably higher during certain times of the day and year. Many types of mold require a humidity of only 50% to commence growth, and air-conditioning systems are a common source of mold in many households. One of the causal factors in this respect is the fact that HVAC systems do not operate continuously, which can result in a fluctuation in humidity levels. The EPA warns that if an HVAC system is turned off before occupants perform tasks such as mopping, the humidity levels can suddenly surge and cause moisture and mold problems. If an HVAC system is improperly programed (which is a common problem), then the air-conditioning system may cycle off when the air is cooled but before it has had time to dry sufficiently, causing moisture problems. HVAC maintenance issues can be a further source of mold-related problems, such as when excessive moisture accumulates on the air-conditioning coils, resulting in the growth of mold. This mold can subsequently be blown through the air-conditioning ducts and released into the surrounding air. Holes or gaps in air ducts can also lead to the formation of condensation, which creates a perfect breeding ground for mold to grow. Such mold can cause numerous health-related problems, including respiratory problems, skin rashes, and allergic reactions [17].

In a study designed to examine the efficiency of various filters used for trapping microorganisms, Al-Abdalall and Al-Abkari [20] isolated the bacteria and fungi colonizing air-conditioning systems in different types of buildings during each of the four season s, in the provinces of Dammam and Qatif in eastern Saudi Arabia, and determined the respective frequency distributions. The air-conditioning systems were found to be contaminated by different types of bacterial and fungal species. Specifically, the isolated bacteria included *Serratia*

Figure 9.
Ducts should be cleaned periodically.

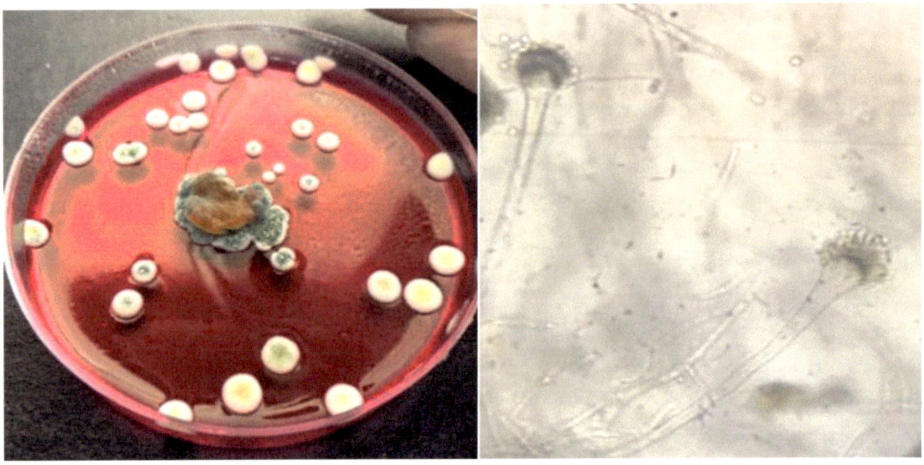

Figure 10.
Microorganisms collected by swabbing an air-conditioning duct and using the swabs to inoculate a fungal growth medium. The left-hand panel shows Aspergillus fumigatus growth on inoculated medium. The right-hand panel shows the mycelial growth and conidiophores of A. fumigatus viewed under a compound microscope.

liquefaciens, *Bacillus pumilus*, *Bacillus cereus*, *Bacillus subtilis*, *Staphylococcus lentus*, and *Oligella ureolytica,* whereas the common fungal taxa included *Cryptococcus laurentii*, *Aspergillus niger*, *Aspergillus flavus*, *Cladosporium* sp., and *Rhizoctonia* sp. Ironically, the findings of the study indicated that buildings that were in good condition were those likely to have the highest levels of microbial contamination (**Figure 10**).

Among the microparticles suspended in air, there is an abundance of biological material, including fungal spores, pollen grains, bacteria, and viruses. Air-conditioning systems can readily become polluted by these biological contaminants, which disperse throughout indoor areas and raise the risk of infection among the occupants [48]. The amounts of bacteria and fungi harbored by these systems tend to differ according to location, and numbers and frequencies also show seasonal differences [49]. Furthermore, differences in the number of microorganisms isolated and the distribution of different types of isolates can also depend on the type of filter used and the frequency of cleaning. In this regard, [20] found *Cladosporium* sp. to be a dominant contaminant and also identified *Alternaria* sp., *Aspergillus flavus*, *Aspergillus niger*, and *Rhizoctonia* sp. with frequencies of 24.16, 12.96, 12.8, 8.29, and 4.96%, respectively. Similar results were obtained by [50–53]. Consistently, Al-Suwaine et al. [54] mentioned that *Aspergillus* and *Cladosporium* spp. were the common isolates detected in closed systems in Riyadh, KSA, whereas other fungal genera, including *Fusarium* and *Rhizopus*, were isolated in low frequency, similar to findings of [51, 55].

6.3 HVAC systems and indoor pollution

Heating, ventilating, and air-conditioning (HVAC) systems function by drawing in air through a network of intake ducts, cooling the air, and then releasing the cooled air back into the home through return ducts. The constant recirculation of air in HVAC systems means that pollutants are continuously blown through indoor areas. In the eastern region of Saudi Arabia, given that the summer is generally very hot and humid, many properties are at risk from the growth of mold within air ducts.

7. The microbial colonization of traditional filters

7.1 The relationship between microbes and traditional filters

We have previously examined the nature of the relationship between filters and airborne microbes, using small pieces (1 × 1 cm) of traditional filters (sponge, polyester, and HEPA) These materials were sterilized with alcohol, then dried, and subsequently moistened with glucose yeast extract medium. We then prepared suspensions of the examined fungal strains (*Aspergillus niger*, *Aspergillus flavus*, *Cladosporium* sp., and *Rhizoctonia* sp.). These were retained in sterilized petri dishes, whereas other groups were prepared for carbon-free sources. They were monitored for 1 to 3 months and thereafter examined under a microscope. Heavy growth of mycelium was observed. The fungal filaments are looped around filter fibers also assembled into the filter cavities, forming a tangled knot of fungal mycelium and filters fibers (**Figure 11**).

Figures 12–15 show that the microscopic structures of the filters shown in **Figure 11** have been colonized by *Aspergillus niger* in (**Figure 12**), *Aspergillus flavus* in (**Figure 13**), *Rhizoctonia* in (**Figure 14**) and *Cladosporium* in (**Figure 15**) which indicate that these fungi use the filters as a support for fungal mycelium.

Microorganisms can exploit various parts of air-conditioning systems, including filters, as sheltered sites, which are conducive to rapid grow and reproduction [16, 17, 56]. The high levels of humidity in air-conditioning systems [16, 57] and the accumulated dust in the filters and other parts of these systems provide an environment that is suitable for the growth of a range of different microbes.

Microorganisms can secrete a diverse array of extracellular enzymes to exploit the various available filter materials, such as cellulose, as sources of nutrition [56, 58].

Kuehn [57] pointed out that moisture promotes fungal growth in filter tissues and can also favor bacterial reproduction leading to subsequent transmission to and dispersal within indoor environments. Such moisture often originates from the drops of condensate that form air-conditioning towers [59].

Maus [60] have suggested that the spores of some bacteria and fungi trapped within air filters can retain their viability and reproduce under the prevailing environmental conditions. These microorganisms can be dispersed through purification and air-conditioning systems and be inhaled by workers and residents in buildings [48].

Microbiological particles constitute one of the most important sources of air pollution that determine the purity of the air. It is known that atmospheric air is a carrier of disease-causing organisms, including fungal spores and microbial

Figure 11.
Structure of filters observed under an optical compound microscope. The right-hand panel shows sponge filter cavities of different sizes. These openings are wider than those of other filters. The middle panel shows a polyester filter, which is characterized as a network installation with narrower openings than the sponge filter that are regular in shape. The left-hand panel shows a HEPA filter, characterized by complex knit and numerous narrow openings that increase efficiency by preventing the passage of fine particles.

cells, the concentrations of which vary widely according to environmental condition, particularly the nature of the internal environment and the various activities of humans who reside or work therein [61]. In order to achieve the desired level of microbial contamination control in air-conditioning systems and develop suitable air purification techniques, it is generally necessary to conduct extensive studies [17].

Hamada and Fujita [62] noted that the contamination of filters tends to be very low during the first year after installation, reaching 257 cells/m^3 of room air, whereas by the sixth year of use, they found that the number of contaminating cells had increase threefold to 692.

Durand [63] demonstrated that species of *Aspergillus* and *Penicillium* are the fungi most commonly isolated from the filters of air-conditioning systems, whereas species of *Cladosporium* and other types tend to be detected at relatively low rates.

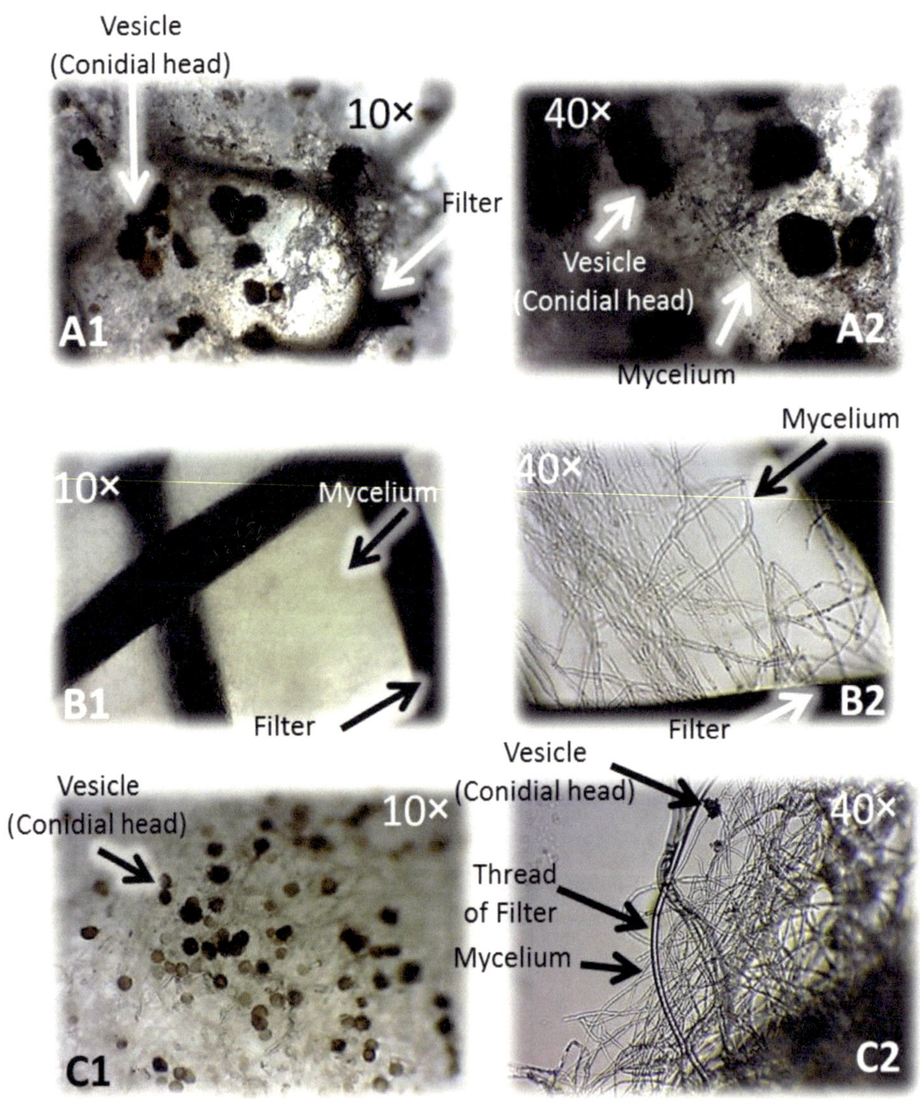

Figure 12.
The mycelial growth and conidiophores of Aspergillus niger on the studied filters: (a) sponge, (b) polyester, and (c) HEPA.

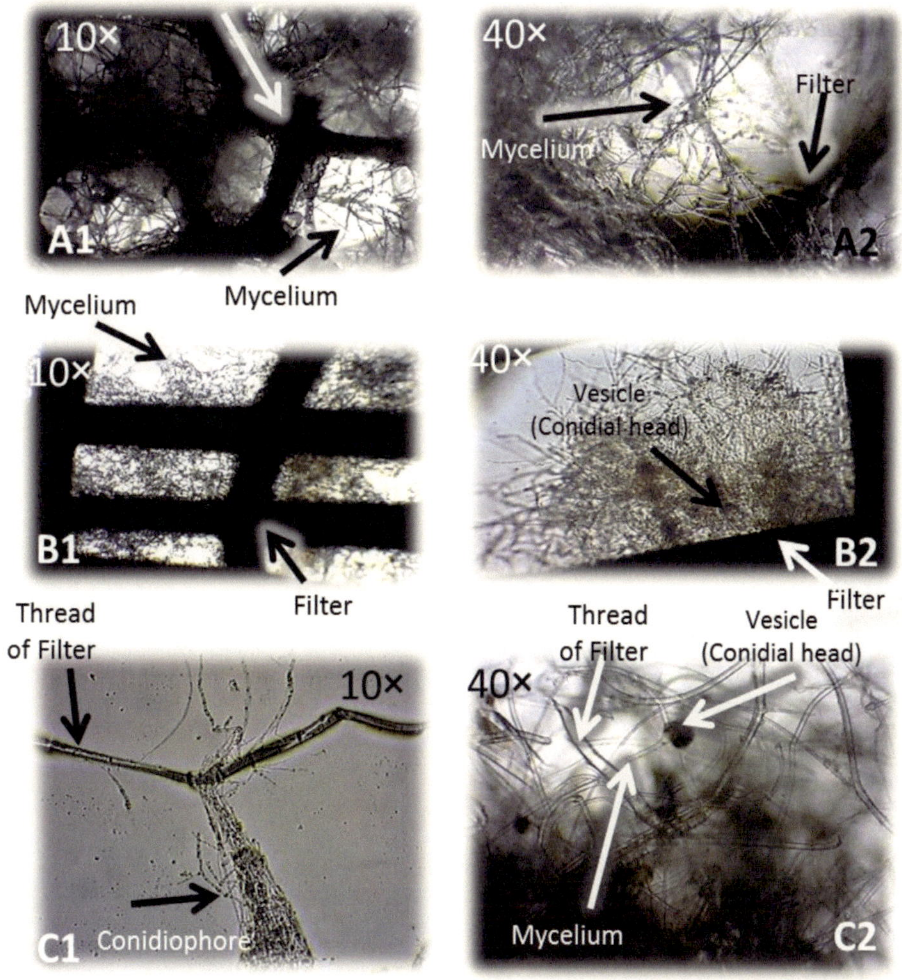

Figure 13.
The mycelial growth and conidiophores of Aspergillus flavus on the studied filters: (a) sponge, (b) polyester, and (c) HEPA.

For bacteria, species of *Actinomycetes* and *Bacillus* (cocci and Gram-negative types) tend to be the most commonly isolated.

Al-Abkari [17] recommended that air-conditioning filters should be cleaned regularly and that regular maintenance is necessary to prevent an accumulation of contaminants and to remove the suspended dust. In this regard, spongy filters can be readily washed and cleaned with detergents, and be reused after cleaning. In contrast, HEPA filters, which consist of interlocking fibers, are very difficult to wash and clean and should thus be replaced on a regular basis. Furthermore, it has been demonstrated that the number of microbial colonies (bacteria and fungi) growing on culture dishes that were exposed to air that had passed through different filters increased after 30 min and then decreased after 60 min. This indicates an inverse relationship between the period of operation of the air-conditioning system and the quantity of air that had been purified [64].

Al-Abkari [17] examined the extent of microbial growth on the most common types of filter used in the eastern region of Saudi Arabia (sponge, polyester, and HEPA) and found that sponge filters harbored the highest microbial moist mass, reaching 0.177 and 0.257 gm for bacteria and fungi, respectively.

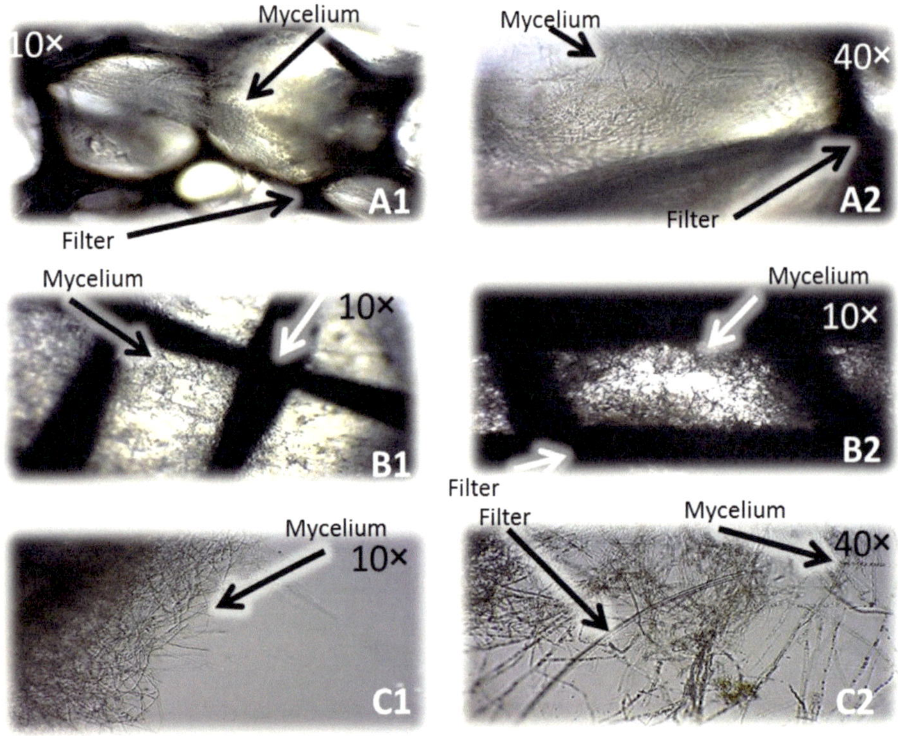

Figure 14.
The mycelial growth of Rhizoctonia sp. on the studied filters: (a) sponge, (b) polyester, and (c) HEPA.

Comparatively, the bacterial mass recorded on polyester and HEPA reached 0.024 and 0.037 gm, respectively, and the corresponding fungal mass on these filters was 0.072 and 0.047 gm, respectively. Douwes [65] isolated polysaccharide compounds known to be excreted by various fungi that grow on dust in residential homes, and the detection of these compounds is accordingly considered to be a good indicator of the presence of these fungi.

Moray and Williams [66] performed direct microscopic observations of the porous soft filters typically used in air-conditioning systems and accordingly identified pollen grains, cellulose fibers, synthetic fibers, decayed plant leaves, hairs, parts of insects, dust, mites, and numerous organic compounds, all of which can provide a refuge for microbes.

7.2 Microbial colonization of the types of filters commonly used in air-conditioning systems

Microorganisms that are captured by filters can thrive on the filters and can potentially be released into the air, thereby resulting in sick building syndrome [67]. Furthermore, it has been determined that the number of microbes found in indoor air is less than that colonizing the surface of filters used in air-conditioning systems [68]. Foarde [69] and Kowalski [68] examined the efficiency of these systems and provided solutions for the HEPA filters. In addition, they noted that the tested filter samples trapped *Bacillus subtilis*, with efficiencies ranging from 19 to 100%, whereas in contrast the efficiency in trapping viruses was low, ranging from 0.7 to 20%.

Al-Abkari [17] examined the ability of microorganisms (bacteria and fungi) to degrade various types of filter commonly used in air-conditioning systems,

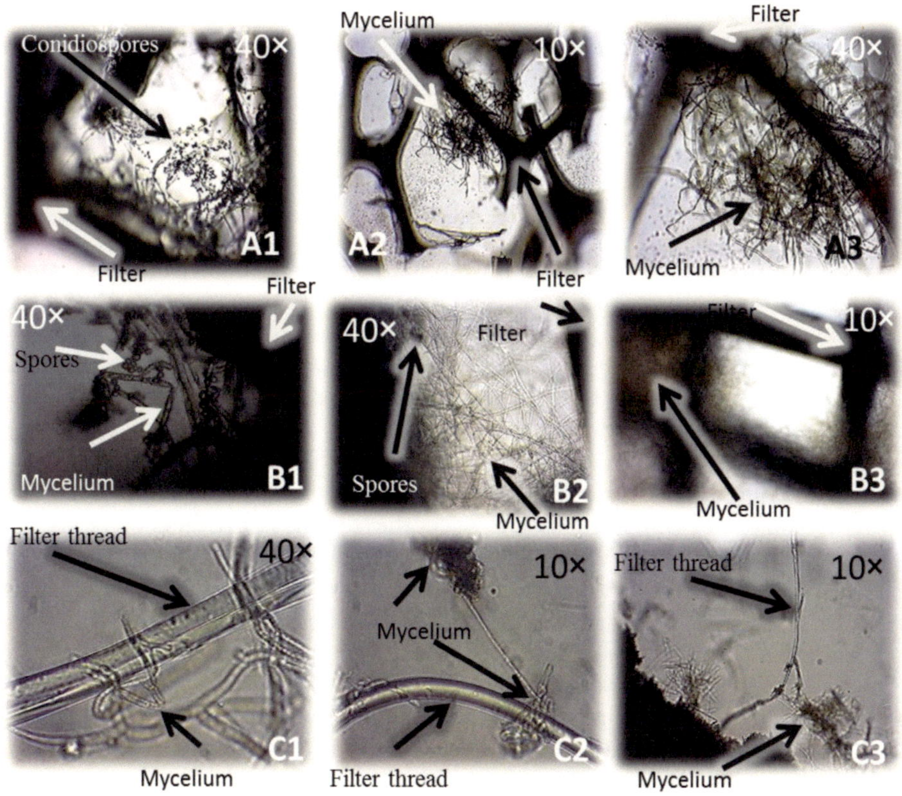

Figure 15.
The mycelial growth and conidia of Cladosporium sp. on the studied filters: (a) sponge, (b) polyester, and (c) HEPA.

namely, sponge, polyester, HEPA, and the environmental conditions, such as dust, temperature, and moisture, which enable these organisms to take refuge, grow, and reproduce. The results indicated that the growth of bacterial strains was dependent on the filter media containing a carbon source. The average of bacterial moist mass loading on different filters was found to be positively related to the length of incubation period (1, 2, and 3 months), with weights reaching 0.061, 0.09, and 0.101 g after incubations for 1, 2, and 3 months, respectively. Furthermore, it was found that the average microbial mass detected on sponge filters (0.177 g) was larger than that on either polyester (0.024 g) or HEPA (0.037 g).

Generally, it was observed that the average of moist weights of bacterial mass on all filters increases with an increase in the length of the incubation period, with recorded averages of (0.134, 0.169, and 0.228 g) and (0.019, 0.024, and 0.03 g) and (0.031, 0.035, and 0.046 g) for sponge, polyester, and HEPA filters, respectively. In contrast, it was found that the moist mass of microbial growth on culture medium lacking a carbon source remained essentially constant with increasing incubation time, with values of 0.023, 0.023, and 0.028 g; 0.03, 0.035, and 0.039 g; and 0.163, 0.171, and 0.162 g) for sponge, polyester, and HEPA filters, respectively.

With regard to the growth of fungal strains, when these were grown in a medium containing a carbon source, the average moist fungal mass loading on different filters showed a positive relationship with incubation period (1, 2, and 3 months), with weights reaching 0.87, 0.118, and 0.142 g, respectively. Similar to bacterial growth, the average weight of fungal biomass growing on sponge filters (0.257 g) was larger than that on polyester (0.072 g) and HEPA (0.047 g). The weight of fungal mass on polyester and HEPA filters was 0.05, 0.082, and 0.085 g and 0.022, 0.04, and 0.078 g, respectively. Notably, however, fungal growth on sponge filters increased with increasing incubation time, reaching 0.181 and 0.324 g, following incubation for 1 and 2 months, respectively, whereas after incubation for 3 months, it had decreased to 0.265 g.

8. Conclusion

Microbial pollution is one of the most fundamental indoor environmental quality problems in buildings. Therefore, this chapter has presented several solutions for indoor air quality monitoring in an effort to enhance the healthcare by describing the potential impact of HVAC systems on the indoor air quality. The principles of air filtration and traditional air filter types were presented.

Also, the chapter covered the filtration technology and the indoor air quality topics. Subsequently, the air duct cleaning devices, mold formation, and HVAC systems and indoor pollution were illustrated. Moreover, this chapter provided the ASHRAE standards, which was used to select the suitable HVAC filters. The six most common designs of HVAC filters were briefly described. This chapter was followed by the modern filters. All advanced air filters, UV lights, activated carbon, deodorizing, antibacterial, electrostatic and HEPA filters, microbial filtration efficiency of HEPA filters were discussed extensively.

A brief description of the microbial colonization on the commonly used traditional filter types for air-conditioning systems were provided, followed by a detailed explanation of the relationship between traditional filters and microbe's formation.

Even though there is highly development in designing advanced filters to overcome microbial pollution, we are still facing indoor air pollution problems. The most challenging step is providing an affordable construction, easy to install, made of environmentally friendly and long-term materials, available in different designs and able to avoid the existence of microbial growth.

Author details

Amira Hassan Al-abdalall[1]*, Sarah Abdullah Al-dakheel[2]
and Hmidah Abdulhadi Al-Abkari[1]

1 Department of Biology, Imam Abdulrahman Bin Faisal University, Dammam, Kingdom of Saudi Arabia

2 Department of Physics, Imam Abdulrahman Bin Faisal University, Dammam, Kingdom of Saudi Arabia

*Address all correspondence to: aalabdalall@iau.edu.sa

References

[1] Hayleeyesus SF, Manaye AM. Microbiological quality of indoor air in university libraries. Asian Pacific Journal of Tropical Biomedicine. 2014;**4**:S312-S317

[2] Harrison RM, Jones AM, Biggins PDE, Pomeroy N, Cox CS, Kidd SP, et al. Climate factors influencing bacterial count in background air samples. The International Journal of Biometeorology. 2005;**49**:167-178

[3] Maron PA, Lejon DPH, Carvalho E, Bizet K, Lemanceau P, Ranjard L, et al. Assessing genetic structure and diversity of airborne bacterial communities by DNA fingerprinting and 16S rDNA clone library. Atmospheric Environment. 2005;**39**:3687-3695

[4] Baxter DM. A regional comparison of mold spore concentrations outdoors and inside "clean" and "mold contaminated" Southern California buildings. Journal of Occupational and Environmental Hygiene. 2005;**2**(1):8-18

[5] Jones AM, Harrison RM. The effects of meteorological factors on atmospheric bioaerosol concentrations—A review. Science of the Total Environment. 2004;**326**(1-3):151-180

[6] Griffin DW. Atmospheric movement of microorganisms in clouds of desert dust and implications for human health. Clinical Microbilogy Reviews. 2007;**20**:459-477

[7] Yang CS. Biological Contamination in the HVAC System. Cherry Hill, New Jersey: P&K Microbiology Services, Inc.; 2000

[8] Ljaljević M, Vukojević J, Stupar M. Fungal colonization of air-conditioning systems. Archives of Biological Science (Belgrade). 2008;**49**(3-4):123-128

[9] Hamady M, Lozupone C, Knight R. Fast UniFrac: Facilitating high-throughput phylogenetic analyses of microbial communities including analysis of pyrosequencing and PhyloChip data. ISME Journal. 2009;**4**:17-24

[10] Frank DN, Feazel LM, Bessesen MT, Price CS, Janoff EN, et al. The human nasal microbiota and *Staphylococcus aureus* carriage. PLoS One. 2010;**5**:e10598

[11] Lazarevic V, Whiteson K, Hernandez D, François P, Schrenzel J. Study of inter- and intra-individual variations in the salivary microbiota. BMC Genomics. 2010;**11**:523

[12] Cephas KD, Kim J, Mathai RA, Barry KA, Dowd SE, et al. Comparative analysis of salivary bacterial microbiome diversity in edentulous infants and their mothers or primary care givers using pyrosequencing. PLoS One. 2011;**6**:e23503

[13] Nasidze I, Li J, Schroeder R, Creasey JL, Li M, et al. High diversity of the saliva microbiome in Batwa Pygmies. PLoS One. 2011;**6**:e23352

[14] Eckmanns T, Ruden H, Gastmeier P. The influence of high efficiency particulate air filtration on mortality and fungal infection among highly immunosuppressed patients: A systematic review. Journal of Infectious Diseases. 2006;**193**(10):1408-1418

[15] Lee BU, Yun SH, Ji J, Bae GN. Inactivation of *S. epidermidis* , *B. subtilis, and E. coli* bacteria bioaerosols deposited on a filter utilizing air borne silver nano particles. Journal of Microbiology and Biotechnology. 2008;**18**:176-182

[16] Yassin MF, Almouqatea S. Assessment of airborne bacteria and fungi in an indoor and outdoor environment. International Journal of Environmental Science and Technology. 2010;**7**(3):535-544

[17] Al-Abkari HA. Studies on microbial contamination in air conditioning systems in the Eastern region of Saudi Arabia Kingdom and their control [M.Sc. thesis in Microbiology]. Dammam, Saudi Arabia: University of Dammam; 2014

[18] Lee YH, Lee BU. In activation of air borne *E. coli* and *B. subtilis* bioaerosols utilizing thermal energy. Journal of Microbiology and Biotechnology. 2006;**16**:1684-1689

[19] Wakefield J. Allergies: The new lore of Spores. Environmental Health Perspectives. 2006;**114**(10):A576

[20] Alabdalall AH, AL-Abkari HA. Microbial contamination of heating, ventilation and air conditioning (HVAC) systems and the efficiency of various filter types in trapping microorganisms in Eastern Saudi Arabia. The Asian International Journal of Life Sciences. 2017;**26**(2):283-302

[21] Huang R, Agranovski I, Pyankov O, Grinshpun S. Removal of viable bioaerosol particles with a low-efficiency HVAC filter enhanced by continuous emission of unipolar air ions. Indoor Air. 2008;**18**(2):106-112

[22] ASHRAE. Method of Testing General Ventilation Air-Cleaning Devices for Removal Efficiency by Particle Size. ANSI/ASHRAE Standard 52.2; 2017. ISSN: 1041-2336

[23] Qian J, Hospodsky D, Yamamoto N, Nazaroff WW, Peccia J. Size-resolve demission rate so fair borne bacteria and fungi in an occupied class room. Indoor Air. 2012;**22**(4):339-351. DOI: 10.1111/j.16000668.2012.00769. xPMID:22257156

[24] Xu Z. Fundamentals of Air Cleaning Technology and Its Application in Cleanrooms. Springer Heidelberg New York Dordrecht London; 2014. DOI: 10.1007/978-3-642-39374-7

[25] Komaladewi AA, Khoiruddin K, Surata IW, Subagia ID, Wenten IG. Recent advances in antimicrobial air filter. E3S Web of Conferences. 2018;**67**:03016. DOI: 10.1051/e3sconf/20186703016 3rd i-TREC 2018

[26] Kim KY, Kim CN. Airborne microbiological characteristics in public buildings of Korea. Building and Environment. 2007;**42**:2188-2196

[27] Chuaybamroong P, Chotigawin R, Supothina S, Sribenjalux P, Larpkiattaworn S, Wu CY. Efficacy of photocatalytic HEPA filter on microorganism removal. Indoor Air. 2010;**20**:246-254

[28] Wamer WG, Yin JJ, Wei RR. Oxidative damage to nucleic acids photosensitized by titanium dioxide. Free Radical Biology and Medicine. 1997;**23**:851-858

[29] ASHRAE. HVAC Design Manual for Hospitals and Clinics. Georgia: American Society of Heating, Refrigerating and Air-Conditioning Engineers, Inc; 2003

[30] Goswami TK, Hingorani SK, Greist H, Goswami DY, Block SS. Photocatalytic system to destroy bioaerosols in air. Journal of Advanced Oxidation Technologies. 1999;**4**:185-188

[31] Jankowska E, Reponen T, Willeke K, Grinshpun SA, Choi K-J. Collection of fungal spores on air filters and spore reentrainment from filters into air. Journal of Aerosol Science. 2000;**31**:969-978

[32] Goswami DY, Trivedi DM, Block SS. Photocatalytic disinfection of indoor air. Transactions of the ASME. 1997;**119**: 92-96

[33] Thai Meteorological Department. Climate of Thailand. 2009. Available from: http://www. tmd.go.th/info/info. php?FileID=56. [Accessed: 25 September 2009]

[34] Goswami DY. Decontamination of ventilation system using photocatalytic air cleaning technology. Journal of Solar Energy Engineering. 2003;**125**:359-365

[35] Li Y, Wu CY. Role of moisture in adsorption, photocatalytic oxidation, and reemission of elemental mercury on a SiO_2-TiO_2 nanocomposite. Environmental Science and Technology. 2006;**40**:6444-6448

[36] Li FB, Li XZ, Ao CH, Lee SC, Hou MF. Enhanced photocatalytic degradation of VOCs using Ln_3+TiO_2 catalysts for indoor air purification. Chemosphere. 2005;**59**:787-800

[37] Peccia J, Werth HM, Miller S, Hernandez M. Effects of relative humidity on the ultraviolet induced inactivation of airborne bacteria. Aerosol Science and Technology. 2001;**35**:728-740

[38] Podgórski A, Bałazy A, Gradoń L. Application of nanofibers to improve the filtration efficiency of the most penetrating aerosol particles in fibrous filters. Chemical Engineering Science. 2006;**61**:6804-6815

[39] John W, Reischl G. Measurements of the filtration efficiencies of selected filter types. Atmospheric Environment. 1978;**12**:2015-2019

[40] Setekleiv A, Eddie S, Hallvard F. Operation and dynamic behavior of wire mesh pads. Chemical Engineering Journal. 2012;**68**:624-639

[41] Ahn Y, Park S, Kim G, Hwang Y, Lee C, Shin H, et al. Development of high efficiency nanofilters made of nanofibers. Current Applied Physics. 2006;**6**:1030-1035

[42] Jamriska M, Martin D, Morawska L. Investigation of the filtration efficiency of HEPA and ULPA filters in submicron particle size range. Clean Air Environment. 1997;**31**:31-37

[43] Hanley J, Ensor D, Smith D, Sparks L. Fractional aerosol filtration efficiency of induct ventilation air cleaners. Indoor Air. 1994;**4**:169-178

[44] Brincat JP, Sardella D, Muscat A, Decelis S, Grima JN, Valdramidis V, et al. A review of the state-of-the-art in air filtration technologies as may be applied to cold storage warehouses. Trends in Food Science and Technology. 2016;**50**:175-185

[45] Wang N, Raza A, Si Y, Yu J, Sun G, Ding B. Tortuously structured polyvinyl chloride/polyurethane fibrous membranes for high-efficiency fine particulate filtration. Journal of Colloid and Interface Science. 2013;**398**:240-246

[46] Abe K. Fungal index and contamination in air conditioners when cooled. Journal of the Society of Indoor Environment Japan. 1998;**1**:41-50

[47] Kujundzic E, Angenent LT, Zander DA, Henderson DE, Miller SL, Hernandez M. Effects of ceiling-mounted HEPA-UV air filters on airborne bacteria concentrations in an indoor therapy pool building. Journal of the Air and Waste Management Association. 2005;**55**:210-218

[48] Mendell MJ, Lei-Gomez Q, Mirer AG, Seppänen O, Brunner G. Risk factors in heating, ventilating, and air-conditioning systems for occupant symptoms in US office buildings:

The US EPA BASE study. Indoor Air. 2008;**18**:301-316

[49] Quintero E, Rivera-Mariani F, Bolaños-Rosero B. Analysis of environmental factors and their effects on fungal spores in the atmosphere of a tropical urban area (San Juan, Puerto Rico). Aerobiologia. 2010;**26**(2):113-124

[50] Griffin DW, Kellogg CA, Shinn A. Dust in the wind: Long range transport of dust in the atmosphere and its implications for global public and ecosystem health. Global Change Human Health. 2001;**2**:20-33

[51] Gorny R, Dutkiewicz J. Bacterial and fungal aerosols in indoor environment in central and eastern European countries. Annals of Agricultural and Environmental Medicine. 2002;**9**:17-13

[52] Ejdys E, Michalak J, Szewczyk KM. Yeast-like fungi isolated from indoor air in school buildings and the surrounding outdoor air. Acta Mycologica. 2009;**44**(1):97-107

[53] Al-Abdalall AHA. Populations and distribution of filamentous fungi and bacteria in Some educational buildings in Dammam, eastern Saudi Arabia. Scientific International Journal of Food, Agriculture and Environment-JFAE. 2011;**9**(3&4):886-891

[54] Al-Suwaine AS, Hasnain SM, Bahkali AH. Viable airborne fungi in Riyadh, Saudi Arabia. Aerobiologia. 1999;**15**(2):121-130

[55] Aydogdu H, Asan A, Otkun T, Ture M. Monitoring of fungi and bacteria in the indoor air of primary schools in Edirne City, Turkey. Indoor and Built Environment. 2005;**14**(5):411-425

[56] Vitel C. The quality of the air in our buildings. Indoor and Built Environment. 2001;**10**:266-270

[57] Kuehn TH, Pui DYH, Vesley D, Berg CD, Peloquin M. Matching filtration to health requirements. ASHRAE Transactions. 1991;**97**(2):164-169

[58] Flannigan B, Miller JD. Health implications of fungi in in-door environments—An overview. In: Samson RA, Flannigan B, Flannigan ME, Verhoeff AP, Adan OCG, Hoekstra ES, editors. Health Implications of Fungi in Indoor Environments; Air Quality Monographs. Vol. 2. NY: Elsevier. 1994. pp. 3-28

[59] Simmons RB, Crow SA. Fungal colonization of air filters for use in heating, ventilating, and air conditioning (HVAC) systems. Journal of Industrial Microbiology. 1995;**14**:41-45

[60] Maus R, Goppelsröder A, Umhauer H. Survival of bacterial and mold spores in air filter media. Atmospheric Environment. 2001;**35**:105-113

[61] Bonadonna L, Marconi A. Aerosol A comparison of two air samplers for recovery of indoor bioaerosols. Aerobiologia. 2006;**10**:153-156

[62] Hamada N, Fujita T. Effect of air-conditioner on fungal contamination. Atmospheric Environment. 2002;**36**:5443-5448

[63] Durand KTH, Muilenberg ML, Burge HA, Seixas NS. Effect of sampling time on the culturability of airborne fungi and bacteria sampled by filtration. The Annals of Occupational Hygiene. 2002;**46**(1):113-118

[64] Jo W, Lee J. Airborne fungal and bacterial levels associated with the use of automobile air conditioners or heaters, room air conditioners, and humidifiers. Archives of Environmental and Occupational Health. 2008;**63**(3):101-107

[65] Douwes J, van der Sluis B, Doekes G, et al. Fungal extracellular polysaccharides in house dust as a marker for exposure to fungi: Relations with culturable fungi, reported home dampness, and respiratory symptoms. The Journal of Allergy and Clinical Immunology. 1999;**103**:494-500

[66] Moray P, Williams C. Porous insulation in buildings: A potential source of microorganisms. In: Proceedings of the 5th International Conference on Indoor Air Quality and Climate. Indoor Air, 1990;4:529-533

[67] Kelly-Wintenberg K, Sherman DM, Tsai PPY, Gadri RB, Karakaya F, Chen Z, et al. Air filter sterilization using a one atmosphere uniform glow discharge plasma (the volfilter). Transactions on Plasma Science. 2000;**28**(1):64-71

[68] Kowalski WJ, Bahnfleth WP, Witham DL, Severin BF, Whittam TS. Mathematical modeling of UVGI for air disinfection. Quantitative Microbiology. 2000;**2**(3):249-270

[69] Foarde KK, Hanley JT, Veeck AC. Efficacy of antimicrobial filter treatments. ASHRAE Journal of Air Filters. 2000;**42**(12):52-58

Chapter 11

Designing Spiral Plate Heat Exchangers to Extend Its Service and Enhance the Thermal and Hydraulic Performance

Lázaro Canizalez Dávalos, Edilberto Murrieta Luna, Mario Alberto Rodríguez Ángeles and Víctor J. Cruz Delgado

Abstract

Spiral plate heat exchangers are well suitable for handling fluids with features linked to fouling, high viscosities, fluids with suspended fragments of solids and process streams with tough heat transfer targets. Correlations to describe the thermal and hydraulic performance are a function of the geometrical configuration of the equipment. The present work shows procedures to design spiral plate heat exchangers as a function of the fluid arrangements, government flow, as well whether the thermal equipment is for condensing, cooling or heating duty. An additional study looking for determining the geometrical variables that allow to enhance and improve the thermal and hydraulic performance is presented. Moreover, computational fluid dynamics to validate the thermal and hydraulic method is performed.

Keywords: design, thermal and hydraulic performance, spiral plate heat exchanger, single- and two-phase flow, computational fluid dynamics

1. Introduction

A spiral heat exchanger is assembled by two metallic plates separated by studs. They are welded on the plate surface. The objective is to maintain a constant spacing between plates at the time the plates are rolled up. Thus, the first turn represents one flow section; at this point, the second flow section initiates. Fluids pass in spiral plate heat exchangers by two arrangements, cross-flow and spiral flow, and these configurations are shown in **Figure 1**. Since these patterns were stated, manufacture companies and researchers have proposed more understandings regarding designs, thermal and hydraulic analyses and ways to provide more worth to these heat exchangers.

Spiral plate heat exchangers have important industrial applications; particularly, they are suitable for dirty fluids and viscous fluids. However, most of the correlations and methods explain the single-phase liquid-liquid, and as a consequence, this is not sufficiently to describe heat transfer and hydraulic behaviour, i.e., two-phase flow in spiral heat exchangers liquid–gas and liquid-vapour. Sathiyan et al. [1, 2] presented a study to evaluate a new equation to approximate the Nusselt number of

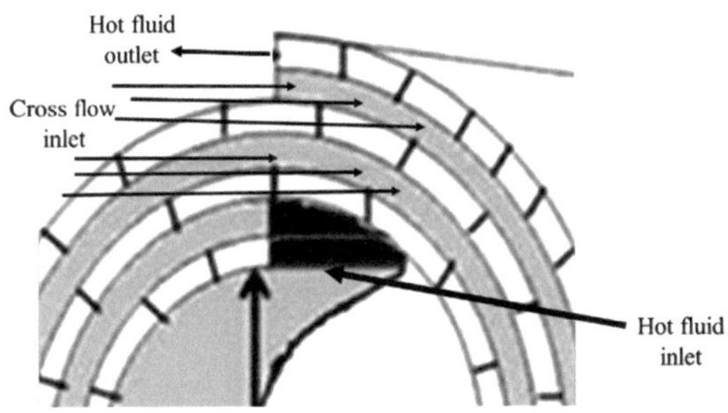

Figure 1.
Cross-flow and spiral flow arrangement.

an immiscible mixture using a countercurrent spiral heat exchanger for two-phase flow. The new correlation was based on the experimental data, and the results were agreement with the theoretical correlation. Khorshidi and Heidari [3] fabricated a spiral heat exchanger geometry to study the performance, the examination showed that spiral heat exchanger is an excellent option to transfer heat especially from fouling fluids and also a computational fluid dynamic was presented to determine a previous design. Ramachandran et al. [4] determined the heat transfer behaviour for a system of two fluids by implementing a countercurrent spiral plate heat exchanger; the data were obtained from varying mass fraction inlets and demonstrated efficient results between the experiment and the correlation. Maruyama et al. [5] measured the thermal effectiveness of a cross-flow spiral plate heat exchanger; the aim was to convert radiation energy from a combustion chamber. Wang [6] analysed the thermal performance of a spiral plate heat exchanger used as an adsorber in a refrigeration process, the flows were configured to follow a spiral trajectory and the spiral exchanger resulted appropriated for a refrigeration system. Bahiraei et al. [7, 8] presented a study to evaluate the thermal and hydraulic performance of a spiral plate heat exchanger under a turbulent flow of a nanofluid. The experimental procedure was to determine the effects due to the spiral geometry varying the flow rate to define the optimal operational conditions.

The thermal and hydraulic concept is an innovative tool to achieve designs for heat exchangers and applies to all types of heat exchangers. Previously, researches have been using this procedure [9, 10]. Compact and conventional exchangers are employed to develop two duties heating and cooling. Nevertheless, they behave differently from each other due to their geometrical configuration, effectiveness, outlet temperature, pressure drop and heat transfer area [11–15].

The current study is organised to describe four main purposes: 1. To present two design methods by a thermal and hydraulic procedure. The first design is for a cooler using a cross-flow arrangement (liquid-gas) without phase change, to evaluate if a spiral plate heat exchanger can take part as a radiator of the cooling system car. The second approach is to size a vertical spiral heat exchanger condenser for a cryogenic operation. 2. To extend the operational activities of spiral plate heat exchangers. 3. To improve the spiral thermal and hydraulic performance by modifying the spacing plate. 4. A numeric analysis applying computational fluid dynamics to validate the method.

2. Empirical thermal and hydraulic model

The method to design spiral plate heat exchangers includes two main equations, the film heat transfer coefficient and the pressure drop, which both are functions of the fluid properties, heat load, geometrical standard parameters, flow section area and metal construction characteristics. Thermal and hydraulic model is a key relationship which reduces the calculations to sizing and performs heat exchangers. Commonly, five iterations are needed to reach the balance between pressure drop and heat transfer. Finally, the heat transfer area is determined. The present procedure primarily introduces a hydraulic equation which is a function of pressure drop, the spacing between plates, flow rate and spiral plate length. The correlation is solved iteratively for the length; if the calculated length does not satisfy the pressure drop design, the spacing plate could adjust to maximise the use of permissible pressure drop.

2.1 Hydraulic equations

The hydraulic equations were presented by Minton [16]. These correlations are a function of a flow along the spiral channel which is separated by studs to give support to the plates. Factor 1.5 in Eq. 1 (**Table 1**) supposes 17 studs per square foot. Every stud has a diameter of 0.3125, and, then, in every 0.118 in^2, a stud is installed [16]. Eq. 2 (**Table 1**) has the same approach; however, the pressure drop is negligible because the fluid flows across the plate width, and value close to zero represents minor influence even by installing the studs [16]. Calculating the Reynolds number and the critical Reynolds number values is possible to select the correct equation to describe the hydraulic operation of the spiral plate heat exchangers. The equations are developed for the three flow regimes: laminar, transition and turbulent. The hydraulic correlations used in this study are presented in **Table 1**.

2.2 Thermal equations

The thermal model equations were introduced by Minton, although Sander [17] proposed the heat transfer correlation carlier.

Eq. (3) describes the heat transfer coefficient for a liquid fluid flowing by the spiral side. Similarly, Minton presented 11 mechanisms to determine the heat transfer coefficient as a function of flow configuration, type of service (condensing and heating) and a vertical nucleate boiling. Eq. (4) is for gas fluid where the Reynolds number is higher than 10,000. Even when this number is above critical Reynolds number, gases have low heat capacity, and they have poor heat transfer coefficient values (**Table 2**).

Flow configuration	Empirical pressure drop correlation
For spiral flow without phase change $Re > Re_c$	(1) $\Delta P = 0.001\frac{L}{s}\left[\frac{F}{d_s H}\right]^2\left[\frac{1.3\mu^{1/3}}{(d_s+0.125)}\left(\frac{H}{F}\right)^{1/3} + 1.5 + \frac{16}{L}\right]$
For axial flow without phase change $Re > 10,000$	(2) $\Delta P = \frac{4x10^{-5}}{s{d_s}^2}\left[\frac{F}{L}\right]^{1.8}\left[0.0115\mu^{0.2}\left(\frac{H}{d_s}\right) + 1 + 0.03H\right]$

Table 1.
Correlations for pressure drop [16].

Flow configuration	Empirical heat transfer coefficient correlation
For spiral flow without phase change (liquid fluid) $Re > Re_c$	(3) $h = \left(1 + 3.54\frac{D_e}{D_H}\right)0.023cGRe^{-0.2}Pr^{-2/3}$
For axial flow without phase change (gas fluid) $Re > 10,000$	(4) $h = 0.0144cG^{0.8}D_e^{-0.2}$

Table 2.
Correlations for heat transfer coefficient [16].

The Reynolds number and critical Reynolds number are represented by Eqs. (5) and (6):

$$Re = 10,000\left(\frac{F}{H\mu}\right) \tag{5}$$

$$Re_c = 20,000\left(\frac{D_e}{D_H}\right)^{0.32} \tag{6}$$

where F is the flow rate in lb./hr., H is the plate width in inches, μ is the viscosity in cp, De is the equivalent diameter in ft. and D_H is the spiral diameter in ft.

Eq. (7) describes the equivalent diameter:

$$D_{e=}\left[\frac{(d_s H)^{0.625}{}_{1.3}}{(d_s+H)^{0.25}}\right] \tag{7}$$

where ds is the channel spacing between plates.

2.3 Geometric additional equations

Auxiliary equations are needed to complete the thermal and hydraulic spiral plate model, for instance, the spiral outside diameter is determined by Eq. (8):

$$D_s = [15.36L(d_{sc} + d_{sh} + 2x)]^{0.5} \tag{8}$$

L is the plate length; d_{sc} and d_{sh} are the cold and cold spacing cannel, respectively; and x is the plate thickness.

The total heat transfer area is defined by the two spiral plates and is shown in Eq. (9):

$$A = H(2L) \tag{9}$$

Dongwu presented an entire description to calculate the number of spiral turns. This equation is based on plate length L, spiral semicircles, plate spacing, core diameter d and plate thickness. These geometric values are presented in Eq. (10). Due to a constant spacing of the two rolled passages, the number of turn equation gives an effective accuracy [18]:

$$N = \frac{-\left(d - \frac{t}{2}\right) + \sqrt{\left(d - \frac{t}{2}\right)^2 + \frac{4tL}{\pi}}}{2t} \tag{10}$$

where d is the core diameter of the first turn at the centre of the spiral heat exchanger and t is composed by Eq. (11):

$$t = d_{sh} + d_{sc} + 2x \tag{11}$$

The improvement of the spiral heat exchangers depends on geometrical variables. For instance, the spacing plate can increase the pressure drop if the separation between the spiral plates decreases. That means the heat exchanger needs more pumping energy, besides this, improves the heat transfer coefficient and the thermal effectiveness. The variation of the spacing channel allows to enhance the heat transfer and optimise the thermal and hydraulic performance.

Table 3 shows some recommended spacing plate values, and the values depend on the calculated plate width.

Plate spacing (m)	Plate width (m)
0.00476	0.101
	0.152
	0.304
	0.304
0.00635	0.457
	0.457
	0.609
	0.609
	0.762
	0.914
	1.219
0.00793	1.524
0.00952	1.778
0.0127	For more than 1.778 m
0.0158	
0.0190	
0.0254	

Table 3.
Recommended spacing plate values for standard plate widths [16].

2.4 Thermal performance

The validation of the method was carried out by two forms: the calculation of the NTU method and a numerical simulation with the commercial software for computational fluid dynamics ANSYS Fluent. The number of transfer units was implemented to calculate the thermal effectiveness of the spiral plate exchanger, as shown in Eq. (12):

$$NTU = \frac{UA}{CP_{min}} \tag{12}$$

where U is the overall heat transfer coefficient, A is the total heat transfer area and CP_{min} is the minimum CP for the stream with minimum heat capacity times the flow rate.

The overall heat transfer coefficient was calculated by the following equation:

$$U = \frac{1}{\left(\frac{1}{h_h} + \frac{x}{kA} + \frac{1}{h_c}\right)} \tag{13}$$

The CP ratio was set by Eq. (14) based on the hot stream $CP_h = F_h c_{ph}$ and the cold stream $CP_c = F_c c_{pc}$:

$$R = \frac{CP_{min}}{CP_{max}} \tag{14}$$

Bes and Roetzel [12, 19] reported an analytical equation to calculate thermal effectiveness. Correlation 15 contains the number of transfer units, the number of turns and the CP ratio:

$$\varepsilon = \frac{1 - e^{(R-1)NTU}}{1 - Re^{(R-1)NTU}} \tag{15}$$

The next expressions were used to calculate the hot outlet temperature and the cold outlet temperature. These equations depend on the CP ratio, effectiveness and maximum temperature difference. The procedure to calculate the hot temperature at the exit of the spiral flow is as follows:

If the $CP_h > CP_c$, then the correct form of the equation is

$$T_{hout} = T_{hin} - \varepsilon(T_{hin} - t_{cin}) \tag{16}$$

If the $CP_h < CP_c$, then the equation takes the next form:

$$T_{hout} = T_{hin} - \varepsilon R(T_{hin} - t_{cin}) \tag{17}$$

Next, the procedure to calculate the cold temperature at the exit of the axial flow is shown in Eqs. (18) and (19).

If the $CP_h < CP_c$, then the equation takes the next form:

$$t_{cout} = t_{cin} + \varepsilon(T_{hin} - t_{cin}) \tag{18}$$

If the $CP_h > CP_c$, then the correct equation is

$$t_{cout} = t_{cin} + \varepsilon R(T_{hin} - t_{cin}) \tag{19}$$

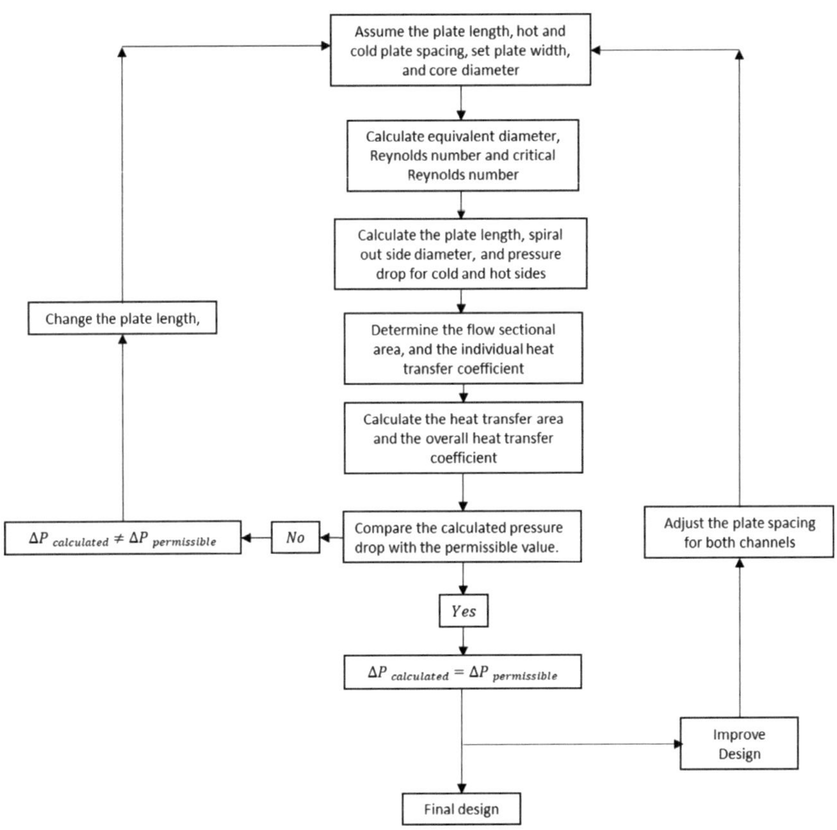

Figure 2.
Flow chart to design the cross-flow spiral heat exchanger.

	Water		Air
Total heat transfer area (m^2)		11.57	
Thermal effectiveness		0.84	
Outlet temperature (°C)	80.95		86.75
Mass flow (kg/h)	4200		5200
Inlet temperatures (°C)	98		20
Height (m)		0.41	
Length (m)		0.54	
Width (m)		0.028	

Table 4.
Car radiator design (car radiator) [20].

Figure 2 shows the sequential steps to design the spiral plate heat exchanger. A visual basic programming code was used to achieve these calculations.

2.5 Computational fluid dynamics (CFD)

The numerical simulation was performed by the software ANSYS Fluent. The mathematical model κ-ε solved the three balances: mass, energy and momentum.

Conservation of mass:

$$\nabla \cdot (\rho v) = 0\nabla \tag{20}$$

Conservation of momentum:

$$\nabla \cdot (\rho v v) = -\nabla P + \nabla \cdot (\mu \nabla v) \tag{21}$$

Conservation of energy:

$$\nabla \cdot (\rho v c p T) = \nabla \cdot (k \nabla T) \tag{22}$$

where k is the thermal conductivity, ρ is the density, μ is the dynamic viscosity, cp is the specific heat, v is the velocity, T is the temperature and P is the pressure (**Table 4**).

Figure 3.
Mesh of the spiral plate heat exchanger.

	Water	Air
Flow (kg/h)	4200	5200
Inlet temperatures (°C)	98	20
Viscosity (cp)	0.3	0.2
Maximum pressure drop (kPa)	6.89	6.89
Thermal conductivity (W/m K)	0.6	0.0259
Heat capacity (J/Kg K)	4270.53	544.28
Thermal conductivity (aluminium, W/m K)	205	
Plate width (m)	0.152	0.152
Core diameter (m)	0.0508	
Specific gravity	0.97	1
Channel spacing (m)	0.0043	0.0088
Plate thickness (m)	0.0787	

Table 5.
Second case study.

Figure 3 shows the virtual spiral plate heat exchanger designed by the software workbench, the mesh was structured with 97,250 control volumes and the geometrical features (plate width, plate spacing, plate length, thickness, etc.) are shown in **Tables** 5 and 8.

The boundary conditions such as the inlet velocity, inlet temperatures, fluid properties, flow rates and metal properties were considered from the results shown in **Tables** 5 and 8.

3. Case study

In order to demonstrate the rating and the design method, a case study was proposed. The example consists of two streams, water (hot stream) and air (cold stream), where the hot fluid flows by a spiral channel and the cold fluid flows through a cross-flow arrangement. **Table 5** explains the operational conditions for the first case study. This data was taken from a normal operation of a car radiator where they proposed to design an equipment to cool down a hot stream by 17 degrees. **Table 4** shows the operational data for the car radiator.

Table 5 summarizes the operational conditions. The plate width is set by a numeric value of 0.1524 m. The initial plate spacing for the hot stream is 0.0043 m and for the cold stream is 0.0088 m, and the core diameter is 0.0508 m.

The second case study consists of designing a condenser for a cryogenic process. The data were taken from a case study to design a compact heat exchanger [13].

The hot temperature must cool down by 9°C to transfer a latent heat to the cold fluid. The cold temperature is 15°C. The initial core spiral diameter was 0.5 m. A plate width of 0.076 m was assumed to find the optimal plate length for the spacing plates of 0.00254 and 0.00635 m for cold and hot stream, respectively.

4. Results and discussions

The new spiral was sized by considering the same instal space than the actual radiator. The features were 0.4 m of height, 0.15 m of depth and 0.4 m of width.

A suggested car cooling system performance must cool down the hot fluid 10°C or maximum 20°C. The results were achieved by implementing the thermal and hydraulic model, where one of the fundamental variables to measure was the hot outlet temperature. The difference between hot inlet and hot outlet was 6.4°C. The thermal and hydraulic method is an option to design heat exchangers. One of its goals is to find the lowest heat transfer area because the permissible pressure drop was fixed as a parameter to use completely. Then, from this value, the method seeks for some geometrical configuration to satisfy the required pressure drop. The method determined the heat transfer area, and the numeric value was of 2.04 m^2 (**Tables 6** and 7).

The permissible pressure drop was set by 1 psi for the two passages. The simulation calculated a value of 0.975 psi for the hot channel. The value for cold channel was 0.00013 psi because the cold channel has a length of 0.15 m. This flow section was considered as an open channel. This hydraulic behaviour was determined by the spiral diameter of 0.387 m, a plate length of 6.7 m, 10 spiral turns and the plate

	Water	R134a
Flow (kg/s)	1	0.38
Inlet temperatures (°C)	15	42.14
T_{out} (C)	31.33	33.65
Viscosity (cp)	1.1	0.175
Maximum pressure drop (kPa)	150	950
Thermal conductivity (k, W/m K)	0.61	0.0763
Heat capacity (J/kg K)	4528.02	1563.76
Thermal conductivity of steel (W/m K)	13	
Plate width (m)	0.0762	
Core diameter (m)	0.508	
Specific gravity	0.97	
Channel spacing (m)	0.00254	0.00635
Plate thickness (m)	0.003175	

Table 6.
Second case study [13].

	Water		Air
T_{out} (°C)	6.72		60.7
Pressure drop (kPa)	0.975		0.00068
Area (m^2)		2.04	
Diameter (m)		0.378	
HTC (W/m^2 K)	12227.62		305
U (W/m^2 K)		52.41	
Effectiveness		0.52	
Number of rounds		10	
Plate length (m)	6.7		6.7

Table 7.
Results for case study 1.

spacing for both channels. To achieve a pressure drop close to the ideal value, it should be necessary to increase the plate length, but the spiral diameter would be higher than the delimited value of 0.4 m. The thermal performance remained with no significantly variations.

For the improvement of the spiral design, only the plate spacings for the cold stream and the hot stream were reduced to 4 and 4.2 mm, respectively. The results are shown in **Table 8**.

The results in **Table 8** show that the thermal and hydraulic performance was improved, by decreasing the plate gap, because the flow distribution along the channel enhances the heat transfer, and the hot outlet temperature increased the difference by 9.4°C. The available pressure drop for the hot stream was used fully, while the cold pressure drop presented the variation of 0.00072 psi. The heat transfer area continued as 2.04 m, the number of turns raised by 1.5 rounds and the spiral outside diameter was modified to 0.32 and 0.06 m lower than the first case study.

The results for the second design were compared with the compact heat exchanger design, and they are shown in **Table 9**.

The results are similar regarding the hot outlet temperature; however, there was a significant difference between the heat transfer areas. The compact heat exchanger was designed to use fins to increase the thermal effectiveness.

	Water		Air
T_{out} (°C)	88.6		79.4
Maximum pressure drop (kPa)	6.89		0.004826
Area (m^2)		2.04	
Diameter (m)		0.32	
HTC (W/m^2 K)	12577.35		628.41
U (W/m^2 K)		105.37	
Effectiveness		0.76	
Number of rounds		11.5	
Plate length (m)	6.7		6.7

Table 8.
Results for an improvement design.

	R134a		Water	Compact HE
T_{out} (°C)	31.72		16.7	32.85
Pressure drop (kPa)	0.875		5.38	84.5
Area (m^2)		0.0929		1.46
Diameter (m)		0.120		
HTC (W/m^2 K)	69098.78		18091.41	
U (W/m^2 K)		3185.5		
Effectiveness		0.38		
Number of rounds		2.43		
Plate length (m)		0.609		

Table 9.
Results for a condenser design.

The compact heat exchanger was implemented to handle three stages: superheated phase, condensed phase and subcooling phase. The spiral plate heat exchanger has the same purpose, which is to condensate the refrigerant and cooling until the target temperature is reached. Though the compact exchanger has more heat transfer area than spiral exchanger, the compact used completely its maximum area to achieve the duty, but the spiral heat exchanger achieved the service using less heat transfer area. This feature is demonstrated by a pressure drop for R134a, because the maximum pressure drop was fixed to 6.89 kPa. If the pressure drop increased, the result will be a larger spiral exchanger. The hot outlet temperature could be minor than 31.72°C. These results validate the use of spiral plate heat exchangers as a part of a cryogenic process.

Spiral plate heat exchangers have a potential participation in the cryogenic process. Heat exchangers are 30% approximately of the total cost of the cryogenic plant [14, 15]. Only two types of heat exchangers are considered for cryogenic applications: tubes (concentric tubes and coil wounded tubes) and plates (perforated plates and plate-fin) [14]. The principal duty that must fill the heat exchangers in the refrigeration process is the high thermal efficiency. Then, spiral plate heat exchangers are capable of dealing with that situation. The thermal and hydraulic model presented in this work showed a suitable precision since the procedure does not support the traditional design such as shell and tube heat exchanger. It is possible to size spiral plate exchangers even with poor flow distribution and axial thermal behaviour and think forward to design a multi-stream spiral plate heat exchanger for cryogenic challenges.

The numerical and computational results were collected by colours which they represent a profile of the measured variables, the inlet and outlet temperatures mainly. Colour red means the hottest temperature and colour blue represents the coldest temperature. **Figure 4** describes the temperature profile for the water and the air and the arrangement of the inlet and the outlet streams. The hot liquid (water) enters at the centre of the spiral moving out along the plate, and the cold stream (air) crosses the plate width.

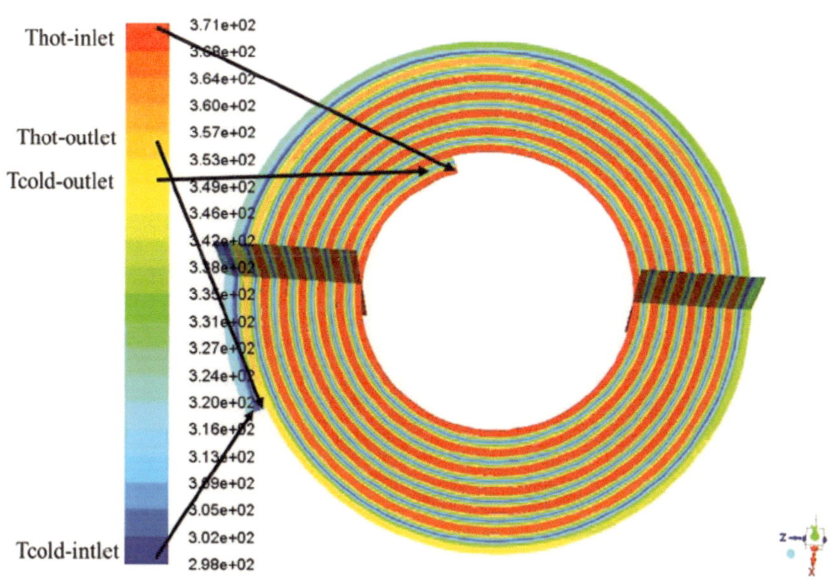

Figure 4.
The temperature profile for the cold and the hot streams.

	CFD		Method	
	Water	Air	Water	Air
T_{in}	98	20	98	20
T_{out}	86.85	75.85	88.6	79.4
Pressure drop	17.23	60	6.89	0.004826

Table 10.
Thermal and hydraulic performance.

The numerical study determined an appreciable accuracy between the method design and the computational simulation. **Table 10** shows the approximation of the outlet temperatures. The hydraulic performance was measured by calculating the outlet pressure for both streams. The results determined that the maximum pressure drop was observed at the hot section. The design method calculated a pressure drop of 6.89 kPa, and the numerical result was 17.23 kPa, because the spiral flow section has the less wide spacing between the plates. The minimum pressure drop was expected for the cold stream. The method reported 0.004826 kPa. The simulation determined a value of 3.4 kPa, due to the cross-flow section that has a spacing wider than the spiral section and furthermore because the channel is open.

5. Conclusions

This work presented two new methods to design cross-flow spiral plate heat exchangers: the first design was to compare the thermal behaviour of the spiral plate heat exchanger versus a car radiator. The radiator is an option to remove the excess of heat; however, this device needs to increase the heat transfer area by installing fins to dissipate the heat, specifically when a gas phase is involved. The constant spiral movement promotes an effective heat transfer, even when a laminar regimen is observed. An additional thermal performance can be proposed by decreasing the spacing plates. The spiral outer diameter will be reduced, and the flow velocity will increase; nevertheless, more energy to pumping flows will be necessary. The studs have not represented a constraint for the flows, but more studs could be installed with a higher diameter to increase heat transfer area and optimise the subcooling zone in spiral plate condensers. The spiral plate heat exchanger has many industrial applications, and this study contributed to expanding the usages by implementing new simple methodologies for cross-flow arrangement particularly for a condenser and for a cooler.

Computational fluid dynamics is a robust instrument to simulate and validate the empirical methods. The results between CFD simulation and the design method approve the accuracy of the method. It allows to extend the service of the spiral plate heat exchangers, as a part of the industrial process, cooling systems, heat networks and recover energy.

Conflict of interest

All authors have contributed in (1) the proposal and design or analysis and interpretation of the data, (2) drafting the article or revising it critically for important intellectual content and (3) the approval of the final version.

This manuscript has not been submitted to, nor is under review at, another journal or other publishing institutions.

DOI: http://dx.doi.org/10.5772/intechopen.85345

Nomenclature

A_c	free flow area (m^2)
A_p	plate area (m^2)
A	heat transfer area (m^2)
C	core diameter (m)
C_p	heat capacity (J/Kg K)
D_e	equivalent diameter (m)
D_h	helix diameter (m)
d_s	spacing plate (m)
F	flow (kg/hr)
h	film heat transfer coefficient (W/m^2 K)
H	plate width (m)
k	thermal conductivity (W/m K)
L	plate length (m)
Pr	Prandtl number
q	heat load (W)
Re	Reynolds number
Re_c	critical Reynolds number
s	specific gravity
T	temperature (°C)
U	overall heat transfer coefficient (W/m^2 K)
G	fluid velocity (kg/hr. m^2)
x	plate thickness (m)

Greek symbols

ΔP	pressure drop (kPa)
μ	viscosity (cp)

Subindices

h	hot side
c	cold side
in	inlet
out	outlet
f	film fluid properties
b	bulk fluid properties

Author details

Lázaro Canizalez Dávalos[1*], Edilberto Murrieta Luna[2], Mario Alberto Rodríguez Ángeles[2] and Víctor J. Cruz Delgado[3]

1 University Autonomous of Zacatecas, “Francisco García Salinas”, Zacatecas, Mexico

2 Metallurgy Engineering Department, Polytechnic University of Juventino Rosas, Santa Cruz de Juventino Rosas, Mexico

3 CONACYT – Unity of Materials, Center of Scientific Research of Yucatán, Merida, Mexico

*Address all correspondence to: lazarocada@uaz.edu.mx

References

[1] Sathiyan S, Rangarajan M, Ramachandran S. An experimental study of spiral-plate heat exchanger for nitrobenzene-water two-phase system. Bulgarian Chemical Communications. 2010;**42**(3):205-209

[2] Rajavel R, Saravanan K. An experimental study of spiral plate heat exchanger for electrolytes. Journal of the University of Chemical Technology and Metallurgy. 2008;**43**(2):255-260

[3] Khorshidi J, Heidari S. Design and construction of a spiral heat exchanger. Advances in Chemical Engineering and Science. 2016;**6**:201-208. DOI: 10.4236/aces.2016.62021

[4] Ramachandran S, Kalaichelvi P, Sundaram S. Heat transfer studies in a spiral plate heat exchanger for water–palm oil two phase system. Brazilian Journal of Chemical Engineering. 2008;**25**(3). DOI: 10.1590/S0104-66322008000300006

[5] Maruyama S, Aoki T, Igarashi K, Sakai S. Development of a high efficiency radiation converter using a spiral heat exchanger. Energy. 2005;**30**:359-371. DOI: 10.1016/j.energy.2004.04.021

[6] Wang RZ, Wu JY, Xu YX, Teng Y, Shi W. Experiment on a continuous heat regenerative adsorption refrigerator using spiral plate heat exchanger as adsorbers. Applied Thermal Engineering. 1998;**18**(1, 2):13-23. DOI: 10.1016/S1359-4311(97)00038-0

[7] Bahiraei M, Ahmadi AA. Thermohydraulic performance analysis of a spiral heat exchanger operated with water-alumina nanofluid: Effects of geometry and adding nanoparticles. Energy Conversion and Management. 2018;**170**:62-72. DOI: 10.1016/j.enconman.2018.05.061

[8] Bahiraei M, Kiani Salmi H, Safaei MR. Effect of employing a new biological nanofluid containing functionalized graphene nanoplatelets on thermal and hydraulic characteristics of a spiral heat exchanger. Energy Conversion and Management. 2019;**180**:72-82. DOI: 10.1016/j.enconman.2018.10.098

[9] Picón-Núñez M, Canizalez-Dávalos L, Medina-Flores JM. Alternative sizing methodology for compact heat exchangers of the spiral type. Heat Transfer Engineering. 2009;**30**(9):744-750. DOI: 10.1080/01457630802678508

[10] Picón-Núñez M, Canizalez-Dávalos L, Martínez-Rodriguez G, Polley GT. Shortcut design approach for spiral heat exchangers. IChemE Food and Bioproducts Processing. 2007;**85**(C4):322-327. DOI: 10.1205/fbp07073

[11] Nguyen D-K, San J-Y. Effect of solid heat conduction on heat transfer performance of a spiral heat exchanger. Applied Thermal Engineering. 2015;**76**:400-409. DOI: 10.1016/j.applthermaleng.2014.11.021

[12] Bes TH, Roetzel W. Distribution of heat flux density in spiral heat exchangers. International Journal of Heat and Mass Transfer. 1992;**35**(6):1331-1347. DOI: 10.1016/0017-9310(92)90026-O

[13] Capata R, Zangrillo E. Preliminary design of compact condenser in an organic rankine cycle system for the low grade waste heat recovery. Energies. 2014;**7**:8008-8035. DOI: 10.3390/en7128008

[14] Pacio JC, Dorao CA. Review on heat thermal hydraulic models for cryogenic applications. Cryogenics. 2011;**51**:

366-379. DOI: 10.1016/j.cryogenics. 2011.04.005

[15] Fredheim AO, Heiersted RS. Possibilities for cost reductions in base-load. In: EUROGAS96. In: Proceedings from the European Applied Research Conference on Natural Gas. 1996. pp. 101-114

[16] Minton PE. Designing Spiral-Plate Heat Exchangers. Union Carbide; 1970. pp. 127-136

[17] Sander J. Rosenblads AB. Patenter, Stockholm, Sweden, 1955; unpublished

[18] Dongwu W. Geometric calculations for the spiral heat exchanger. Chemical Engineering and Technology. 2003;**26** (5). DOI: 10.1002/CEAT.200390091

[19] Bes TH, Roetzel W. Thermal theory of the spiral heat exchanger. International Journal of Heat and Mass Transfer. 1993;**36**(3):765-773. DOI: 10.1016/0017-9310(93)80052-V

[20] Carlos Alberto Romero Piedrahita, Yamid Alberto Carranza Sánchez. Rating of thermal capacity of a heat radiator. Scientia et Technica Año XIII, No. 35, Agosto de 2007. ISSN 0122-1701

Chapter 12

Air Cooling in Steam Plant Condenser Using Refrigeration System for Improving Vacuum Pump Performance

Ahmed Hegazy and Abraham Engeda

Abstract

Modifying the steam plant condenser by incorporating a vapor compression refrigeration system, which is utilized to chill the water used for cooling the condenser air cooler, is proposed in this work. Hence a considerable drop in air temperature is obtained, and more steam associated with air is condensed resulting in reducing both steam lost and vacuum pump power. A thermodynamic analysis is developed for working out the performance of the proposed hybrid system. The results obtained using this analysis showed that for a condensate subcooling of 3°C the total work saving due to this modification has maximums of 38.5, 33.9 and 28.9% and occur at temperature reduction below the condensate temperature of 12, 10, and 8°C when the temperature of steam admitted to the condenser is 20, 30, and 40°C, respectively. The savings in steam lost through the vacuum pump in these cases amount to 87.7, 84, and 79.2%, respectively. It is recommended to select the decrease in condenser air temperature in the range of 4°C higher than those decreases, at which the minimum total work comes about, as the saving in steam lost becomes fairly greater (5–7%), while the maximum increase in minimum total work is 3%.

Keywords: steam plant condenser, vapor compression refrigeration system, condenser air cooler, vacuum pump, power saving

1. Introduction

The primary purpose of the steam condenser in a steam plant is to convert the turbine exhaust steam into water for reuse in the boiler and to create and maintain vacuum at the turbine exhaust. This is achieved by cold water circulation through tube bundles and condensing the steam on their outer surfaces. Referable to the large reduction in the specific volume on changing from steam to condensate, a vacuum is formed within the condenser. Getting to this vacuum continuously as minimum as possible is very important for the steam plant to attain the highest possible power from the turbine. A rise in the condenser pressure can be produced due to the existence of air, ammonia, and other noncondensable gases [1]. This rise in pressure in the condenser results in lowering the useful turbine power and in turn the plant efficiency degenerates [2–5].

The presence of air lowers the partial pressure of steam and therefore brings down the saturation temperature of steam, which leads to increasing evaporation enthalpy (latent heat), and therefore, more cooling water will be required in the condenser. In addition, this gives rise to increasing condensate subcooling, which is undesirable because it means that the excess heat removed for this purpose serves no useful process and entails providing additional boiler firing [6]. Thus, continuous removal of the air and noncondensable gases from the condenser is very critical for maintaining the plant efficiency possibly high. Beside this task, the air extraction system prevents air blanketing of condenser tubing that could dramatically reduce the heat transfer and stop the condensing process. Also, it reduces the condensate-dissolved oxygen levels that could lead to corrosion of boiler tubing.

The air and noncondensables are extracted from the bottom of the condenser, where the temperature is the lowest, by using vacuum pump. The air and noncondensables sucked by the vacuum pump contain a relatively big amount of noncondensed steam; each kilogram of air can contain more than 2 kg of water vapor [7]. This results in increasing the parasitic work of the pump. In condenser design, it is considered reducing the subcooling of the condensate possibly to the lowest degree to minimize heat removal. For reducing the amount of steam mixed with the air sucked by the air pump, the steam condenser has a portion of tubes near the air pump suction screened off and the condenser tubes in this portion contain the coldest water, and the number of cooling tubes is increased. Hence, the temperature of the air drops below that of the outlet condensate, and more steam associated with the air is condensed so that the mass flow rate of the air and steam mixture drawn by the air pump is reduced and, in turn the power of the dry pump is lowered.

The air extraction system (vacuum pump) can be centrifugal compressor, steam jet ejector air ejector, or water ring pump [8, 9]. A hybrid system of these components is preferred as they offer several advantages toward minimizing venting equipment size and save parasitic consumption [7]. The power required for operating these equipment sorts is dependent on both the rates of air and the noncondensables and noncondensed steam sucked by the vacuum pump. The temperature decrease of the air in the condenser air cooler is limited when the normal cooling water is used. For obtaining considerable drop in air temperature it is proposed in this work to modify the condenser by incorporating a refrigeration system into the condenser. This system is utilized to chill the cooling water, which is used for cooling the air in the condenser rather than the normal cooling water. Of course the use of the refrigerant directly to cool the air exploiting the air cooler as an evaporator for the refrigeration system and circulating the cold liquid refrigerant of this system through the tubes of the air cooler will be more efficient than using the chilled water. However, this configuration calls for redesigning the air cooler to be adapted for receiving and evaporating the refrigerant, which will be expensive for already existing power plants. The latest configuration can be taken into consideration for the design of the new plants.

The concept of using dry ammonia for cooling steam plant condenser is described in Ref. [10]. This concept states that the condenser cooling water is replaced by liquid ammonia, which evaporates as it acquires heat from the condensing steam. The heat of condensation absorbed by the ammonia is rejected in an air cooled condenser of a refrigeration machine into the surrounding atmosphere. This concept was tested and well documented [11–17], with the participation of several major equipment vendors (Baltimore Air Coil, the Trane Company, Curtiss-Wright, CB&I, and Union Carbide). In addition, the concept of using vapor compression refrigeration system (VCRS) combined with steam plant condenser was studied in detail in [14–19]. The results of these studies showed significant rise in net power as well as efficiency of the steam plant. However, for realizing this combined system, a relatively huge VCRS is required, which leads to material increase of the combined system cost. In the current study, a relatively small VCRS is required as it serves to cool only the air contained in the air cooler of steam plant condenser. This leads to lessening the temperature of the air and water vapor mixture in the air cooler, resulting in condensing some of the water vapor. Hence, the mass flow rate of the air and water vapor mixture extracted from the steam condenser is decreased and consequently the power consumed by the vacuum pump lessens. Meanwhile, water vapor leaving the steam condenser and lost to the surroundings is reduced. In general, this contributes to contracting both the power consumed in the auxiliary systems of the steam plant and the steam consumption.

2. Conception of the modified steam plant condenser

The steam plant condenser is typically a shell and tubes heat exchanger. **Figure 1** depicts a section through a steam plant condenser. The steam leaving the plant turbine enters the condenser at its top with temperature T_{ci} and flows around the outer surface of the cooling tubes in which cooling water is circulated. Heat is transferred from the steam into the cooling water where steam condenses and the condensate is extracted out of the condenser at the bottom with temperature T_{ce}. A segment of the condenser (called air cooler) is shielded for cooling the air associated with the condensate and leaked from the environment. This segment is possibly located near the inlet of cooling water, and the number of the cooling tubes is increased in it, so that the air associated with some water vapor collected in this section is cooled to a temperature T_{vpi}, which is lower than that of the condensate. This will lead to decreasing the mass rate of water vapor, the air pulled by the vacuum pump connected to this segment. As a result, the energy used up by the vacuum pump for extracting the air and its accompanying water vapor to the environment are decreased. In the current study, it is suggested to circulate chilled

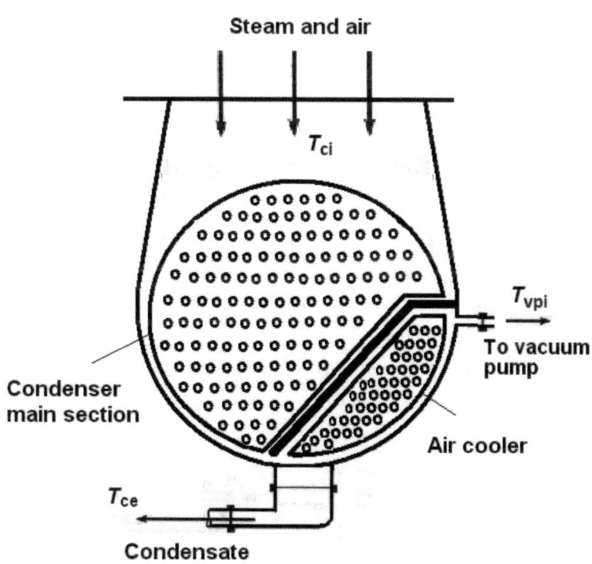

Figure 1.
A schematic section through steam plant condenser.

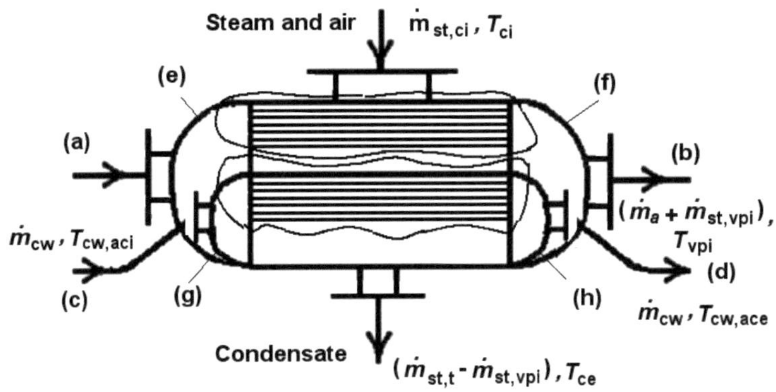

Figure 2.
A configuration of the modified steam plant condenser. a, normal cooling water in; b, normal cooling water out; c, chilled water in; d, chilled water out; e–h, headers.

water within the cooling pipes of the cooling air section, so that the temperature of the air and accompanying steam is further reduced.

Figure 2 illustrates a configuration for the modified steam plant condenser that enables the use of chilled water to lower the temperature of the air cooler section. The cooling water (a) coming from a water body (river, sea, ...) or a cooling tower is fed into the header (e) where it is distributed to the most of the cooling water tubes of the condenser. It helps to cool the steam exhausting out of the plant turbine. This cooling water is called here normal cooling water. While being heated up, it exits the cooling tubes into the header (f). The warmer normal cooling water (b) is extracted from the header (f) and pumped back to the water body or the cooling tower. The chilled water (c) is fed to the header (g) and it flows through the cooling tubes of the air cooling part of the steam condenser. This chilled water acts as a coolant for air along with some water vapor collected in the air cooler. This leads to cooling the mixture of the air and water vapor to a temperature lower than that of the condensate. The temperature of the chilled water is raised as it flows through the cooling tubes due to heat transfer from the mixture of air and steam in the air cooler. As a result, some of the steam contained in the mixture is condensed and flows down the condenser to the hot well. The warm chilled water goes into the header (h), from which it is transmitted backward to a chiller. Finally, the rest of the air and water vapor mixture in the air cooler is sucked by a vacuum pump and expelled into the atmosphere. The vacuum pump is not shown in **Figure 2**.

3. Description of the studied combined system

In the current study, a vapor compression refrigeration system (VCRS) is used to chill the cooling water used for cooling the air cooler of the above described steam condenser. Hence, **Figure 3** shows a schematic of the studied combined system. The refrigeration cycle is made up of a two-stage compression system (c) and (d) (these are also designated by I and II, respectively); a refrigerant condenser (e); two throttling valves (f) and (h); a flash tank (g) for intercooling and flash gas removal; a liquid-line/suction-line heat exchanger (LLSL-HE) (b); and an evaporator (a), which acts at the same time as a heat exchanger for chilling the cooling water used in the air cooler. In this system, the low-pressure refrigerant leaving the evaporator (a) is heated in the liquid suction heat exchanger (b) as it

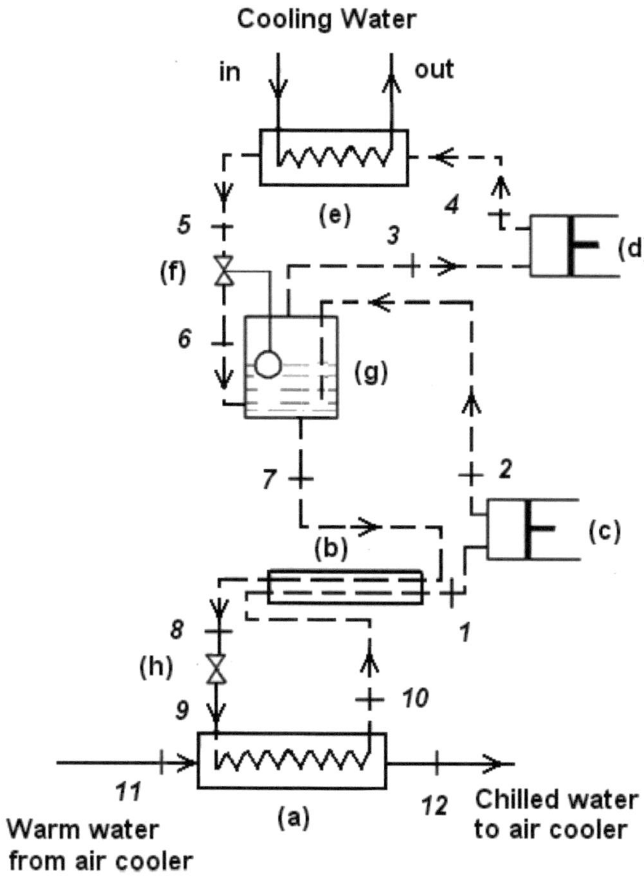

Figure 3.
A schematic of the combined proposed steam plant condenser and vapor compression refrigeration system. _____, water; _ _ _, refrigerant; a, refrigerant evaporator/HE; b, liquid suction HE; c, refrigerant compressor I; d, refrigerant compressor II; e, refrigerant condenser; f, throttle valve I; g, flash chamber; h, throttle valve II.

absorbs heat from the higher-temperature liquid refrigerant coming from the flash chamber (g). It is drawn by the first-stage compressor (c) where its pressure is raised to the intermediate pressure of the flash chamber. It is sent to the flash chamber (g) in which it is mixed with the refrigerant coming out of the throttle valve (f) and it gets cooled. The refrigerant vapor is separated from the liquid refrigerant and it is drawn by the second-stage compressor (d) and its pressure is elevated to the condenser pressure. The high-pressure refrigerant leaving the compressor (d) flows through the refrigerant condenser (e) where it is condensed and it streams further to the throttle valve (f) where its pressure is reduced to the intermediate pressure of the flash chamber, and it is fed into the flash chamber. The liquid refrigerant leaving the flash chamber is fed to the liquid suction HE (b). Here it is cooled as it gives a portion of its sensible heat to the refrigerant vapor departing the evaporator (a). The cooled liquid refrigerant is reduced in pressure to the evaporator pressure on running through the throttle valve (h). The refrigerant is then run to the evaporator/water chiller to cool down the warm water coming from the air cooler. The low-pressure vapor refrigerant leaving the evaporator/water chiller (a), completes the refrigeration cycle. This refrigeration cycle has been selected among the available refrigeration cycles due to its simplicity and relatively high efficiency.

4. Thermodynamic analysis of the studied combined system

For the following analysis of steam and air flow through the plant condenser, it is assumed that both fluids are ideal gases. Since the steam and air flow inside the condenser are at relatively very low pressure, they can be considered behaving to acceptable accuracy as ideal gases. Referring to **Figures 1** and 2, the steam exhausting the plant turbine (it is usually slightly wet) enters the condenser with temperature T_{ci} and dryness fraction x_{ci}. Knowing the temperature T_{ci}, both the saturation pressure $p_{s,ci}$ (it equals the partial pressure of the steam) and specific volume $v_{s,ci}$ of the saturated steam at condenser inlet can be fixed. Given the steam mass flow rate $\dot{m}_{st,t}$ entering the turbine, the mass flow rate $\dot{m}_{st,ci}$ of the steam at condenser entry can be calculated by:

$$\dot{m}_{st,ci} = x_{ci}\,\dot{m}_{st,t} \tag{1}$$

The volume flow rate $\dot{V}_{st,ci}$ at inlet to the condenser, by Dalton's law, is equal to the volume flow rate $\dot{V}_{a,ci}$ of the associated air [20] and is obtained from the following equation:

$$\dot{V}_{st,ci} = \dot{V}_{a,ci} = \dot{m}_{st,ci}\,v_{s,ci} \tag{2}$$

The partial pressure $p_{a,ci}$ of the air at entry to the condenser is calculated by:

$$p_{a,ci} = \frac{\dot{m}_a R_a (T_{ci} + 273.15)}{\dot{V}_{a,ci}} \tag{3}$$

Inserting Eqs. (1) and (2) into Eq. (3) and designating the mass ratio $\dot{m}_a/\dot{m}_{st,t}$ by β, it follows that:

$$p_{a,ci} = \frac{\beta R_a (T_{ci} + 273.15)}{x_{ci}\,v_{s,ci}} \tag{4}$$

The total pressure $p_{c,t}$ at condenser entry is equal to the sum of the partial pressure $p_{a,ci}$ of the air and the saturation pressure $p_{s,ci}$ of the steam entering the condenser. It is taken constant throughout the condenser, since the velocity of steam flow is small. Hence, the total absolute pressure $p_{c,t}$ inside the condenser is given as:

$$p_{c,t} = p_{s,ci} + p_{a,ci} \tag{5}$$

The condensate temperature T_{ce} at condenser outlet is usually ΔT_{ce} lower than the steam temperature T_{ci} at condenser inlet (i.e., $T_{ce} = T_{ci}$-ΔT_{ce}). Knowing the temperature T_{ce}, the saturation pressure $p_{s,ce}$ and specific volume $v_{s,ce}$ of the steam corresponding to this temperature can be determined. If the condenser is not screened, then the partial pressure $p_{a,ce}$ of the air leaving the steam condenser with condensate is given as:

$$p_{a,ce} = p_{c,t} - p_{s,ce} \tag{6}$$

The volume flow rate $\dot{V}_{a,ce}$ of the air to be dealt by the vacuum pump at the exit of the steam condenser can be determined by:

$$\dot{V}_{a,ce} = \frac{\dot{m}_a R_a (T_{ce} + 273.15)}{p_{a,ce}} \tag{7}$$

Hence, the mass flow rate $\dot{m}_{\text{st, ce}}$ of steam associated with the air is assessed by:

$$\dot{m}_{\text{st, ce}} = \frac{\dot{V}_{\text{a, ce}}}{v_{\text{s, ce}}} \tag{8}$$

Dividing both sides of Eq. (8) by $\dot{m}_{\text{st, t}}$ and inserting Eq. (7) into Eq. (8), it follows that:

$$\gamma = \frac{\dot{m}_{\text{st, ce}}}{\dot{m}_{\text{st, t}}} = \frac{\beta R_a \left(T_{\text{ce}} + 273.15\right)}{p_{\text{a, ce}} v_{\text{s, ce}}} \tag{9}$$

In the case of screening the steam condenser and cooling the air, the temperature T_{vpi} of the air and water vapor mixture sucked by the vacuum pump is ΔT_{vpi} lower than the condensate temperature T_{ce} at steam condenser outlet (i.e., $T_{\text{vpi}} = T_{\text{ce}} - \Delta T_{\text{vpi}}$). Knowing the temperature T_{vpi}, the corresponding saturation pressure $p_{\text{s,vpi}}$ and specific volume $v_{\text{s,vpi}}$ can be fixed. The partial pressure $p_{\text{a,vpi}}$ and volume flow rate $\dot{V}_{\text{a, vpi}}$ of the air and mass ratio δ $\left(\dot{m}_{\text{st, vpi}}/\dot{m}_{st,t}\right)$ at inlet of the vacuum pump can be determined using Eqs. (6), (7) and (9), respectively, by replacing the subscripts ce by vpi and γ by δ. Accordingly, it follows that $p_{\text{a,vpi}}$, $\dot{V}_{\text{a, vpi}}$, and δ are given by:

$$p_{\text{a, vpi}} = p_{c,t} - p_{\text{s, vpi}} \tag{10}$$

$$\dot{V}_{\text{a, vpi}} = \frac{\dot{m}_a R_a \left(T_{\text{vpi}} + 273.15\right)}{p_{\text{a, vpi}}} \tag{11}$$

and

$$\delta = \frac{\dot{m}_{\text{st, vp}}}{\dot{m}_{\text{st, t}}} = \frac{\beta R_a \left(T_{\text{vpi}} + 273.15\right)}{p_{\text{a, vpi}} v_{\text{s, vpi}}} \tag{12}$$

Regarding the VCRS, the p-h diagram of its cycle is illustrated in **Figure 4**. The numerals of **Figure 4** correspond to the points 1–10 given in **Figure 3**. It is to be considered here that the refrigerant condenser is cooled exactly as it is conducted with the steam plant condenser. Consequently, it is assumed here that the refrigerant leaving the refrigerant condenser has a temperature equal to that of the condensate in the steam plant condenser (i.e., $T_5 = T_{\text{ce}}$). Considering this fact and

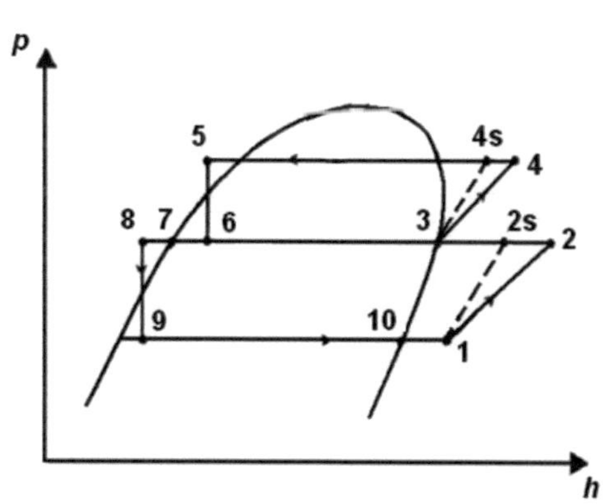

Figure 4.
p-h *diagram of the VCRS cycle of the studied refrigeration system.*

knowing the subcooling $\Delta T_{rc,sub}$ of the refrigerant condenser, the pressure of the refrigerant in the condenser (e) can be fixed. The evaporator temperature $T_9 = T_{10}$ is defined according to the temperature required by the chilled water at inlet to the air cooler of the steam condenser. T_9 as well as T_{10} is ΔT_e less than the temperature $T_{cw,aci}$ (i.e., $T_9 = T_{10} = T_{cw,aci} - \Delta T_e$) of the chilled water entering the steam condenser air cooler. Hence, the pressure of the refrigerant in the evaporator is the saturated one corresponding to the temperature T_9/T_{10}. Knowing the refrigerant pressures in the evaporator (a) and condenser (e), the adiabatic efficiencies $\eta_{rco,I}$ and $\eta_{rco,II}$ and mechanical efficiencies $\eta_{m,rco,I}$ and $\eta_{m,rco,II}$ of the compressor I (c) and II (d), respectively, and the effectiveness ε_{LLSL} of the LLSL-HE, the states of the different points of the VCRS cycle can be determined as explained in any refrigeration text book (e.g., [21]).

For obtaining the value of the mass ratio ζ $(\dot{m}_{cw}/\dot{m}_{st,t})$ of the chilled water flow rate through the air cooling segment and steam flow rate crossing the turbine, **Figure 5** shows the mass flow rates of air mixture and chilled water, and their temperatures through this segment. An energy balance for the air cooler leads to the following equation:

$$\dot{m}_a c_{p,a} \left(T_{ce} - T_{vpi}\right) + \left(\dot{m}_{st,aci} - \dot{m}_{st,vpi}\right) LH \left[at\ saturation\ pressure\, p_{s,vpi}\right] = \dot{m}_{cw} C_w \left(T_{cw,ace} - T_{cw,aci}\right) \tag{13}$$

It is to be noticed here that the sensible heat of the condensed steam in the air cooler was neglected as it is relatively very small, and $\dot{m}_{st,aci}$ is equal to $\dot{m}_{st,,ce}$.

Solving Eq. (13) to get $\dot{m}_{cw}/\dot{m}_a$ and multiplying both sides of the resulted equation by $\dot{m}_a/\dot{m}_{st,t}$, the ratio ζ $(\dot{m}_{cw}/\dot{m}_{st,t})$ is obtained as:

$$\zeta = \frac{\dot{m}_{cw}}{\dot{m}_{st,t}} = \frac{\beta c_{p,a}\left(T_{ce} - T_{vpi}\right) + (\gamma - \delta) x LH \left[at\ saturation\ pressure\, p_{s,vpi}\right]}{c_w \left(T_{cw,ace} - T_{cw,aci}\right)} \tag{14}$$

For determining the mass ratio $\dot{m}_{r,I}/\dot{m}_{cw}$ of the refrigerant and cooling water, a heat balance is performed for the evaporator/water HE (a), which yields to:

$$\dot{m}_{r,I}\left(h_{10} - h_9\right) = \dot{m}_{cw} C_w \left(T_{cw,ace} - T_{cw,aci}\right) \tag{15}$$

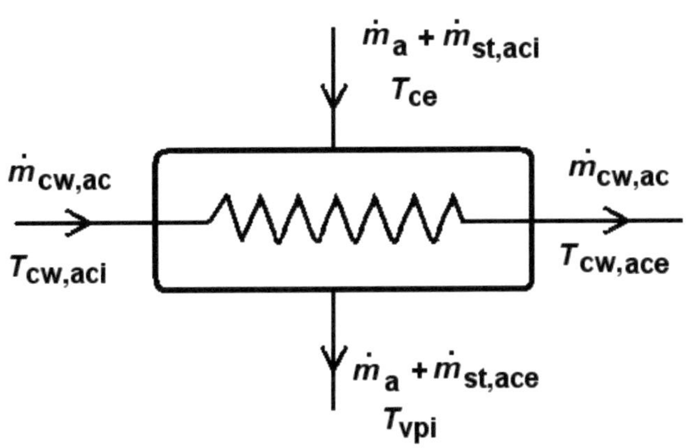

Figure 5.
Mass flow and temperature of air, steam, and chilled water through the air-cooling segment.

From which it follows that:

$$\frac{\dot{m}_{\mathrm{r,I}}}{\dot{m}_{\mathrm{cw}}} = \frac{C_w\left(T_{\mathrm{cw,ace}} - T_{\mathrm{cw,aci}}\right)}{(h_{10} - h_9)} \tag{16}$$

Multiplying both sides of Eq. (16) by the mass ratio $\dot{m}_{cw}/\dot{m}_{\mathrm{st,t}}$, the mass ratio ξ_I $(\dot{m}_{\mathrm{r,I}}/\dot{m}_{\mathrm{st,t}})$ is worked out as:

$$\xi_I = \frac{\dot{m}_{\mathrm{r,I}}}{\dot{m}_{\mathrm{st,t}}} = \zeta\frac{C_w\left(T_{\mathrm{cw,ace}} - T_{\mathrm{cw,aci}}\right)}{(h_{10} - h_9)} \tag{17}$$

The mass ratio ζ is obtained from Eq. (14). A heat balance of the flash chamber (**Figure 3**) yields:

$$\dot{m}_{\mathrm{r,II}}h_6 + \dot{m}_{\mathrm{r,I}}h_2 = \dot{m}_{\mathrm{r,I}}h_7 + \dot{m}_{\mathrm{r,II}}h_3 \tag{18}$$

It follows from Eq. (18) that:

$$\frac{\dot{m}_{r,II}}{\dot{m}_{r,I}} = \frac{(h_2 - h_7)}{(h_3 - h_6)} \tag{19}$$

Multiplying both sides of Eq. (19) by $\dot{m}_{r,II}/\dot{m}_{r,I}$, it follows that the mass ratio ξ_{II} $(\dot{m}_{r,II}/\dot{m}_{st,t})$ of the refrigerant flow rate through the compressor II to the steam flow rate of the steam turbine is given by:

$$\xi_{II} = \frac{\dot{m}_{\mathrm{r,II}}}{\dot{m}_{\mathrm{st,t}}} = \frac{\dot{m}_{\mathrm{r,II}}}{\dot{m}_{\mathrm{r,I}}}x\frac{\dot{m}_{\mathrm{r,I}}}{\dot{m}_{\mathrm{st,t}}} = \xi_I\frac{\dot{m}_{\mathrm{r,II}}}{\dot{m}_{\mathrm{r,I}}} = \xi_I\frac{(h_2 - h_7)}{(h_3 - h_6)} \tag{20}$$

From Eqs. (17) and (20), the mass ratio ξ_t can be found as:

$$\xi_t = \frac{\dot{m}_{\mathrm{r,t}}}{\dot{m}_{\mathrm{st,t}}} = \xi_I x \xi_{II} = \zeta\frac{C_w\left(t_{\mathrm{cw,ace}} - t_{\mathrm{cw,aci}}\right)}{(h_{10} - h_9)}\left[1 + \frac{(h_2 - h_7)}{(h_3 - h_6)}\right] \tag{21}$$

$$\xi_t = \frac{\dot{m}_{\mathrm{r,t}}}{\dot{m}_{\mathrm{st,t}}} \tag{22}$$

The characteristic parameters describing the performance of the refrigeration cycle are given as:

$$q_{\mathrm{rc}} = \xi_{II}(h_4 - h_5) \tag{23}$$

$$w_{\mathrm{rco,I}} = \frac{\xi_I}{\eta_{\mathrm{m,co,I}}}(h_2 - h_1) = \frac{\xi_I}{\eta_{\mathrm{m,rco,I}}\eta_{\mathrm{i,rco,I}}}(h_{2s} - h_1) \tag{24}$$

$$w_{\mathrm{rco,II}} = \frac{\xi_{II}}{\eta_{\mathrm{m,rco,II}}}(h_4 - h_3) = \frac{\xi_{II}}{\eta_{\mathrm{m,rco,I\ I}}\eta_{\mathrm{i,rco,II}}}(h_{4s} - h_3) \tag{25}$$

$$w_{\mathrm{rco,t}} = (w_{\mathrm{rco,I}} + w_{\mathrm{rco,II}}) \tag{26}$$

$$q_e = \xi_{II}(h_{10} - h_9) \tag{27}$$

$$\mathrm{COP} = \xi_{II}(h_{10} - h_9)/\left[\xi_I(h_{2s} - h_1)/\eta_{\mathrm{co,I}}\eta_{\mathrm{m,co,I}} + \xi_{II}(h_{4s} - h_3)/\eta_{\mathrm{co,I}}\eta_{\mathrm{m,co,II}}\right] \tag{28}$$

The hybrid system proposed in the current work can lead to two benefits, the first benefit is a decrease in the mass flow rate of steam lost in venting the air from the SPC, and the second one is a reduction in mass flow rate of air and steam mixture drawn by the vacuum pump and hence the pump power is lowered.

However, a new energy consumption comes out, which is the total work ($w_{rco,t}$) of the two compressors employed in the VCRS. To be capable to judge the goodness of the proposed hybrid system, both the energies used for venting the air from SPC and for operating the compressors of the refrigeration system should be determinable. The total work $w_{rco,t}$ of the two refrigerant compressors can be determined using Eqs. (24)–(26). As for the work consumed for venting process, there are some vacuum pumps that can be utilized for this function, among which centrifugal compressor is the most effective and efficient instrument for performing this task. It is selected here only for the sake of judgment of the hybrid system goodness. In venting the air and steam mixture out of the steam condenser without air cooling, the specific vacuum pump (compressor) work $w_{vp,wac}$ referred to each kilogram of steam flow through the steam turbine can be expressed by aid of any thermodynamics text book (e.g., [20]) as:

$$w_{vp,wac} = \frac{k_m R_m T_{ce} (\beta + \gamma)}{(k_m - 1)\, \eta_{m,vp} \eta_{i,vp}} \left[\left(\frac{p_{atm}}{p_t} \right)^{\left(\frac{k_m - 1}{k_m}\right)} - 1 \right] \quad (29)$$

where k_m and R_m are the isentropic exponent and gas constant, respectively, of the air steam mixture. They are given by [20]:

$$k_m = \frac{\beta C_{p,a} + \gamma C_{p,st}}{\beta C_{v,a} + \gamma C_{v,st}} \quad (30)$$

$$R_m = \frac{\beta R_a + \gamma R_{st}}{\beta + \gamma} \quad (31)$$

When the steam condenser is fitted with air cooler, the air extracting compressor work w_{vp} is given by:

$$w_{vp,ac} = \frac{k_m R_m T_{vpi} (\beta + \delta)}{(k_m - 1)\, \eta_{m,vp} \eta_{i,vp}} \left[\left(\frac{p_{atm}}{p_t} \right)^{\left(\frac{k_m - 1}{k_m}\right)} - 1 \right] \quad (32)$$

k_m and R_m are calculated in this case by aid of Eqs. (30) and (31), respectively, by replacing γ by δ.

The total specific work w_t ($w_{rco,t} + w_{vp,ac}$) employed for cooling and driving out steam condenser air is calculated by summing up Eqs. (26) and (32).

5. Results and discussion

The thermodynamic analysis developed in Section 4 for predicting the condensation rate in the steam plant condenser due to air cooling is first validated with the experimental data of reference [22]. In this work, experiments were conducted in a 2-m-long square cross-sectional channel (0.34 m × 0.34 m) to study the heat and mass transfer in the condensation of water vapor from humid air. The air flowing inside the channel was cooled by cold water flowing outside and adjoining only one side (0.34 m × 2 m) of the channel. Experimental data were obtained from five tests at various operating conditions as shown in **Table 1**. The thermodynamic analysis in Section 4 was slightly modified to be adapted for using the data given in **Table 1** for computing the condensation rate. For solving the equations in this analysis and finding out the rate of condensation, the commercial computer package EES [23] was used. The thermal properties of the humid air at different conditions were found using the built-in functions available in the EES package. The condensation

Test	Inlet humid air temperature $T_{ha,in}$ (°C)	Velocity of humid air v_{ha} (m/s)	Inlet relative humidity of humid air φ	Average cooling flux $q_{c,av}$ (kW/m^2)
1	82.66	1.46	100	7.3
2	80.61	2.02	100	9.0
3	79.13	2.52	97.83	10
4	78.73	3.01	87.35	11.1
5	75.02	3.59	96.55	12.5

Table 1.
Experimental conditions from Ref. [22].

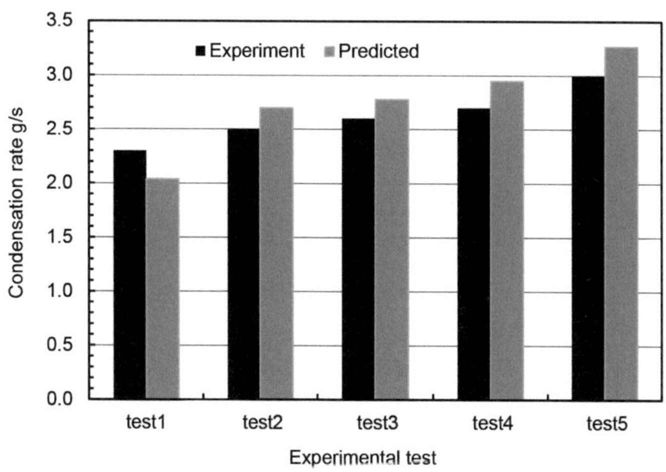

Figure 6.
Comparison of the predicted and experimental values of condensation rate for all the tests of Ref. [22].

rate computed from the present model is displayed in **Figure 6**, which shows satisfactory agreement with the experimental data of Ref. [22] since the maximal discrepancy does not exceed 10%.

The thermodynamic analysis developed in Section 4 for performance prediction of the combined system proposed in this work necessitates knowing some basic design and operational data, which is listed in **Table 2**. It is to be noticed here that the values of isentropic and mechanical efficiencies of the compressors involved in this study have been selected close to the practical values of compressors in use in industry [24]. The results presented hereafter are based on these data. All parameter values given in **Table 2** will be kept unchanged except for the case where the effect of a specific parameter is to be examined; it is handled as a variable.

The refrigerant of the refrigeration system is selected to be ammonia. The physical properties of air, water, steam, and ammonia needed for computation are predicted using the built-in functions of the commercial computing package EES [23], which is used for solving the equations of the analysis of Section 4.

In **Figure 7**, the mass ratio γ is plotted versus the temperature T_{ci} for values of ΔT_{ce} of 1, 3, and 5°C. It is seen from **Figure 7** that γ runs linearly with very low rate with T_{ci}. This can be explained as follows: since β is constant, the mass rate of steam condensed because cooling the air depends mainly on ΔT_{ce} and it is very little dependent on T_{ci}. Of course the condensed steam rate in the air cooler is a pit higher at higher T_{ci}. For constant β and T_{ci}, the amount of steam associated with the

Parameter	Value
Mass ratio (β)	0.0003
Compressor isentropic efficiency ($\eta_{i,rco,I}$, $\eta_{i,rco,II}$, $\eta_{i,vp}$)	0.85
Compressor mechanical efficiency ($\eta_{m,rco,I}$, $\eta_{m,rco,II}$, $\eta_{m,vp}$)	0.75
Dryness fraction of steam entering the SPC (x_{ci})	0.9
Effectiveness of the liquid suction HE (ε_{LLSL})	0.8
Temperature difference ($T_{ci} - T_5$), °C	0
Temperature difference ($T_{cw,ace} - T_{cw,aci}$), °C	4
Temperature difference ($T_{cw,aci} - T_{10}$), °C	4
Temperature difference ($T_{cw,ace} - T_{vpi}$), °C	4
Temperature difference ΔT_{ce}, °C	3
Subcooling of the refrigerant condenser ($\Delta T_{rc,sub}$), °C	3

Table 2.
Basic design and operational data of the proposed combined steam plant condenser and refrigeration system.

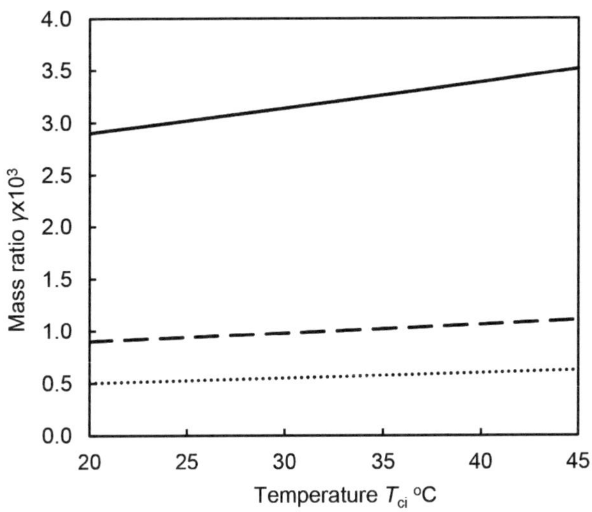

Figure 7.
Effect of condensate temperature on the extracted steam mass rate by the vacuum pump without air cooler.
———, ΔT_{ce} = 1°C; — — —, ΔT_{ce} = 3°C; ...., ΔT_{ce} = 5°C.

condenser air and, in turn, the rate of condensed steam decrease progressively with ΔT_{ce}. This accounts for the remarkable drop in γ with an increase in ΔT_{ce} as shown in **Figure 7**. In contrast, the amount of steam associated with the condenser air and consecutively the rate of condensed steam for constant β and ΔT_{ce} are almost unvarying with T_{ci}. This explains the very low rate of increase in γ with rising T_{ci}.

It is worth noting here that in **Figures 8–14**, which will be displayed in this section, the small values of temperature difference ΔT_{vpi} close to zero represent the case in which there is virtually no air cooler is employed. These values are not practical as the refrigeration system will be useless. Yet these values are included in these images just for illumination. In **Figure 8**, the mass ratio δ is drawn against the temperature difference ΔT_{vpi} for temperature T_{ci} of 20, 30, and 40°C. It is seen from **Figure 8** that δ declines with an increase in ΔT_{vpi} where the rate of declination is relatively high at small values of ΔT_{vpi} and it decreases progressively with ΔT_{vpi}

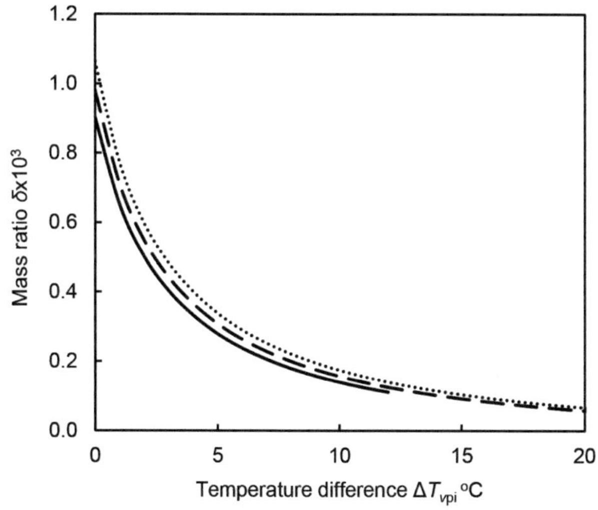

Figure 8.
Dependence of the mass of steam associated with air on the air temperature at vacuum pump entrance. ———, T_{ci} = 20°C; — — —, T_{ci} = 30°C; ………, T_{ci} = 40°C.

and becomes immaterially small for values of ΔT_{vpi} greater than 12°C. This is brought about due to the large drop in steam content in the air and likewise the rate of steam condensed as the temperature T_{vpi} falls down. It is also seen from **Figure 8** that T_{ci} has almost negligible effect on δ as the amount of steam mixed with the condenser air and in turn the rate of condensed steam for constant β and ΔT_{vpi} are almost invariable with T_{ci}.

Figure 9 illustrates the saving percentage $(\gamma - \delta) \times 100/\gamma$ in the steam amounts to be condensed in the air cooler and not sucked by the vacuum pump as steam. It is clear from **Figure 9** that this saving has a reversed trend to that of the mass ratio δ, it is equal to zero at $\Delta T_{vpi} = 0$, and it increases steeply with ΔT_{vpi}. The rate of increase in this saving with ΔT_{vpi} falls increasingly with the rise in ΔT_{vpi} where it becomes inconsiderably small at ΔT_{vpi} of 12°C.

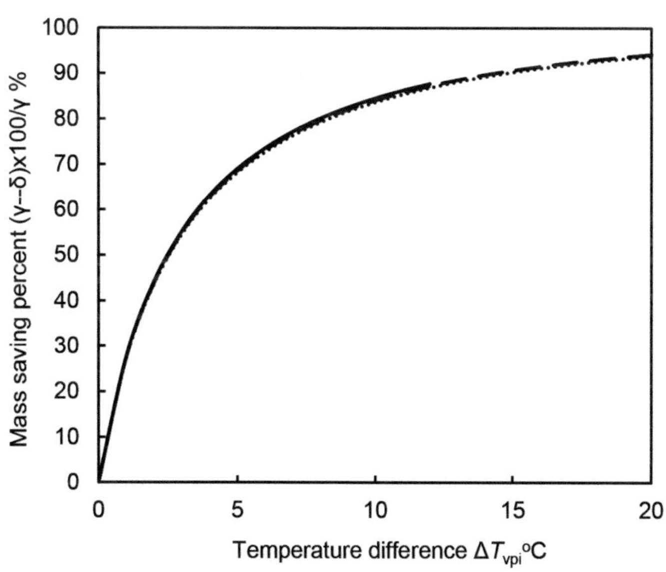

Figure 9.
Saving percentage in steam mass associated with air when using air cooler. ———, T_{ci} = 20°C; — — —, T_{ci} = 30°C; ……… T_{ci} = 40°C.

It follows from **Figures 8** and **9** that the amount of steam to be condensed in the air cooler is relatively high at small values of ΔT_{vpi} and the rate of increase in this amount falls progressively with ΔT_{vpi}. Therefore, the amount of the cooling chilled water and in turn the refrigerant needed for chilling the cooling water takes the same trend of the percentage saving $(\gamma - \delta) \times 100/\gamma$ (see **Figure 9**). **Figures 10** and **11** show the mass ratios ζ and ξ_t, respectively, as a function of ΔT_{vpi}.

In **Figure 12**, the coefficient of performance COP of the refrigeration system is plotted versus the temperature difference ΔT_{vpi} for T_{ci} of 20, 30, and 40°C. **Figure 12** discloses distinctly that COP decreases sharply with ΔT_{vpi}. This is ascribed mainly to the falling value of the evaporator temperature of the

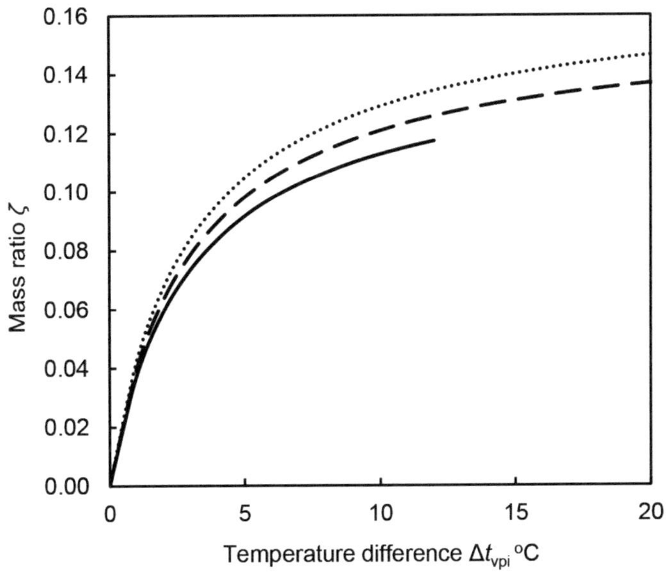

Figure 10.
Relationship between the mass of chilled cooling water required for steam condenser air cooler and the air temperature at vacuum pump entrance. ———, T_{ci} = 20°C; — — —, T_{ci} = 30°C; ………, T_{ci} = 40°C.

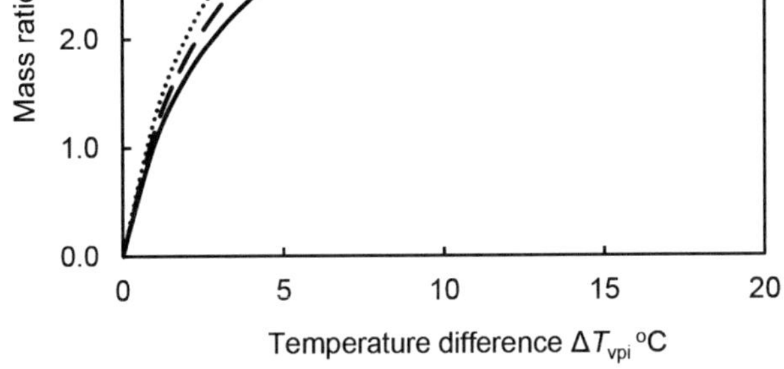

Figure 11.
Effect of the air temperature at vacuum pump entrance on refrigerant mass rate when using air cooler. ———, T_{ci} = 20°C; — — —, T_{ci} = 30°C; ………, T_{ci} = 40°C.

refrigeration system and the increasing temperature difference between the refrigerant condenser and the evaporator. The temperature T_{ci} has almost a negligible effect on COP as the temperature difference between refrigerant condenser and evaporator alters with ΔT_{vpi} and not with T_{ci}.

The specific works $w_{vp,wac}$, $w_{rco,t}$, $w_{vp,ac}$ as well as w_t are plotted in the diagrams of **Figure 13** versus ΔT_{vpi} for temperature T_{ci} of 20, 30, and 40°C. Although $w_{vp,wac}$ is independent of ΔT_{vpi}, it is represented on the diagrams of **Figure 13** as horizontal lines for the sake of comparison. It can be seen from **Figure 13** that $w_{vp,wac}$ decreases with an increase in T_{ci}, which is caused mainly due to the drop in pressure ratio of the vacuum pump. The specific work $w_{vp,ac}$ is reduced steeply with ΔT_{vpi} until a value of ΔT_{vpi} around 12°C; then, the rate of decrease in $w_{vp,ac}$ diminishes remarkably. This can be interpreted as follows: on the one hand, at low temperature difference ΔT_{vpi}, the amount of steam mixed with air is relatively high as can be seen from **Figures 8** and **9**, which results in relatively high mass rate of the mixture of air and water vapor flowing through the vacuum pump, and therefore, greater pump work $w_{vp,ac}$ is obtained. As ΔT_{vpi} is raised, the amount of steam flowing with air dwindles and so the mass flow rate through the pump declines, which gives rise to decreasing the pump work $w_{vp,ac}$. On the other hand, the pressure ratio through the vacuum pump is raised with ΔT_{vpi} and hence the work $w_{vp,ac}$ is increased. However, the increase in $w_{vp,ac}$ is relatively small at low values of ΔT_{vpi} compared to the work decrease due to the drop in pump mass flow rate. Therefore, the pump work $w_{vp,ac}$ falls off with relatively high rate, as ΔT_{vpi} grows. As the mass flow induced by the pump declines and the pressure ratio through the pump grows with increasing ΔT_{vpi}, the effect of the former parameter diminishes, while the effect of the latter parameter grows up and the net result is a considerable drop in the rate of decrease in $w_{vp,ac}$. The specific work $w_{rco,t}$ rises almost linearly with ΔT_{vpi}. This is attributed mainly to the declination of the refrigeration system COP (see **Figure 12**). The rate of increase in $w_{rco,t}$ is almost independent of the temperature T_{ci}.

The saving percentage $(w_{vp,wac} - w_t) \times 100/w_{vp,wac}$ in work by using refrigeration system for cooling the air contained in the steam condenser is plotted in **Figure 14** versus ΔT_{vpi} for T_{ci} of 20, 30, and 40°C. It is seen from **Figures 13** and **14**

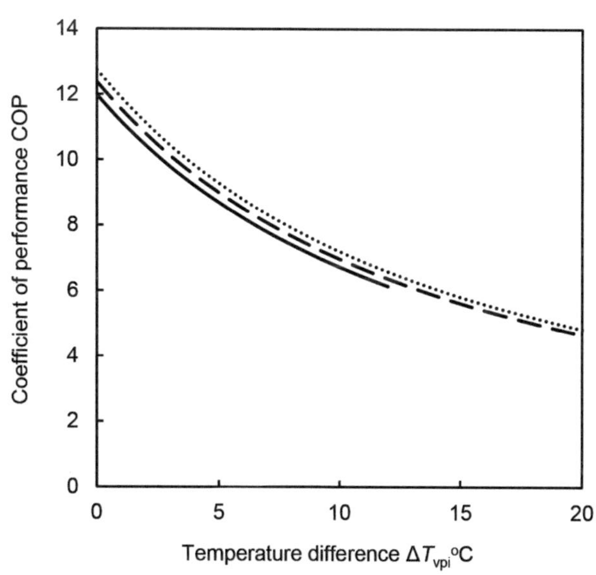

Figure 12.
Effect of the air temperature at vacuum pump entrance on coefficient of performance of the refrigeration system. ———, T_{ci} = 20°C; — — —, T_{ci} = 30°C; ...., T_{ci} = 40°C.

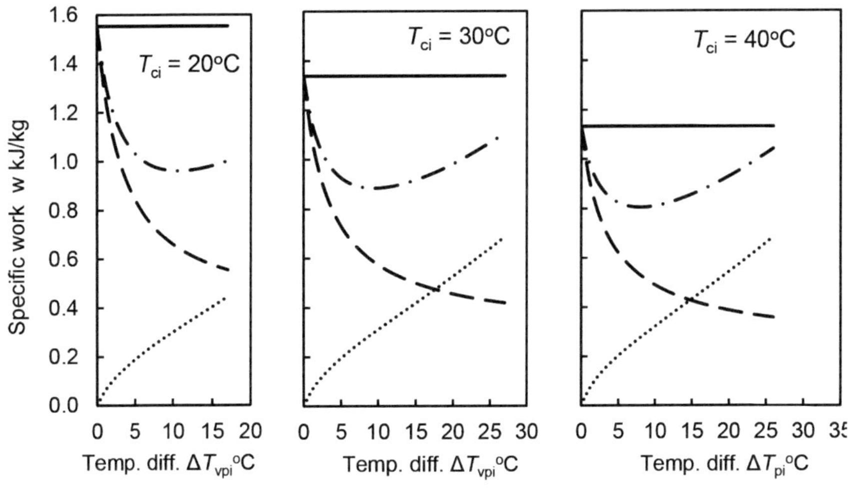

Figure 13.
Specific work dependence on the air temperature at vacuum pump entrance. ———, $w_{vp,wac}$; — — —, $w_{vp,ac}$; ...., $w_{rco,t}$; —. —, w_t.

that the sum of the specific works $w_{vp,ac}$ and $w_{rco,t}$ (i.e., the total specific work w_t) has a minimal value (maximum saving in the total work sum w_t). This minimum value depends on the value of T_{ci} and it is less than the corresponding $w_{vp,wac}$ according to the temperature T_{ci} by 38.5, 33.9, and 28.9% at ΔT_{vpi} of 12, 10, and 8°C for T_{ci} of 20, 30, and 40°C, respectively. Also, it is seen from **Figures 13** and **14** that w_t is maximally higher than the minimal value corresponding to T_{ci} by 3% when ΔT_{vpi} is 4°C higher than its value at which the minimal total specific work occurs. On the contrary, the saving in the steam lost increases depending on T_{ci} in the range of 5–7%. Therefore, it is more advantageous to choose values for ΔT_{vpi} higher than those at which the minimal total work occurs in the range of 4°C, as it results in fairly less lost steam rate to be drawn by the vacuum pump and inconsiderable increase in the total work. Higher values than 4°C cause inconsiderably small increase in saving the lost steam, but the total work is significantly raised.

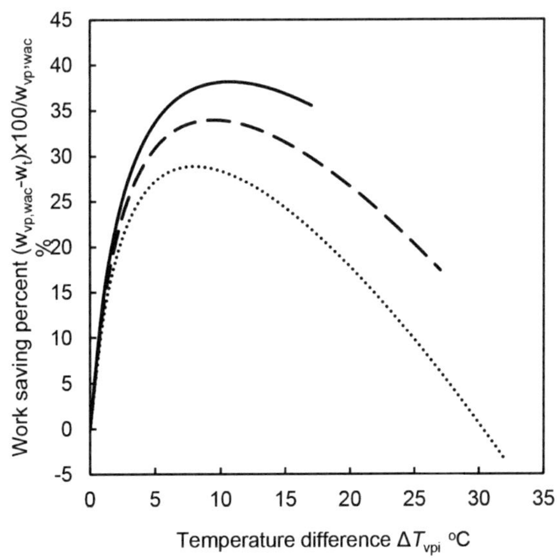

Figure 14.
Saving percent in specific work due to refrigeration cooling of steam condenser air cooler. ———, T_{ci} = 20°C; — — —, T_{ci} = 30°C; .... T_{ci} = 40°C.

It is to be mentioned here that the parameters γ, δ, ζ, ξ_t, $w_{\mathrm{vp,wac}}$, $w_{\mathrm{vp,ac}}$, $w_{\mathrm{rco,t}}$, and w_t are displayed in **Figures 7, 8, 10, 11,** and **13** for the value for β of 0.0003. These parameters are directly proportional to the mass ratio β. This is explained as follows: the mass of steam mixed with each kilogram of air and condensed and in turn the amount of cooling normal/chilled water used for cooling a kilogram of air and condensing the steam as well as the amount of refrigerant utilized for chilling the cooling water are dependent only on the initial and final air temperatures of the cooling process. Therefore, the parameters mentioned above are directly proportional to the mass ratio β. On the contrary to that, the saving percentages $(\gamma - \delta) \times 100/\gamma$ and $(w_t - w_e) \times 100/w_e$ in the amount of steam to be condensed and total work, respectively, are independent of β.

6. Conclusions

The current work is concerned with the use of vapor compression refrigeration system (VCRS) for chilling cooling water used with the air cooler of the steam plant condenser (SPC). A thermodynamic analysis is developed for working out the performance of the hybrid system of VCRS and SPC. The results obtained using this analysis showed that subcooling of the SPC condensate can cause considerable reduction in steam rate associated with the air induced by the vacuum pump. However, this is possibly avoided as it represents heat loss in the condensate heat content, which should be compensated in the plant boiler. In addition, the results of this work led to drawing the following conclusions for condensate subcooling of 3°C, which represents a reasonable and practical subcooling of the condenser condensate.

1. Temperature reductions of the condenser air of 5, 10, and 15°C below the condensate temperature result in reducing steam rate lost in venting air from the condenser relative to the loss when using no air cooler, by around 69, 85, and 90%, respectively.

2. The total work saving when using chilled water for cooling the air in the condenser air cooler from that in case of no air cooling is applied, has maximums of 38.5, 33.9, and 28.9% and occurs at temperature decrease below the condensate temperature of 12, 10, and 8°C when the temperature of steam admitted to the condenser is 20, 30, and 40°C, respectively. In these cases, the savings in steam lost in venting process amount to 87.7, 84, and 79.2%, respectively.

3. Selecting the reduction in condenser air temperature in the range of 4°C higher than that temperature reduction, at which the minimum total work occurs, is very advantageous where the saving in steam lost becomes fairly greater while the saving in total work is slightly lower than the minimum total works; in the range of 5–7 and maximally 3%, respectively.

Nomenclature

COP	coefficient of performance of the refrigeration system
C_p	specific heat capacity at constant pressure (kJ/kg K)
C_w	specific heat capacity of water (kJ/kg K)
h	specific enthalpy (kJ/kg)
k	isentropic exponent

LH	specific latent heat (kJ/kg)
$\dot{m}$	mass flow rate (kg/s)
p	pressure (kPa)
q	relative heat transfer rate (kW/kg)
R	gas constant (kJ/kg K)
T	temperature (°C)
v	specific volume (m^3/kg)
V	air velocity (m/s)
$\dot{V}$	volume flow rate (m^3/s)
x	dryness fraction (–)
w	specific work (kJ/kg)

Greek letters

β	mass ratio $\dot{m}_a/\dot{m}_{\mathrm{st,t}}$ (–)
γ	mass ratio $\dot{m}_{st,ce}/\dot{m}_{\mathrm{st,t}}$ (–)
δ	mass ratio $\dot{m}_{st,vp}/\dot{m}_{\mathrm{st,t}}$ (–)
ΔT_{ce}	temperature difference of steam at condenser inlet and exit (K/°C)
ΔT_{e}	temperature difference of chilled water at air cooler inlet and saturated refrigerant in VCRS evaporator (K/°C)
$\Delta T_{\mathrm{rc,sub}}$	subcooling in the refrigerant condenser
ΔT_{vpi}	temperature difference of condensate in hot well and air entering the vacuum pump (K/°C)
ε	heat exchanger effectiveness (–)
ζ	mass ratio $\dot{m}_{\mathrm{cw}}/\dot{m}_{\mathrm{st,t}}$ (–)
η_i	isentropic efficiency of the compressor (–)
η_m	mechanical efficiency of the compressor (–)
ξ_I, ξ_{II}, ξ_t	mass ratios $\dot{m}_{\mathrm{r,I}}/\dot{m}_{\mathrm{st,t}}$, $\dot{m}_{\mathrm{r,II}}/\dot{m}_{\mathrm{st,t}}$, and $(\dot{m}_{\mathrm{r,I}}+\dot{m}_{\mathrm{r,II}})/\dot{m}_{\mathrm{st,t}}$, respectively (–)
φ	relative humidity (–)

Subscripts

a	air
ac	with air cooler
ace	air cooler exit for chilled water
aci	air cooler inlet for chilled water
atm	atmospheric
av.	average
c	steam condenser, cooling
ce	steam condenser outlet
ci	steam condenser inlet
cw	chilled water
e	evaporator
ha	humid air
i	isentropic
in	inlet
m	air and steam mixture, mechanical
r	refrigerant
rc	refrigerant condenser
rco	refrigerant compressor
s	saturated steam
st	steam

t	steam turbine, total pressure, total work
vp	vacuum pump
vpi	vacuum pump inlet
w	water
wac	without air cooler
1–10	state numbers of the refrigerant of the refrigeration cycle
I, II	refrigerant compressor no. I and II, respectively

Abbreviations

LLSL-HE	liquid-line/ suction-line heat exchanger
SPC	steam plant condenser
VCRS	vapor compression refrigeration system

Author details

Ahmed Hegazy[1] and Abraham Engeda[2*]

1 Mechanical Engineering Department, Menoufia University, Shibin El Kom, Egypt

2 Turbomachinery Lab, Michigan State University, East Lansing, MI, USA

*Address all correspondence to: engeda@msu.edu

References

[1] Loy Yang B. Plant Description Manuals, Feedwater, Steam and Condensate Systems, LYB PDM. Vol. 6, Sections 1 & 2; 2005

[2] Srivastava NK. Uprating, Renovation & Modernization of Ageing Thermal Power Plant, Report. R&M Engg. NTPC-INDIA; 2010

[3] Banerjee S, Tripathi SN, Yadav M. Thermal Power Plant Performance Improvement Using Quality Initiatives, NTPC Report; 2008

[4] Renovation and Modernization of Thermal Power Plants in India Planning and Implementation Guidelines. 2010

[5] Mule SM, Kulkarni SH, Gulhane NP. A study of effect of air in-leakage on performance of steam surface condenser. International Journal of Mechanical and Production Engineering. 2017;**5**(8):56-62

[6] Rodgers GFC, Mayhew YR. Thermodynamic and Transport Properties of Fluids—SI Units. 5th ed. Cambridge, Massachusetts, USA: Blackwell Publishers Ltd; 1996

[7] Kubik WJ, Spencer E. Improved steam condenser gas removal system. American Society of Mechanical Engineers. Heat Transfer Division. 1988; **104**(2):255-263

[8] Birgenheier DB, Butzbach TL, Bolt DE, Bhatnagar RK, Ojala RE, Aglitz J. Designing steam jet vacuum systems. Chemical Engineering. July 1993;**100**: 116-121

[9] Ranga N. Evacuation systems for steam surface condensers: vacuum pump or steam jet air ejectors? In: ASME 2016 Power Conference. POWERENERGY 2016–59067; June 26–30, 2016; Charlotte, North Carolina, USA

[10] Allemann RT, Johnson BM, Werry EV. Wet-Dry Cooling Demonstration: A Transfer of Technology. EPRI CS-5016. Palo Alto, CA: Electric Power Research Institute; 1987

[11] Allemann RT et al. Wet-Dry Cooling Demonstration: Test Results. EPRI CS-4321. Palo Alto, CA: Electric Power Research Institute; 1986

[12] Allemann RT. Advanced Concepts Test (ACT) Facility Summary Safety Report. EPRI CS-1915. Palo Alto, CA: Electric Power Research Institute; 1981

[13] Allemann RT. Development of an Advanced Concept of Dry/Wet Cooling of PowerGenerating Plants. EPRI CS-1668. Palo Alto, CA: Electric Power Research Institute; 1981

[14] Hegazy A, Rothan Y, Engeda A. Fea sibility of using vapor compression ref rigeration system for cooling steam pla nt condenser. Applied Thermal Engineering. 2016;**106**:570-578

[15] Yousef K, Bolin C, Hegazy A, Engeda A. Experimental investigation of a refrigerant as a coolant of a power plant condenser. International Journal of Air-Conditioning and Refrigeration. 2014;**22**(4):1450024-1-13

[16] Yousef K, Assefa A, Hegazy A, Engeda A. Comparative study of using R-410A, R-407C, R-22, and R-134a as cooling medium in the condenser of a steam power plant. ASME Journal of Engineering for Gas Turbines Power. 2015;**137**:022002-1-8

[17] Hegazy AS. Possible waste heat recovery in the condenser of a regenerative steam cycles. JSME Journal of Thermal Science and Technology. 2007;**2**(1):1-7. Artical ID: 053002

[18] Hegazy AS. Use of cooling thermal storage as a heat sink for steam power plant. JSME Journal of Thermal Science and Technology. 2008;**3**(2):330-341

[19] Hegazy AS. Improving performance of refrigerant cooled steam power plant by using cooling thermal storage. ASME Journal of Engineering for Gas Turbines and Power. 2009;**131**(5):1-7. Artical ID: 053002

[20] Eastop TD, McConkey A. Applied Thermodynamics. UK: Addison Wesley Longman Group; 1998

[21] Whitman WC, Johnson WM, Tomczyk JA. Refrigeration & Air Conditioning Technology. 5th ed. Australia, Canada, Mexico, Singapore, Spain, United Kingdom, United States: Thomson and Delmar Learning; 2005

[22] Ambrosini W, Bucci M, Forgione N, Oriolo F, Paci S, Magnaud JP, et al. Comparison and analysis of the condensation benchmark results. In: The 3rd European Review Meeting on Severe Accident Research (ERMSAR-2008); 23–25 September 2008; Nessebar, Bulgaria

[23] Klein KA, Alvarado FL. EES—Engineering Equation Solver, Version 6.648 ND. Middleton, WI: FChart Software; 2004

[24] Stoecker WF. Industrial Refrigeration Handbook. first ed. New York, San Francisco, Washington, DC: McGraw-Hill Education; 1998

Chapter 13

Energy and Exergy Analysis of Refrigeration Systems

Shaimaa Seyam

Abstract

Refrigeration systems have the priority in design for residential and industrial applications. The chapter includes five major refrigeration systems: vapor-compression refrigeration; ammonia-water absorption refrigeration; gas refrigeration where standard air is the most popular refrigerant; multi-pressure refrigeration including multistage, cascade, and multipurpose refrigeration system; and heat pump systems. Energy and exergy analysis has been presented for most of the systems. The energetic and the exergetic COP for each system are presented. Renewable energy sources are also discussed including geothermal, solar, and wind energy, a with combination with refrigeration systems in different industrial and residential applications. The overall efficiency of the renewable systems is achieved to be more than 50% providing promising solutions for energy use and having a low environmental impact.

Keywords: refrigeration systems, absorption cooling system, heat pump, geothermal energy, solar energy, wind energy

1. Introduction

The primary application of refrigeration system is to transfer heat from a lower temperature region to a higher temperature one. A refrigeration cycle consists of a source at low temperature, a sink at high temperature, and a device to produce the work done to transfer heat from the source to sink. For the complete circulation, the refrigeration cycle should have an expansion device to circulate the refrigerant to the source.

Major refrigeration systems include vapor-compression refrigeration system (VCRS), heat pump (HP), gas refrigeration system (GRS), multi-pressure refrigeration systems (MPRS), and absorption refrigeration system (ARS), as presented in **Figure 1**. These systems are combined with renewable sources, such as geothermal, solar, and wind energy sources.

2. Vapor-compression refrigeration system

The vapor-compression refrigeration cycle (VCRS) is the most widely used cycle for refrigerators, air-conditioning systems, and heat pumps [1, 2]. It consists of four processes, as shown in **Figure 2**:

1-2 Isentropic compression in a compressor

2-3 Constant-pressure heat rejection in a condenser

3-4 Throttling in an expansion device
4-1 Constant-pressure heat absorption in an evaporator

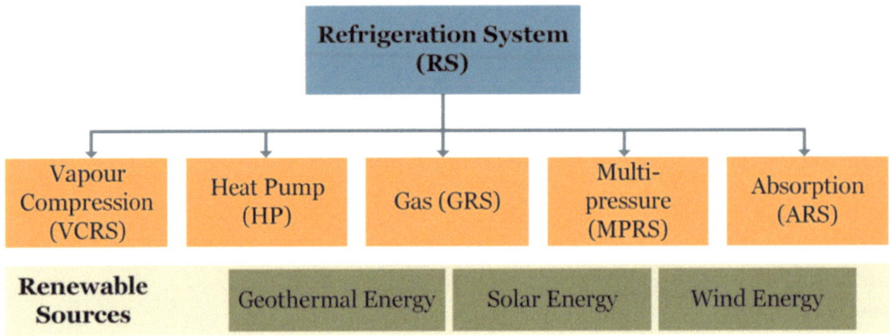

Figure 1.
The classifications of refrigeration systems and renewable sources.

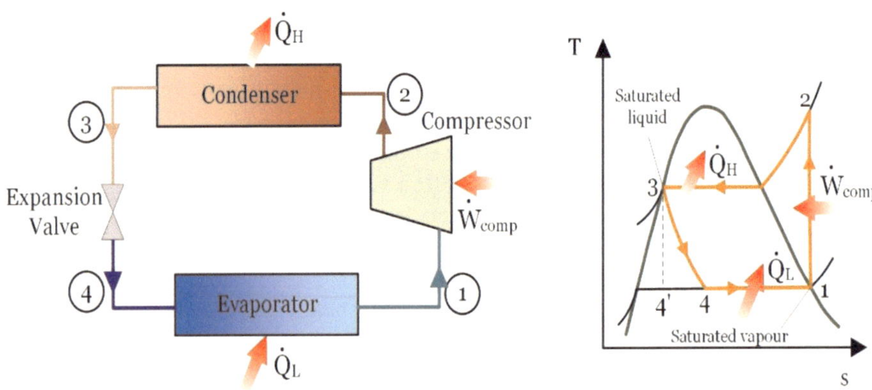

Figure 2.
Schematic and T-s diagram for the ideal VCRS.

The refrigerant enters the compressor from state 1 at saturated vapor to be isentropically compressed from low pressure of state 1 to high pressure and temperature of state 2, which is at the superheated region. Then, the refrigerant of state 2 enters the condenser to reject heat to the warm environment and exits at the saturated liquid as state 3. The refrigerant enters an adiabatic throttling or expansion valve to drop the pressure, which equals the pressure at the compressor inlet of state 1. The refrigerant temperature at state 1 is very low so that it absorbs heat from the refrigerated space at the evaporator and heated to be saturated vapor again. The vapor refrigeration system is a closed cycle where it starts and ends at state 1. This type of refrigeration system can be used for refrigerators, inside the air conditioners as split air conditioners, and separate as in radiant cooling systems [3, 4] and air-to-air systems [1].

2.1 Energy and exergy analysis of vapor-compression RS

The first and second law of Thermodynamics for steady-state flow is applied for each component and the whole system. They include the energy balance equation (EnBE) and exergy balance equation (ExBE) in this order. The energy balance equation considers the heat transfer and work produced or done crossing the control volume of a component or a system, while the exergy balance equation considers the irreversibilities of a process, which are described by the exergy destruction. For the given system of **Figure 2**. The refrigerant mass flow rate is constant through the cycle and denotes as $\dot{m}$ in kg/s, h is the specific enthalpy of each state point in kJ/kg, ex_i is the specific exergy of a state point in kJ/kg and defined as $ex_i = (h_i - h_0) - T_0(s_i - s_0)$, and s is the specific entropy of the refrigerant at each state point, and its unit is kJ/kg.K. The change in kinetic and potential energy is negligible for each component and the entire system. The energy balance equation (EnBE) of the evaporator considers the rate of heat removal by the evaporator, $\dot{Q}_L$ which is released from a low-temperature environment and determined by Eq. (1) [5]. The exergy balance equation (ExBE) of the evaporator is given as Eq. (2). Also, the thermal exergy rate due to the heat transfer from the evaporator is defined as Eq. (3) [6].

EnBE $$\dot{Q}_L = \dot{m}(h_1 - h_4) \tag{1}$$

ExBE $$\dot{m}_4 ex_4 + \dot{E}x_{Q,\,evap} = \dot{m}_1 ex_1 + +\dot{E}x_{des,\,evap} \tag{2}$$

$$\dot{E}x_{Q,\,evap} = \dot{Q}_L\left(\frac{T_0}{T_L} - 1\right) \tag{3}$$

The power input to the compressor, $\dot{W}_{comp}$, can be determined from Eq. (4), the isentropic efficiency of an adiabatic is defined as in Eq. (5) [5], and the exergy destruction of the compressor can be given as Eq. (6) [6]:

EnBE $$\dot{W}_{comp} = \dot{m}(h_2 - h_1) \tag{4}$$

$$\eta_{comp} = \frac{\dot{W}_{in}}{\dot{W}} = \frac{h_{2s} - h_1}{h_2 - h_1} \tag{5}$$

ExBE $$\dot{m}_1 ex_1 + \dot{W}_{in} = \dot{m}_2 ex_2 + \dot{E}x_{des,\,comp} \tag{6}$$

The heat rejection rate from the condenser, $\dot{Q}_H$, to the environment can be written as Eq. (7) [5], while the exergy destruction and thermal exergy rate of the condenser can be given as Eqs. (8) and (9) [6]:

EnBE $$\dot{Q}_H = \dot{m}(h_2 - h_3) \tag{7}$$

ExBE $$\dot{m}_2 ex_2 = \dot{m}_3 ex_3 + \dot{E}x_{Q,\,cond} + \dot{E}x_{des,\,cond} \tag{8}$$

$$\dot{E}x_{Q,\,cond} = \dot{Q}_H\left(1 - \frac{T_0}{T_H}\right) \tag{9}$$

The energy and exergy balance equations for the expansion valve can be expressed as Eqs. (10) and (11), respectively. The expansion valves are considered to be decreasing the pressure adiabatically and isentropically, which means no heat transfer and work done in the throttling process [6]:

EnBE $$\dot{m}h_3 = \dot{m}h_4 \tag{10}$$

ExBE $$\dot{m}_3 ex_3 = \dot{m}_4 ex_4 + \dot{E}x_{des,\,\exp} \tag{11}$$

The energy balance for the entire refrigeration system can be given as [5]:

$$\dot{Q}_H = \dot{Q}_L + \dot{W}_{comp} \tag{12}$$

The coefficient of performance (COP) of the refrigeration system is defined as the ratio of useful energy, which is the rate of heat removal by the evaporator to the required energy, which is the power required to operate the compressor. The COP is given as below [5]:

$$COP = \frac{\dot{Q}_L}{\dot{W}_{comp}} \tag{13}$$

The Carnot or reversible COP is defined as the maximum COP of a refrigeration cycle operating between temperature limits T_L and T_H, which can be given as Eq. (14) [5]:

$$COP_{rev} = T_L/(T_H - T_L) \tag{14}$$

An actual vapor-compression refrigeration cycle differs from the ideal one because of the irreversibilities that occur in various components, such as fluid friction (causes pressure drops) and heat transfer to or from the surroundings. The aim of exergy analysis is to determine the exergy destruction in each component of the system and to determine the exergy efficiency of the entire system. Exergy destruction in a component can be evaluated based on entropy generation and an exergy balance equation using Eq. (15) [6]:

$$\dot{Ex}_{des} = T_0\dot{S}_{gen} \tag{15}$$

where T_0 is the dead-state temperature or environment temperature. In a refrigerator, T_0 usually equals the temperature of the high-temperature medium T_H.

The exergetic coefficient of performance (COP_{ex}) of the refrigeration system is the second-law efficiency of the cycle. It is defined as the ratio of useful exergy rate, which is the thermal exergy of the heat removed by the evaporator, to the required exergy rate, which is the work done by the compressor. The COP_{ex} can be written as [6]:

$$COP_{ex} = \frac{\dot{Ex}_{Q,evap}}{\dot{W}_{comp}} = \frac{\dot{Q}_L\left(\frac{T_0}{T_L} - 1\right)}{\dot{W}_{comp}} \tag{16}$$

by substituting $\dot{W}_{comp} = \dot{Q}_L/COP_{en}$, and the COP_{ex} can be defined as the maximum COP of a refrigeration cycle operating between temperature limits T_L and T_H, which can be given as Eq. (14). Therefore, the second-law efficiency or COP_{ex} can be rewritten as Eq. (17) [6]:

$$COP_{ex} = \frac{\dot{Q}_L\left(\frac{T_0}{T_L} - 1\right)}{\dot{W}_{comp}} = \frac{\dot{Q}_L\left(\frac{T_0}{T_L} - 1\right)}{\dot{Q}_L/COP} = \frac{COP}{T_L/(T_H - T_L)} = \frac{COP}{COP_{rev}} \tag{17}$$

Since $T_0 = T_H$ for a refrigeration cycle, thus, the second-law efficiency is also equal to the ratio of actual and maximum COPs for the cycles, which accounts for all irreversibilities associated within the refrigeration system.

3. Heat pump system

Heat pump system (HP) is similar to VCRS since it consists of a compressor, expansion valve, and outdoor and indoor coils, which operate exchangeably as

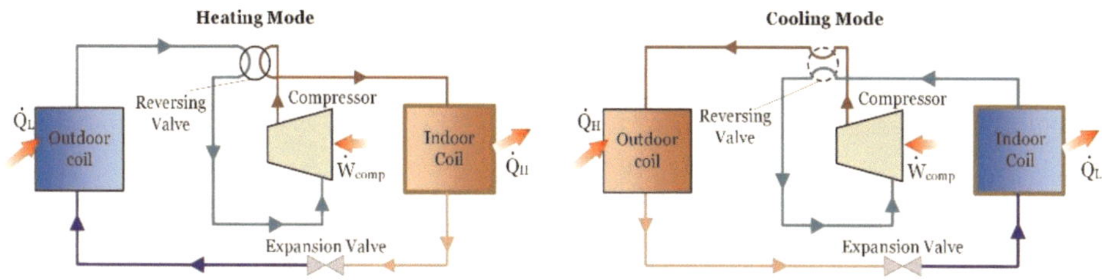

Figure 3.
A heat pump can be used to heat a house in winter and to cool it in summer.

condenser and evaporator. The advantage of HP systems is the ability to provide cooling and heating for the desired space, especially for the long winter season as in Canada and north European countries. This can be achieved by adding a reversing valve, as shown in **Figure 3**. There are two essential modes: heating mode and cooling mode. The condenser and evaporator are exchanging during the cooling and heating season since the reversing valve is switching between two modes according to the weather condition.

The energy source for heat pump can be classified into air-source, water-source, and ground-source. The air-source system uses atmospheric air through the evaporator, while the water-source system uses well water of depth 80 m and operates from 5 to 18°C. The ground-source system uses long piping under the ground since the soil temperature is not affected by climate change. The capacity and efficiency of heat pump drop at low-temperature environment, and therefore, other auxiliary systems, such as heaters or furnaces, are used to provide sufficient heating load for residential buildings.

The COP of a heat pump is defined as the ratio of the heat removed for cooling mode or added for a heating mode of the indoor coil to the compressor power. Therefore, the $COP_{heating}$ and $COP_{cooling}$ are given in Eq. (19). $\dot{Q}_{in}$ can be $\dot{Q}_H$ for heating mode or $\dot{Q}_L$ for cooling mode [6]:

$$COP = \frac{\dot{Q}_{in}}{\dot{W}_{comp}} \tag{18}$$

$$COP_{heating} = \frac{\dot{Q}_H}{\dot{W}_{comp}} \quad and \quad COP_{cooling} = \frac{\dot{Q}_L}{\dot{W}_{comp}} \tag{19}$$

The exergetic COP is defined as the ratio of thermal exergy rate divided by the compressor power. It is also given as the ratio of COP to the reversible COP for both heating and cooling mode. T_{in} can be considered as T_H for heating mode and T_L for cooling mode [6]:

$$COP_{ex} = \frac{\dot{Q}_{in}|1 - T_0/T_{in}|}{\dot{W}_{comp}} = \frac{COP}{COP_{rev}} \tag{20}$$

4. Gas refrigeration system

The VCRS is known as a modified, reverse Rankine cycle, while the gas refrigeration system (GRS) is known as a reverse Brayton cycle using a noncondensing gas such as air. The main advantage of this system is the small size for achieving the desired cooling due to the lighter weight of air than other refrigerants. This system can be used in aircraft cabin cooling.

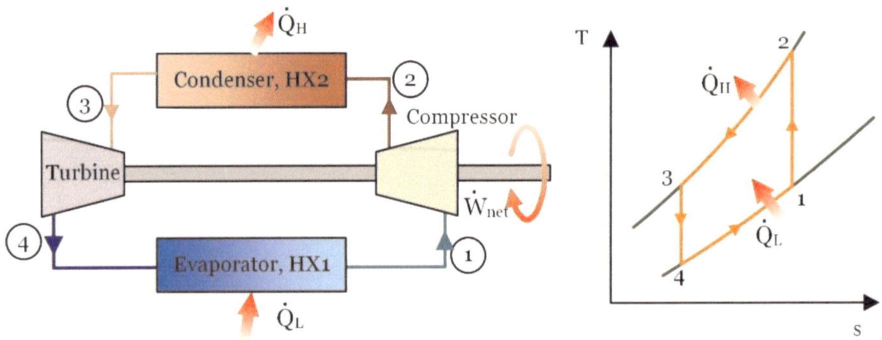

Figure 4.
Simple gas refrigeration cycle and T-s diagram.

As illustrated in **Figure 4**, the major elements of GRS are compressor to raise the pressure of gas from state 1 to 2, a rejecting heat exchanger (condenser), turbine or expander to decrease the gas pressure isentropically, and an absorbing heat exchanger (evaporator) to absorb the heat from the refrigerated space at constant pressure. A regenerator heat exchanger can be added to the system for heat recovery between the hot and cold paths of circulated gas. It can be located between the two heat exchangers. Air is a popular refrigerant of this system since it can be utilized as a refrigerant and air-conditioning medium in smaller equipment units as aircraft cooling systems.

4.1 Energy and exergy analysis of gas refrigeration system

The energy analysis of a gas refrigeration system is similar to that of the vapor refrigeration system except that the gaseous fluid is treated as an ideal gas. Therefore, the enthalpy and entropy equations are written as [5]:

$$\Delta h = (h_e - h_i) = c_p \Delta T = c_p (T_e - T_i) \tag{21}$$

$$\Delta s = (s_e - s_i) = c_p \ln \frac{T_e}{T_i} - R \ln \frac{P_e}{P_i} \tag{22}$$

where the subscripts i and e indicate inlet and exit states, respectively. Therefore, the energy and exergy analysis for each component of **Figure 5** is listed below [5, 6].

Compressor:

EnBE $\quad \dot{m}h_1 + \dot{W}_{Comp} = \dot{m}h_2 \Rightarrow \dot{W}_{Comp} = \dot{m}(h_1 - h_2) = \dot{m}c_p(T_1 - T_2) \quad (23)$

ExBE

$$\dot{Ex}_{des,\,comp} = T_0 \dot{S}_{gen,\,1-2} = \dot{m}T_0(s_2 - s_1)$$
$$= \dot{m}T_0\left(c_p \ln \frac{T_2}{T_1} - R \ln \frac{P_2}{P_1}\right) \tag{24}$$

Heat exchanger 2 (condenser):

EnBE $\quad \dot{m}h_2 = \dot{Q}_H + \dot{m}h_3 \Rightarrow \dot{Q}_H = \dot{m}(h_2 - h_3) = \dot{m}c_p(T_2 - T_3) \quad (25)$

ExBE

$$\dot{Ex}_{des,\,HX2} = T_0 \dot{S}_{gen,\,2-3} = \dot{m}T_0\left(s_3 - s_2 + \frac{q_H}{T_H}\right)$$
$$= \dot{m}T_0\left\{\left(c_p \ln \frac{T_3}{T_2} - R \ln \frac{P_3}{P_2}\right) + \frac{q_H}{T_H}\right\} \tag{26}$$

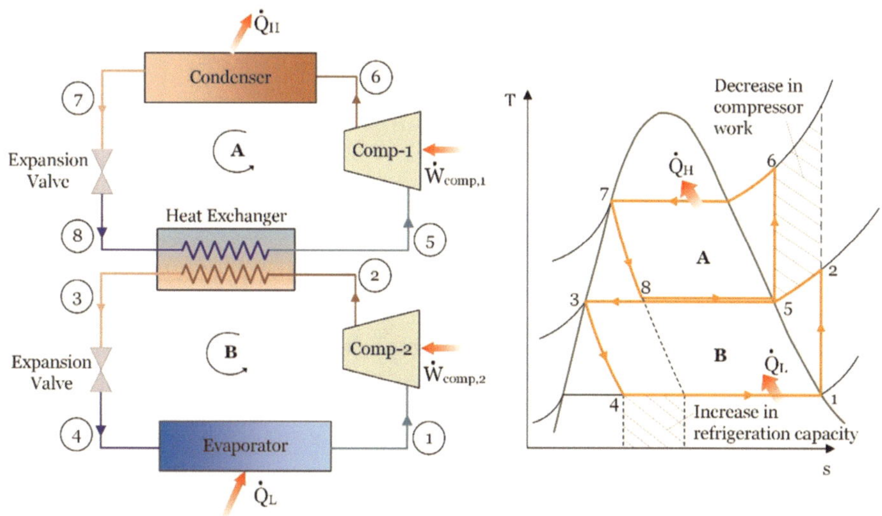

Figure 5.
A two-stage cascade refrigeration system with the same refrigerant in both stages.

Turbine (expander):

EnBE $$\dot{m}h_3 = \dot{W}_{turb} + \dot{m}h_4 \Rightarrow \dot{W}_{turb} = \dot{m}(h_3 - h_4) = \dot{m}c_p(T_3 - T_4) \tag{27}$$

ExBE $$\dot{Ex}_{des,\,turb} = T_0\dot{S}_{gen,\,3-4} = \dot{m}T_0(s_4 - s_3)$$
$$= \dot{m}T_0\left(c_p \ln\frac{T_4}{T_3} - R\ln\frac{P_4}{P_3}\right) \tag{28}$$

Heat exchanger 1 (evaporator):

EnBE $$\dot{m}h_4 + \dot{Q}_L = \dot{m}h_1 \Rightarrow \dot{Q}_L = \dot{m}(h_1 - h_4) = \dot{m}c_p(T_1 - T_4) \tag{29}$$

ExBE $$\dot{Ex}_{des,\,HX1} = T_0\dot{S}_{gen,\,4-1} = \dot{m}T_0\left(s_1 - s_4 + \frac{q_L}{T_L}\right)$$
$$= \dot{m}T_0\left\{\left(c_p \ln\frac{T_1}{T_4} - R\ln\frac{P_1}{P_4}\right) + \frac{q_L}{T_L}\right\} \tag{30}$$

For the entire refrigeration system, the energy balance can be written as:

$$\dot{W}_{comp} + \dot{Q}_L = \dot{W}_{turb} + \dot{Q}_H \tag{31}$$

The net power for the system becomes:

$$\dot{W}_{net} = \dot{W}_{comp} - \dot{W}_{turb} \tag{32}$$

The COP of the gas refrigeration system is given as:

$$COP = \frac{\dot{Q}_L}{\dot{W}_{net}} = \frac{\dot{Q}_L}{\dot{W}_{comp} - \dot{W}_{turb}} \tag{33}$$

The total exergy destruction in the system can be calculated by adding exergy destructions of each component:

$$\dot{Ex}_{des,\,total} = \dot{Ex}_{des,\,turb} + \dot{Ex}_{des,\,comp} + \dot{Ex}_{des,\,HX1} + \dot{Ex}_{des,\,HX2} \tag{34}$$

It can also be expressed as:

$$\dot{Ex}_{des,\,total} = \dot{W}_{net} - \dot{Ex}_{Q,\,HX1} = \dot{W}_{net} - \dot{Q}_L\left(\frac{T_0}{T_L} - 1\right) \tag{35}$$

Thus, the minimum power input to accomplish the required refrigeration load $\dot{Q}_L$ s equal to the thermal exergy rate of the heat exchanger I (evaporator) $\dot{W}_{min} = \dot{Ex}_{Q,\,HX1}$. Consequently, the second-law efficiency or the exergetic COP is defined as [6]:

$$COP_{ex} = \frac{\dot{Ex}_{Q,\,HX1}}{\dot{W}_{net}} = 1 - \frac{\dot{Ex}_{des,\,total}}{\dot{W}_{net}} \tag{36}$$

5. Multi-pressure refrigeration system

The VCRS is the most popular refrigeration cycle because it is simple, inexpensive, and reliable. However, the industrial refrigeration systems should be efficient by providing more refrigeration load. This can be achieved by modifying the simple VCRS into multi-pressure refrigeration systems (MPRS). The MPRS can be classified into cascade RS, multi-compression RS, and multipurpose RS.

5.1 Cascade refrigeration systems

Some industrial applications require low temperature below −70°C with substantially large pressure and temperature difference (−70 to 100°C). VCRS cannot achieve these applications because it can operate within a temperature range of +10 to −30°C. Therefore, a modification of VCRS can be performed by using multiple refrigeration cycles operating in series, the so-called cascade refrigeration systems. The refrigerants of each cycle can be different. The evaporator of the first refrigeration cycle is connected to the condenser of the next refrigeration system forming an interchange heat exchanger between the 2 cycles, as shown in **Figure 5**. Cascade refrigeration systems are mainly used for liquefaction of natural gas, hydrogen, and other gases [7–9]. The major benefit of this system is decreasing the compressor power and increasing the refrigeration load compared with a VCRS with large temperature and pressure difference, as shown in the T-s diagram of cascade system in **Figure 5**. Therefore, reducing system components can be fulfilled in an appropriate way [2].

The net compressor power can be determined by the summation of all compressor power in all cascaded refrigeration system and written as [2]:

$$\dot{W}_{net} = \dot{W}_{comp,\,1} + \dot{W}_{comp,\,2} = \dot{m}_A(h_6 - h_5) + \dot{m}_B(h_2 - h_1) \tag{37}$$

The refrigeration load can be described as:

$$\dot{Q}_L = \dot{m}_B(h_1 - h_4) \tag{38}$$

The heat exchanger that connects the 2 cycles together has an energy balance equation as follows [5]:

$$\dot{m}_B h_2 + \dot{m}_A h_8 = \dot{m}_B h_3 + \dot{m}_A h_5 \tag{39}$$

Therefore, the COP and exergetic COP of the cascade refrigeration system can be explained as the following [5, 6]:

$$COP = \frac{\dot{Q}_L}{\dot{W}_{net}} = \frac{\dot{m}_B(h_1 - h_4)}{\dot{W}_{comp,1} + \dot{W}_{comp,2}} \tag{40}$$

$$COP_{ex} = \frac{\dot{Ex}_{Q,evap}}{\dot{W}_{net}} = \frac{\dot{Q}_L\left(\frac{T_0}{T_L} - 1\right)}{\dot{W}_{comp,1} + \dot{W}_{comp,2}} \tag{41}$$

5.2 Multistage compression refrigeration systems

Similar to the cascade refrigeration system, multistage compression refrigeration system is used for applications below −30°C. This requires a large-pressure-ratio compressor and cannot be performed by one compressor because of the lack of efficiency and performance. Therefore, using multistage compressors connected in series can improve the performance of the refrigeration system by increasing the pressure ratio and increasing the refrigeration load. As shown in **Figure 6**, a two-stage compression refrigeration cycle consists of two compressors, a condenser, an evaporator, a flash intercooler, a mixer, and two throttling valves. The compressors. The upper compressor compresses the total refrigerant mass flow rate in a vapor form from the intermediate pressure of state 9 to the high pressure of state 4. The vapor refrigerant cools down in the condenser to saturated liquid at high pressure of state 5 and then passes through the upper expansion valve to reduce the pressure to intermediate pressure. The wet refrigerant passes through the flash intercooler to split the vapor and liquid phase. The vapor phase at state 3 enters the mixer to mix with the exit superheated refrigerant of the lower compressor at state 2. The liquid phase at state 7 is expanded by the lower throttling valve to state 8, which enters the lower pressure evaporator to absorb heat from the refrigerated space.

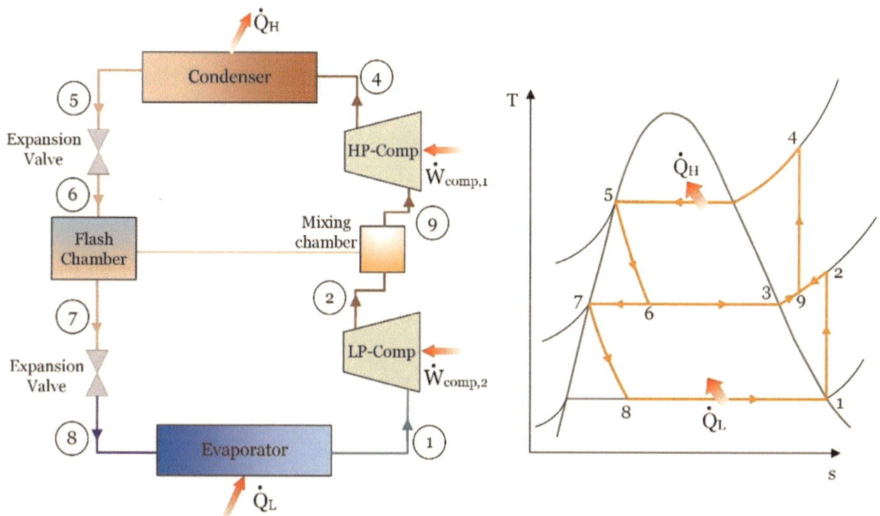

Figure 6.
A two-stage compression refrigeration system with a flash chamber.

The minimum temperature can be achieved by two-stage compression at −65°C, while the three-stage compression can attain about −100°C.

The heat transfer to the evaporator can be written, according to **Figure 6**, as [10]:

$$\dot{Q}_L = \dot{m}(1-x)(h_1 - h_8) \tag{42}$$

where x is the quality ratio of vapor mass to the total refrigerant mass flow rate at the intermediate pressure of the cycle. The term $(1-x)$ refers to the liquid mass ratio of the cycle. The net compressor power of the cycle can be evaluated as [10]:

$$\dot{W}_{net} = \dot{W}_{comp,1} + \dot{W}_{comp,2} = \dot{m}(h_4 - h_9) + \dot{m}(1-x)(h_2 - h_1) \tag{43}$$

Therefore, the COP of this system can be determined as the following [10]:

$$COP = \frac{\dot{Q}_L}{\dot{W}_{net}} = \frac{(1-x)(h_1 - h_8)}{(h_4 - h_9) + (1-x)(h_2 - h_1)} \tag{44}$$

The second efficiency or the exergetic COP can be calculated as [6, 10]:

$$COP = \frac{\dot{E}x_{Q,evap}}{\dot{W}_{net}} = \frac{\dot{Q}_L\left(\frac{T_0}{T_L} - 1\right)}{\dot{W}_{comp,1} + \dot{W}_{comp,2}} \tag{45}$$

5.3 Multipurpose refrigeration systems

Multipurpose refrigeration systems are also considered as a branch of MPRS. This type of system accomplishes different refrigeration loads in one system. Therefore, a modification of VCRS can be done by using multiple evaporators at different low pressure and different refrigerant capacity. Also, this system can be operated using one compressor or multistage compressor.

There are different configurations of multipurpose refrigeration systems [2], as shown in **Figure 7**. Firstly, a system of a single compressor and individual expansion valves consists of two evaporators and single compressor with individual expansion valves for each evaporator and one compressor, as shown in **Figure 7a**. Operation under these conditions means the dropping of pressure from high-pressure evaporators through back pressure valves to ensure the compression of the vapor from the higher temperature evaporators through a pressure ratio. Secondly, a system of a single compressor with multi-expansion valves consists of two evaporators and a compressor with multiple arrangements of expansion valves, as shown in **Figure 7b**.

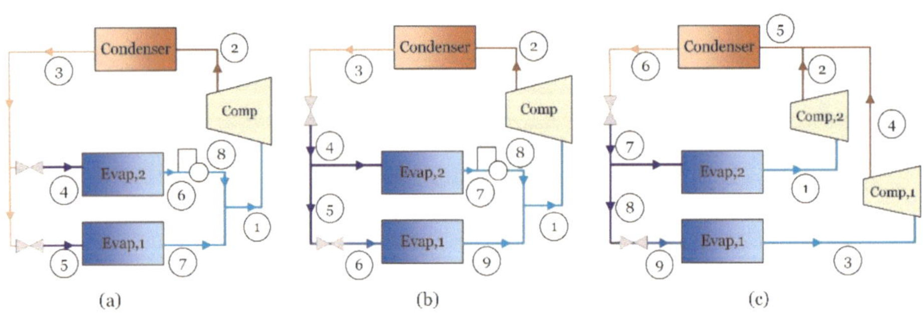

Figure 7.
Multipurpose refrigeration system: (a) two evaporators with individual expansion valve, (b) two evaporators and multi-expansion valve, and (c) individual compressors and multi-expansion valve.

The only advantage of the arrangement is that the flashed vapor at the pressure of the high-temperature evaporator is not allowed to go to the lower-temperature evaporator, thus improving its efficiency. Finally, a system of individual compressors with multi-expansion valves consists of a compressor for each evaporator and multiple arrangements of expansion valves, as shown in **Figure 7c**, to reduce the total power requirement. This amounts to parallel operation of evaporators and is called sectionalizing. There may be a separate condenser for each compressor or a common condenser for the whole plant.

The heat transfer to the evaporators and the net compressor power of the multipurpose refrigeration system despite the system configuration can be evaluated as [2]:

$$\dot{Q}_{evap,\,total} = \sum_{i=1}^{n} \dot{Q}_{evap,\,i} = \sum_{i=1}^{n} \dot{m}_{evap,\,i}\left(h_{evap,\,ex} - h_{evap,\,in}\right)_i \tag{46}$$

$$\dot{W}_{net} = \sum_{k=1}^{m} \dot{W}_{comp,\,k} = \sum_{k=1}^{m} \dot{m}_{comp,\,k}\left(h_{com,\,ex} - h_{com,\,in}\right)_k \tag{47}$$

where i is the number of evaporators from 1 to n, the subscripts *evap, in* and *evap, ex* refer to the inlet and exit states of each evaporator i, k is the number of compressor in the refrigeration system from 1 to m, and the subscripts *comp, in* and *comp, ex* refer to the inlet and exit states of each compressor k.

Therefore, the COP of this system can be determined as the following [5]:

$$COP = \frac{\dot{Q}_{evap,\,total}}{\dot{W}_{net}} \tag{48}$$

The second efficiency or the exergetic COP can be calculated as [6]:

$$COP = \frac{\sum_{i=1}^{n} \dot{Ex}_{Q,\,evap}}{\dot{W}_{net}} = \frac{\sum_{i=1}^{n} \dot{Q}_{evap,\,i}\left(\frac{T_0}{T_{L,\,i}} - 1\right)}{\sum_{k=1}^{m} \dot{W}_{comp,\,k}} \tag{49}$$

6. Absorption refrigeration system

The absorption refrigeration system (ARS) is similar to the VCRS except that the compressor of the vapor-compression system is replaced by three elements: an absorber, a solution pump, and a generator. The ABS medium is a mixture of a refrigerant and absorbent, such as ammonia-water system (NH_3 + H_2O) and water-lithium bromide (LiBr2 + H2O). The solubility of refrigerant (ammonia or lithium bromide) in the absorbent (water) is satisfactory, but the difference in boiling points is significant, which may affect the purity of vaporization. Thus, a purge unit or rectifier is used in the system. The refrigerant concentration in the mixture changes according to the pressure and temperature for each step. The ABS.

As shown in **Figure 8**, the ARS consists of a condenser, an evaporator, an absorber, a regeneration heat exchanger (HX1), heat recovery heat exchanger (HX2), a generator, two expansion valves, and a solution pump. The system includes an analyzer and a rectifier to remove the water vapor that may have formed in the generator. Thus, only ammonia vapor goes to the condenser. This system utilizes the absorbent water to release and absorb ammonia as the refrigerant. Starting from state 3, the strong solution (a high concentration of ammonia

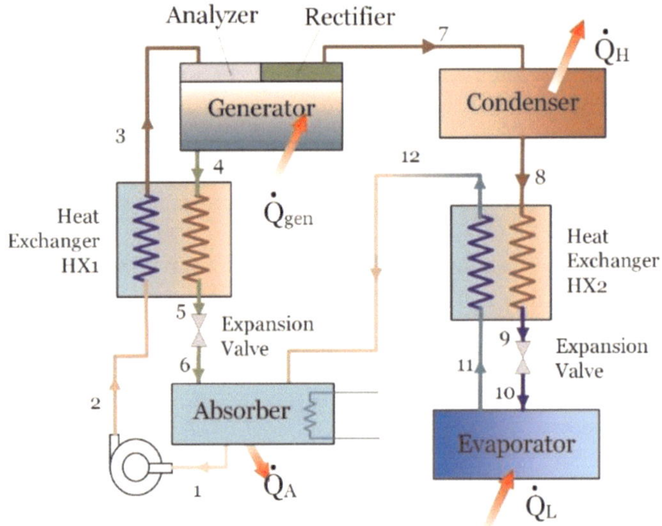

Figure 8.
Ammonia absorption refrigeration cycle.

refrigerant) is heated in the high-pressure generator. This produces refrigerant vapor off the solution at state 7. The hot pure ammonia vapor is cooled in the condenser at state 8 and condenses at state 9 by passing through the HX2 before entering a throttling valve into the low pressure at state 10. Then the refrigerant liquid passes through the evaporator to remove the heat from refrigerated medium and leaves at low-pressure vapor phase of state 11. The pure ammonia is heated by the HX2 to enter the absorber and mixed with the absorbent water. The weak solution (about 24% ammonia concentration) flows down from the generator at state 4 through the regeneration heat exchanger HX1 at state 5 through a throttling valve and enters the absorber at state 6. Therefore, the weak refrigerant is absorbed by the water because of the strong chemical affinity for each other. The absorber is cooled to produce a strong solution at low pressure at state 1. The strong solution is obtained and pumped by a solution pump to the generator passing through HX1, where it is again heated, and the cycle continues. Then, the water absorbs the ammonia in the absorber at the condenser temperature supplied by the circulating water or air, and hence a strong solution (about 38% ammonia concentration) occurs. For ammonia-water ARSs, the most suitable absorber is the film-type absorber because of high heat and mass transfer rates, enhanced overall performance, and large concentration rates [11].

6.1 Energy and exergy analysis of ammonia-water (NH_3-H_2O) ARSs

The energy and exergy analysis for each component is presented according to **Figure 8**. The partial mass balance (PMBE) is also included to determine the concentration mass of ammonia and water in the absorber and generator. That is because the ARS has two fluids as refrigerant and absorbent and their composition at different points is different, particularly in the absorber and generator. The exergy analysis of ammonia-water ARSs is to determine the exergy destruction of each component and to determine the overall exergy efficiency based on the second law of thermodynamics. The exergy analysis (ExBE) for each component is stated below [5, 6]:

Absorber:

EnBE $$\dot{m}_6h_6 + \dot{m}_{12}h_{12} = \dot{m}_1h_1 + \dot{Q}_{Absorber} \quad (50)$$

PMBE $$\dot{m}_{ws}X_{ws} + \dot{m}_r = \dot{m}_{ss}X_{ss} \quad (51)$$

ExBE $$\dot{m}_6ex_6 + \dot{m}_{12}ex_{12} = \dot{m}_1ex_1 + \dot{Ex}_{Q,A} + \dot{Ex}_{des,\,absorber} \quad (52)$$

where $\dot{Q}_{Absorber}$ is the absorber head load in kW; X is the concentration of ammonia (refrigerant); $\dot{m}_{ws}$ is the mass flow rate of the weak solution in kg/s, which equals to $\dot{m}_6$; $\dot{m}_{ss}$ is the mass flow rate of the strong solution in kg/s, which equals to mass flow rate exiting from the absorber at $\dot{m}_1$; and $\dot{m}_r$ is the mass flow rate of pure ammonia (refrigerant) in kg/s, which flows from the generator at state 7 to state 12; $\dot{Ex}_{Q,A}$ is the thermal exergy rate of the absorber due to the heat transfer $\dot{Q}_A$ to the environment, and it is calculated as $\dot{Ex}_{Q,A} = \dot{Q}_A(1 - T_0/T_s)$. Here, state 1 is a saturated liquid at the lowest temperature in the absorber and is determined by the temperature of the available cooling water flow or air flow.

Solution pump:

EnBE $$\dot{m}_1h_1 + \dot{W}_{Pump} = \dot{m}_2h_2 \quad (53)$$

ExBE $$\dot{m}_1ex_1 + \dot{W}_P = \dot{m}_2ex_2 + \dot{Ex}_{des,\,pump} \quad (54)$$

Regeneration heat exchanger (HX1):

EnBE $$\dot{m}_2h_2 + \dot{m}_4h_4 = \dot{m}_3h_3 + \dot{m}_5h_5 \quad (55)$$

ExBE $$\dot{m}_2ex_2 + \dot{m}_4ex_4 = \dot{m}_3ex_3 + \dot{m}_5ex_5 + \dot{Ex}_{des,\,HX1} \quad (56)$$

Generator:

EnBE $$\dot{m}_3h_3 + \dot{Q}_{gen} = \dot{m}_4h_4 + \dot{m}_7h_7 \quad (57)$$

PMBE $$\dot{m}_{ws}X_{ws} + \dot{m}_r = \dot{m}_{ss}X_{ss} \quad (58)$$

ExBE $$\dot{m}_3ex_3 + \dot{Ex}_{Q,gen} = \dot{m}_4ex_4 + \dot{m}_7ex_7 + \dot{Ex}_{des,gen} \quad (59)$$

where $\dot{Q}_{gen}$ is the heat input to the generator in kW; $\dot{m}_{ws} = \dot{m}_4$ and $\dot{m}_{ss} = \dot{m}_3$; $\dot{Ex}_{Q,gen}$ is the thermal exergy rate of the generator due to the heat transfer $\dot{Q}_{gen}$ to the environment, and it is calculated as $\dot{Ex}_{Q,gen} = \dot{Q}_{gen}(1 - T_0/T_s)$.

Condenser:

EnBE $$\dot{m}_7h_7 = \dot{Q}_H + \dot{m}_8h_8 \quad (60)$$

ExBE $$\dot{m}_7ex_7 = \dot{Ex}_{Q,\,cond} + \dot{m}_8ex_8 + \dot{Ex}_{des,\,cond} \quad (61)$$

where $\dot{Ex}_{Q,\,cond}$ is the thermal exergy rate of the condenser due to the heat transfer $\dot{Q}_H$ to warm environment and is calculated as $\dot{Ex}_{Q,\,cond} = \dot{Q}_H(1 - T_0/T_s)$.

Heat recovery heat exchanger (HX2):

EnBE $$\dot{m}_8h_8 + \dot{m}_{11}h_{11} = \dot{m}_9h_9 + \dot{m}_{12}h_{12} \quad (62)$$

ExBE $$\dot{m}_8ex_8 + \dot{m}_{11}ex_{11} = \dot{m}_9ex_9 + \dot{m}_{12}ex_{12} + \dot{Ex}_{des,\,HX2} \quad (63)$$

Expansion valves:

EnBE
$$\dot{m}_5 h_5 = \dot{m}_6 h_6 \Rightarrow h_5 = h_6 \tag{64}$$

$$\dot{m}_9 h_9 = \dot{m}_{10} h_{10} \Rightarrow h_9 = h_{10} \tag{65}$$

ExBE
$$\dot{m}_5 ex_5 = \dot{m}_6 ex_6 + \dot{Ex}_{des,\,EX1} \tag{66}$$

$$\dot{m}_9 ex_9 = \dot{m}_{10} ex_{10} + \dot{Ex}_{des,\,EX1} \tag{67}$$

Evaporator:

EnBE
$$\dot{m}_{10} h_{10} + \dot{Q}_L = \dot{m}_{11} h_{11} \tag{68}$$

ExBE
$$\dot{m}_{10} ex_{10} + \dot{Ex}_{Q,\,evap} = \dot{m}_{11} h_{11} + \dot{Ex}_{des,\,evap} \tag{69}$$

where $\dot{Ex}_{Q,evap}$ is the thermal exergy rate of the evaporator due to the heat transfer $\dot{Q}_L$ from refrigerated space and is calculated as $\dot{Ex}_{Q,evap} = \dot{Q}_L(T_0/T_L - 1)$.

For the entire system, the overall energy balance of the complete system can be written as follows, by considering that there is negligible heat loss to the environment:

$$\dot{W}_P + \dot{Q}_{gen} + \dot{Q}_L = \dot{Q}_A + \dot{Q}_H \tag{70}$$

The COP of the system then becomes:

$$COP = \frac{\dot{Q}_L}{\dot{W}_P + \dot{Q}_{gen}} \cong \frac{\dot{Q}_L}{\dot{Q}_{gen}} \tag{71}$$

where $\dot{W}_P$ is the pumping power requirement, and it is usually neglected in the COP calculation.

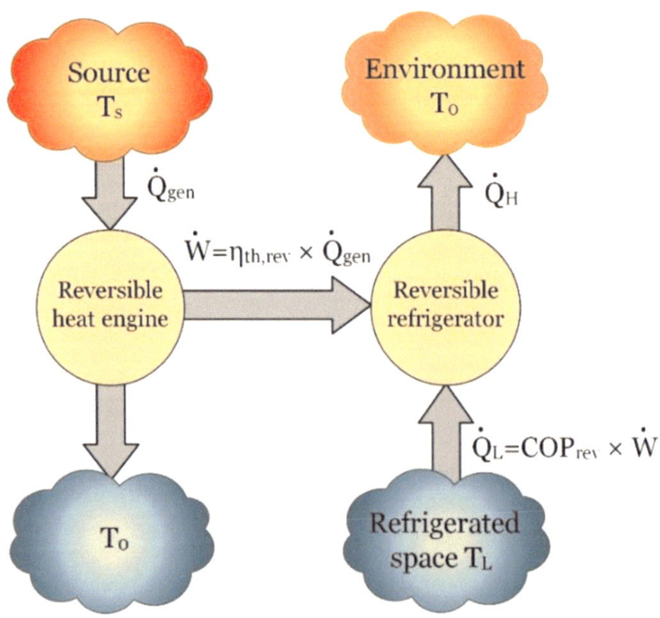

Figure 9.
The maximum COP of an absorption refrigeration system.

The ARS is a heat-driven system, which requires heat pump instead or required power by a compressor. That means the ARS is a combination of a heat pump and a refrigeration cycle without a compressor. Therefore, the maximum (reversible) of an ARS can be achieved by a reversible heat engine and a reversible refrigerator, as shown in **Figure 9**. A reversible heat pump is operating by absorbing the heat from a source at and rejecting heat to an environment of to produce a work output from the heat engine. This work is defined as the reversible efficiency of the heat pump multiplied by the heat absorber from the source, which is the heat transfer from the generator in the ARS. This work output is used by the reversible refrigerator to keep a refrigerated space at T_L while rejecting heat to the environment at T_0. Therefore, the reversible COP of ARS can be obtained by the thermal efficiency of a reversible heat engine and the COP of a reversible refrigerator as in Eq. (72) [10]:

$$COP_{abs,\,rev} = \frac{\dot{Q}_L}{\dot{Q}_{gen}} = \frac{\dot{W}}{\dot{Q}_{gen}}\frac{\dot{Q}_L}{\dot{W}} = \eta_{th,\,rev} COP_{rev} = \left(1 - \frac{T_0}{T_s}\right)\left(\frac{T_L}{T_0 - T_L}\right) \tag{72}$$

The temperature of the heat source is taken as the average temperature of geothermal water. Then the second-law efficiency of this absorption system is determined to be [10]:

$$COP_{ex,\,abs} = \frac{\dot{Ex}_{Q,\,evap}}{\dot{Ex}_{Q,\,gen}} = \frac{\dot{Q}_L(T_0/T_L - 1)}{\dot{Q}_{gen}(1 - T_0/T_s)} = \frac{\dot{Q}_L/\dot{Q}_{gen}}{(1 - T_0/T_s)(T_L/(T_0 - T_L))} \tag{73}$$

$$COP_{ex,\,abs} = \eta_{II,\,abs} = \frac{COP}{COP_{abs,\,rev}} \tag{74}$$

7. Renewable sources for refrigeration system

The refrigeration systems require an input work to release the heat from the refrigerated space to the environment, which is called as a work-driven system. The absorption refrigeration system is based on external heat transfer from an external source, which can be classified as a heat-driven system. For industrial refrigeration systems, energy demand is high and should be provided in a secure and eco-friendly approach to reduce environmental pollution. This can be executed by fossil-based fuels such as oil, natural gas, and coal, which produce substantial carbon monoxide and dioxide emissions that affect global warming and climate change.

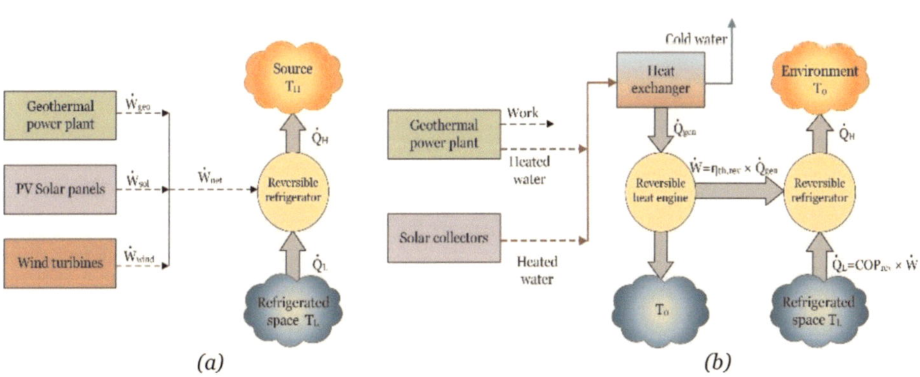

Figure 10.
Renewable sources for refrigeration systems: (a) work-driven and (b) heat-driven source.

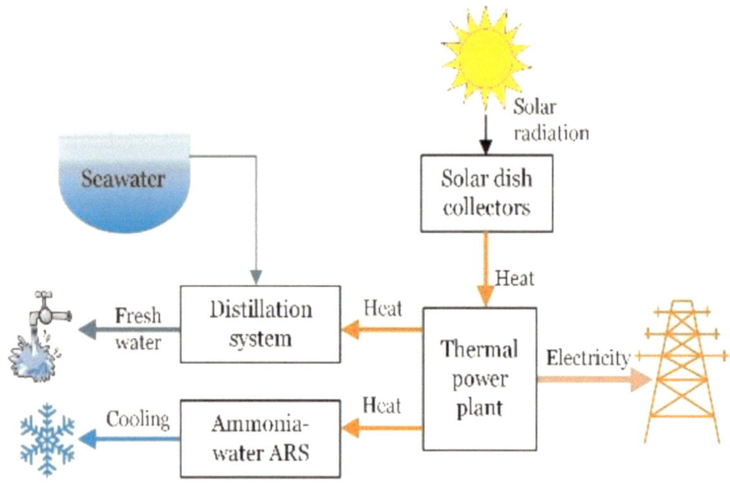

Figure 11.
Schematic diagram of the integrated solar thermal power plant, absorption refrigeration system, and MED cycle (adopted from [12]).

Massive efforts point to renewable sources such as geothermal energy, solar energy, and wind energy, which promise a potential solution to provide the clean energy needed as work or heat to operate the refrigeration. Schematic diagram of **Figure 10** shows possible ways of renewable sources for work-driven and heat-driven refrigeration system.

An integrated system of a concentrated solar power plant integrated with desalination process and absorption refrigeration cycle is utilized to supply power, freshwater, and refrigeration [12]. The system, as shown in **Figure 11**, consists of concentrated solar collectors connected with steam turbine power plant, a multi-effect desalination process with a parallel feed of seawater, and a single-stage ammonia-water absorption refrigeration system. The solar collectors provide thermal energy 21,030 kW to the steam power plant to deliver an electric power of 4632 kW. The refrigeration load from the absorption cooling system is 820.8 kW. The desalination system can also provide 22.79 kg/s freshwater. This cycle has obtained overall energy and exergy efficiencies to be 80.70% and 66.05%, respectively.

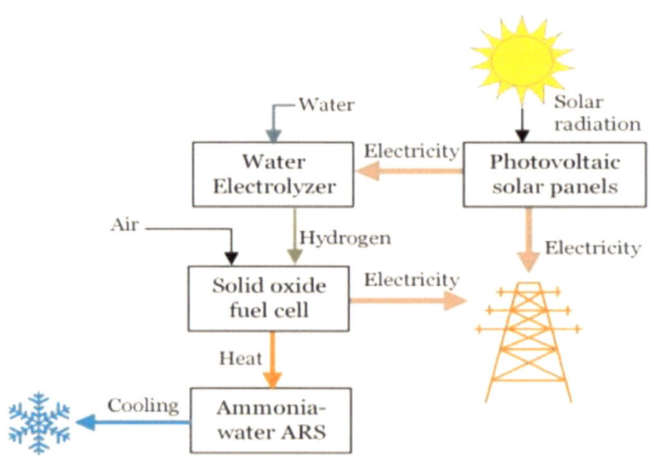

Figure 12.
Schematic of the photovoltaic-fuel cell CHIP system for residential applications (adopted from [13]).

Another example, a small-scale system, is designed to provide an electrical load to residential buildings [13]. This system utilizes, as shown in **Figure 12**, photovoltaic solar system (PV) to provide electrical power. This electric power is used for a water electrolyzer system to split the water electrochemically to produce hydrogen and oxygen gases. The hydrogen gas enters high-temperature solid oxide fuel cells (SOFC) to produce electricity and heat. The heat is transferred to an absorption cooling system by heat recovery generator. The PV system may generate excess electricity more than the demand during off-peak hours. This system is designed for a detached house in Toronto city, Canada. The PV solar system delivers maximum power of 3.35 kW. The water electrolyzer can produce 0.792 and 0.538 kg/day of gaseous hydrogen in summer and winter seasons. The SOFC fuel cell supplies 8.43 kWh per day in summer season. The maximum energy and exergy efficiencies of the photovoltaic system are 17 and 18.3%, respectively, while the maximum total energy and exergy efficiencies are obtained to be 55.7 and 49.0%, respectively.

In a similar study, a hybrid renewable system was designed to produce electricity and clean fuel such as hydrogen gas and provide cooling for a residential building in two locations Egypt and Saudi Arabia in summer season [14]. The cooling loads for a house are 18.06 and 19.3 kW in Egypt and Saudi Arabia, respectively. This system, as shown in **Figure 13**, depends on the photovoltaic solar system and wind turbines to provide excess electricity more than the electric grid. The excess electricity is delivered to a water electrolyzer to produce pure oxygen and hydrogen gases stored in tanks for clean fueling services. Part of the hydrogen gas is used for a proton-exchange membrane (PEM) fuel cell that can produce heat and electricity through an electrochemical process without any mechanical parts. The heat generated from the fuel cell can be utilized by a generator of an ammonia-water ARS to provide cooling. The hybrid renewable system can operate in a significant performance with water mass flow rate of 1.8 kg/s to produce hydrogen with a mass flow rate of 0.2 kg/s and ammonia mass flow rate of about 0.2 kg/s to produce cooling load between 40 and 120 kW more than the design cooling load of one house. The energy and exergy efficiencies are obtained to be about 67 and 68%, respectively. Therefore, this hybrid system can be sufficient for more than one house.

A multigeneration system is designed by [15] and powered by geothermal energy assisted with solar energy to produce five outputs: heating air for residential building, hot domestic water, drying food, refrigeration for industry, and electricity. This multigeneration system, as shown in **Figure 14**, consists of a heat pump

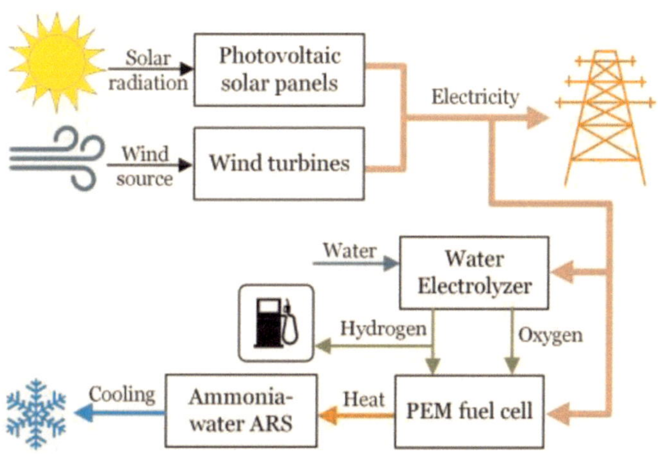

Figure 13.
Schematic diagram of a hybrid renewable system (adopted from [14]).

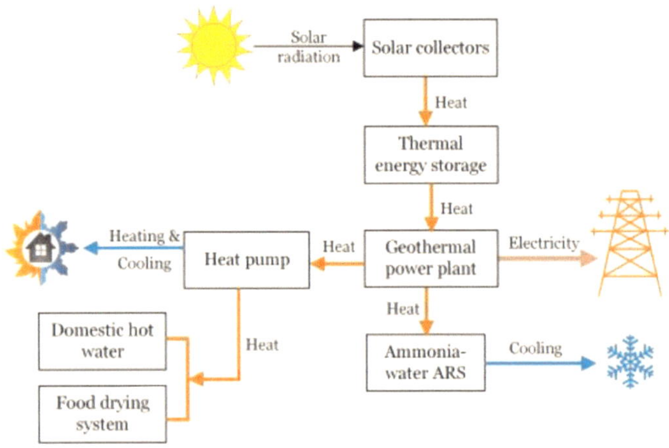

Figure 14.
Schematic diagram of the multigeneration system powered by the solar and geothermal energy (adopted from [15]).

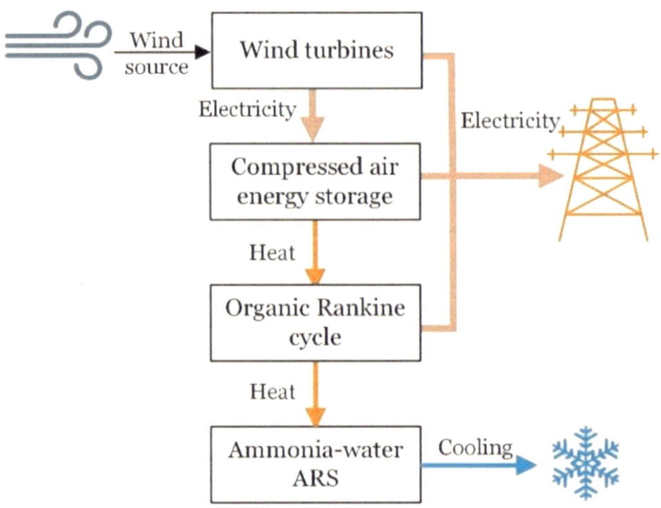

Figure 15.
Schematic diagram of gas power cycle with wind turbine, CAES, ORC, and ARS (adopted from [16]).

system, a single flash geothermal cycle, an absorption cooling system, thermal energy storage connected with auxiliary steam turbine and concentrated solar collectors, hot water system, and drying system. The system has achieved overall energy and exergy efficiencies to be 69.6 and 42.8%, respectively. The first and second steam turbines have the power of 10,043 and 9886 kW. The *COP* and COP_{ex} are 0.678 and 0.253 for the absorption cooling system and 2.029 and 0.1826 for heat pump system, respectively. The refrigeration load is 1787 kW. The overall energy and exergy efficiencies for the whole system are 69.6 and 42.8%, respectively.

A wind system is combined with a refrigeration system, as shown in **Figure 15**. Wind energy is coupled with compressed air energy storage (CAES) systems to store wind energy for long-term usage [16]. The integrated system consists of a combined gas power cycle, including compressors, intercooling heat exchangers, and gas turbine, an organic Rankine power cycle (ORC), and an absorption refrigeration system (ARS). The system objective is to provide electricity, domestic hot water, and cooling load. The system can generate electricity of 33.67 kW provided by wind turbines (83.24 kWh) and fuel combustion (258.97 kWh), cooling load of 2.56 kW, and mass flow rate of hot water of 1.82 ton per day hot. The energy efficiency of the system is achieved to be 53.94%.

8. Conclusion

The refrigeration systems are applied in our life for preserving food, cooling air, and other industrial applications. Most refrigeration systems require external power or external heat to release the heat from the refrigerated space. Many industrial applications involve large cooling energy, which can be operated by multi-pressure refrigeration system, which requires a large amount of external power. The chapter has presented some applications with renewable sources to replace the fossil fuel-driven energy with an environmentally friendly energy source such as geothermal, solar, and wind energy so-called hybrid or integrated systems. In addition to cooling load, the hybrid systems can produce electricity, heating load, and clean fuel such as hydrogen fuel. The absorption refrigeration system is mostly-combined with hybrid system to use the heating load from solar or geothermal energy to produce cooling load.

Author details

Shaimaa Seyam
Faculty of Engineering, Benha University, Benha, Egypt

*Address all correspondence to: shaimaa.seyam@mail.utoronto; shaimaa.abdelhamid@bhit.bu.edu.eg

References

[1] Seyam S. Types of HVAC Systems. HVAC Syst. Rijeka: IntechOpen; 2018. DOI: 10.5772/intechopen.78942

[2] Arora CP. Refrigeration and Air Conditioning. 3rd ed. New Delhi: Tata McGraw-Hill Publishing Company Limited; 2009

[3] Seyam S, Huzayyin A, El-Batsh H, Nada S. Experimental and numerical investigation of the radiant panel heating system using scale room model. Energy and Buildings. 2014;**82**:130-141. DOI: 10.1016/j.enbuild.2014.07.003

[4] Huzayyin A, El-Batsh H, Nada S, Seyam S. Experimental and numerical investigation of a condensation repellent radiant cooling panel system. Journal of Engineering Sciences. 2012;**40**: 1075-1089

[5] Çengel YA, Boles MA. Thermodynamics: A Engineering Approach. 8th ed. USA: McGraw-Hill Education; 2015. DOI: 10.1109/MILCOM.2005.1605829

[6] Dincer I, Rosen MA. Exergy: Energy, Environment and Sustainable Development. 2nd ed. Elsevier; 2013

[7] Aasadnia M, Mehrpooya M. Large-scale liquid hydrogen production methods and approaches: A review. Applied Energy. 2018;**212**:57-83. DOI: 10.1016/j.apenergy.2017.12.033

[8] Mehrpooya M, Ansarinasab H. Advanced exergoeconomic analysis of the multistage mixed refrigerant systems. Energy Conversion and Management. 2015;**103**:705-716. DOI: 10.1016/j.enconman.2015.07.026

[9] Mehrpooya M, Sharifzadeh MMM. Conceptual and basic design of a novel integrated cogeneration power plant energy system. Energy. 2017;**127**: 516-533. DOI: 10.1016/j.energy.2017.03.127

[10] Dinçer I, Kanoglu M. Refrigeration Systems and Applications. 2nd ed. New York: Wiley; 2010

[11] ASHRAE Technical Committee. ASHRAE Handbook: Refrigeration Inch-Pound Edition; 2014

[12] Mehrpooya M, Ghorbani B, Hosseini SS. Thermodynamic and economic evaluation of a novel concentrated solar power system integrated with absorption refrigeration and desalination cycles. Energy Conversion and Management. 2018;**175**: 337-356. DOI: 10.1016/j.enconman.2018.08.109

[13] Hosseini M, Dincer I, Rosen MA. Hybrid solar-fuel cell combined heat and power systems for residential applications: Energy and exergy analyses. Journal of Power Sources. 2013;**221**:372-380. DOI: 10.1016/j.jpowsour.2012.08.047

[14] Seyam S, Al-hamed KHM, Qureshy AMMI, Dincer I. Multi-objective optimization of hydrogen production in hybrid renewable energy systems. IEEE Congress on Evolutionary Computation. 2019:827-834

[15] Ezzat MF, Dincer I. Energy and exergy analyses of a new geothermal-solar energy based system. Solar Energy. 2016;**134**:95-106. DOI: 10.1016/j.solener.2016.04.029

[16] Mohammadi A, Ahmadi MH, Bidi M, Joda F, Valero A, Uson S. Exergy analysis of a combined cooling, heating and power system integrated with wind turbine and compressed air energy storage system. Energy Conversion and Management. 2017;**131**:69-78. DOI: 10.1016/j.enconman.2016.11.003